Quick Reference

NEUROSCIENCE

SECOND EDITION

for Rehabilitation Professionals

The Essential Neurologic Principles Underlying Rehabilitation Practice

Quick Reference

NEUROSCIENCE

SECOND EDITION

for Rehabilitation Professionals

The Essential Neurologic Principles Underlying Rehabilitation Practice

Sharon A. Gutman, PhD, OTR
Associate Professor, Programs in Occupational Therapy
Columbia University
New York, NY

Delivering the best in health care information and education worldwide

www.slackbooks.com

ISBN: 978-1-55642-800-5

All illustrations are by Sharon Gutman, PhD, OTR.

The procedures and practices described in this book should be implemented in a manner consistent with the professional standards set for the circumstances that apply in each specific situation. Every effort has been made to confirm the accuracy of the information presented and to correctly relate generally accepted practices. The authors, editor, and publisher cannot accept responsibility for errors or exclusions or for the outcome of the material presented herein. There is no expressed or implied warranty of this book or information imparted by it. Care has been taken to ensure that drug selection and dosages are in accordance with currently accepted/recommended practice. Due to continuing research, changes in government policy and regulations, and various effects of drug reactions and interactions, it is recommended that the reader carefully review all materials and literature provided for each drug, especially those that are new or not frequently used. Any review or mention of specific companies or products is not intended as an endorsement by the author or publisher.

SLACK Incorporated uses a review process to evaluate submitted material. Prior to publication, educators or clinicians provide important feedback on the content that we publish. We welcome feedback on this work.

Published by: SLACK Incorporated
6900 Grove Road
Thorofare, NJ 08086 USA
Telephone: 856-848-1000
Fax: 856-853-5991
www.slackbooks.com

Contact SLACK Incorporated for more information about other books in this field or about the availability of our books from distributors outside the United States.

Library of Congress Cataloging-in-Publication Data

Gutman, Sharon A.
Quick reference neuroscience for rehabilitation professionals : the essential neurologic principles underlying rehabilitation practice / Sharon A. Gutman. -- 2nd ed.
p. ; cm.
Includes bibliographical references and index.
ISBN-13: 978-1-55642-800-5 (alk. paper)
ISBN-10: 1-55642-800-6 (alk. paper)
1. Neurosciences. 2. Medical rehabilitation. I. Title. II. Title: Neuroscience for rehabilitation professionals.
[DNLM: 1. Nervous System--anatomy & histology--Outlines. 2. Nervous System--physiopathology--Outlines. 3. Rehabilitation--Outlines. WL 18.2 G984q 2008]

RC343.G88 2008
616.8--dc22

2007044583

Printed in the United States of America.

Last digit is print number: 10 9 8 7 6 5 4 3

CONTENTS

About the Author

Sharon A. Gutman, PhD, OTR, is currently an associate professor in Programs in Occupational Therapy at Columbia University. Her clinical expertise lies in the treatment of adults with traumatic brain injury and psychiatric disability. In the last two decades, she has taught courses and written journal articles and books regarding (a) the fundamentals of clinical neuroscience, (b) the neurologic basis of rehabilitation practices, and (c) the neurologic basis of pathological conditions including traumatic brain injury and psychiatric disability.

SECTION 1

Directional Terminology

DIRECTIONAL TERMINOLOGY

Anterior or Ventral

- Refers to the *front* of the organism. Ventral means the "belly" of a four-legged animal.

Posterior or Dorsal

- Refers to the *back* of the organism.

Superior

- Refers to the direction *above*. One structure is above another.

Inferior

- Refers to the direction *below*. One structure is below another.

Rostral

- Refers to the *head* of the organism. Also refers to structures that are *above* others.

Caudal

- Refers to the *tail* of the organism. Also refers to structures that are *below* others.

Medial

- Refers to structures that are *close* to the *midline* of the body.

Lateral

- Refers to structures that are *further* from the *midline* of the body.

PLANES OF THE BRAIN

Midsagittal

- The midsagittal plane *divides* the *left* and *right cerebral hemispheres*. This plane divides the brain in half and runs along the *medial longitudinal fissure*.

Sagittal (also called Parasagittal)

- The sagittal planes run *parallel* to the *midsagittal plane*.

Coronal (also called Frontal or Transverse)

- The coronal planes run *perpendicular* to the *sagittal planes*. Coronal planes divide the *anterior* aspect of the brain from the *posterior* aspect.

Horizontal

- The horizontal planes *divide* the *superior* aspect of the brain from the *inferior* aspect.

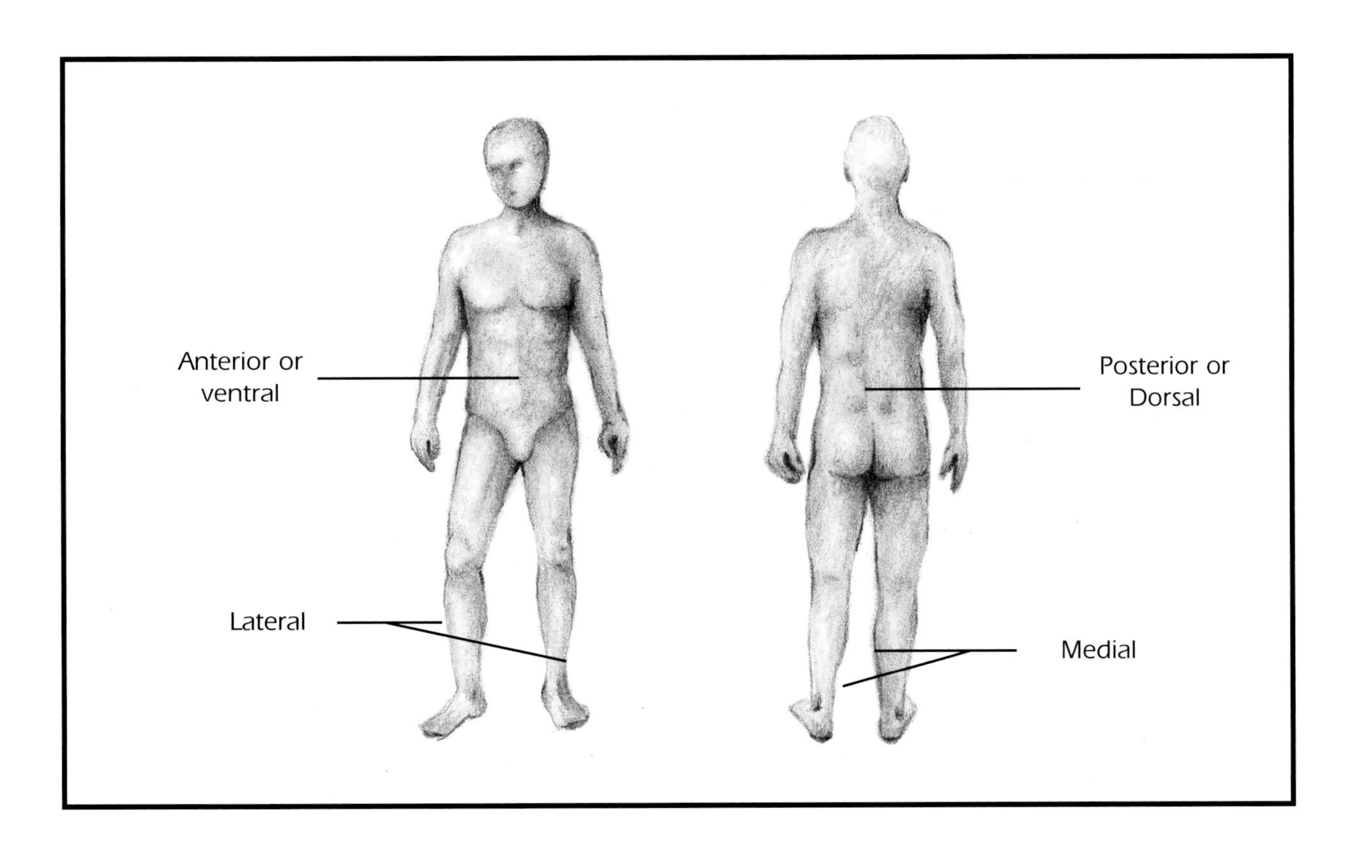
Anterior or
ventral
Posterior or
Dorsal
Lateral
Medial

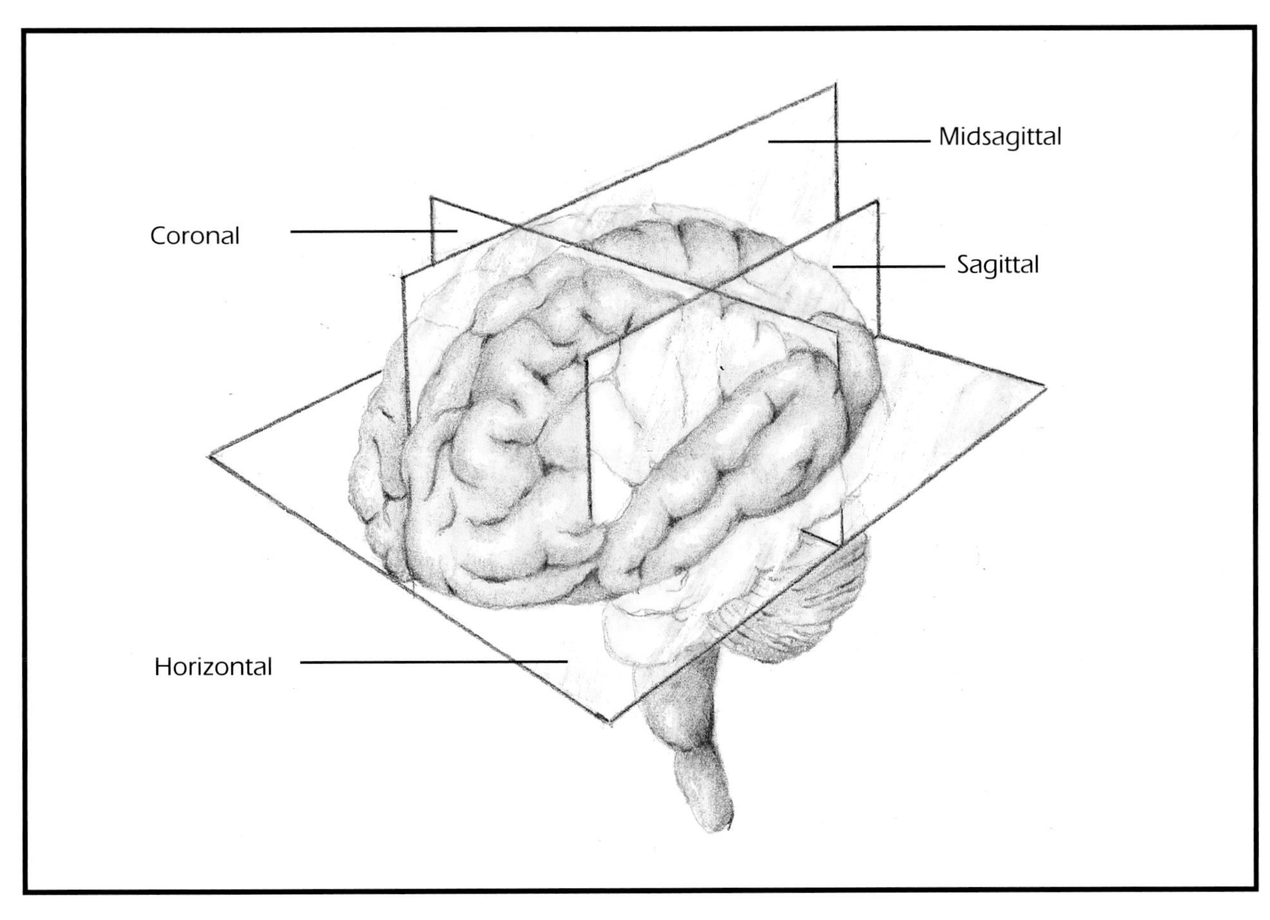
Midsagittal
Coronal
Sagittal
Horizontal

SECTION 2

Division of the Nervous System

- The nervous system is divided into the following:
  - Central Nervous System (CNS)
  - Peripheral Nervous System (PNS)
- The CNS is composed of the following:
  - Brain
  - Spinal cord
- The PNS is composed of the following:
  - Cranial Nerves (CN)
  - Autonomic Nervous System (ANS)
  - Somatic Nervous System (SNS)
- The brain has six major component parts:
  - Cerebral Lobes
  - Cerebellum
  - Basal Ganglia
  - Diencephalon
  - Brainstem
  - Limbic System
- The ANS is composed of the following:
  - Parasympathetic Nervous System
  - Sympathetic Nervous System
- The SNS is responsible for the innervation of skeletal muscles.

Autonomic Nervous System

Innervation of Visceral Muscles and Glands

- ➤ Cardiac Muscle
- ➤ Lungs
- ➤ GI Tract
- ➤ Secretory Glands

Parasympathetic Nervous System

- ➤ Homeostasis
- ➤ Slowing Body Down
- ➤ Decreased Blood Pressure
- ➤ Decreased Heart Rate
- ➤ Peristalsis

Sympathetic Nervous System

- ➤ Arousal
- ➤ Fight/Flight
- ➤ Increased Blood Pressure
- ➤ Increased Heart Rate
- ➤ Cessation of Peristalsis

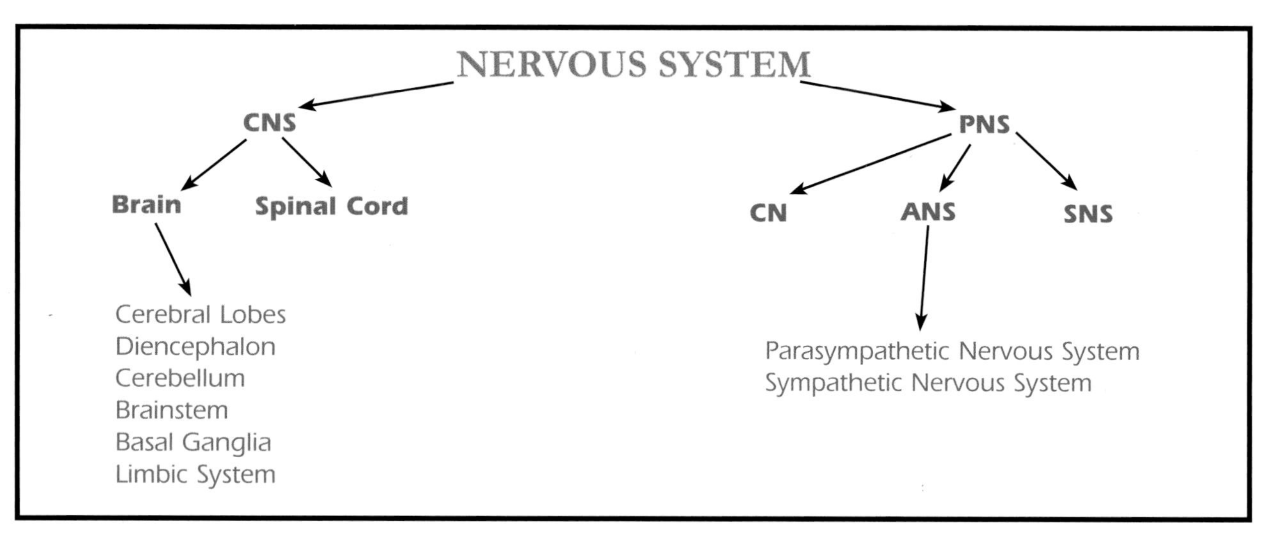
NERVOUS SYSTEM
CNS
PNS
Brain
Spinal Cord
CN
ANS
SNS
Cerebral Lobes
Diencephalon
Cerebellum
Brainstem
Basal Ganglia
Limbic System
Parasympathetic Nervous System
Sympathetic Nervous System

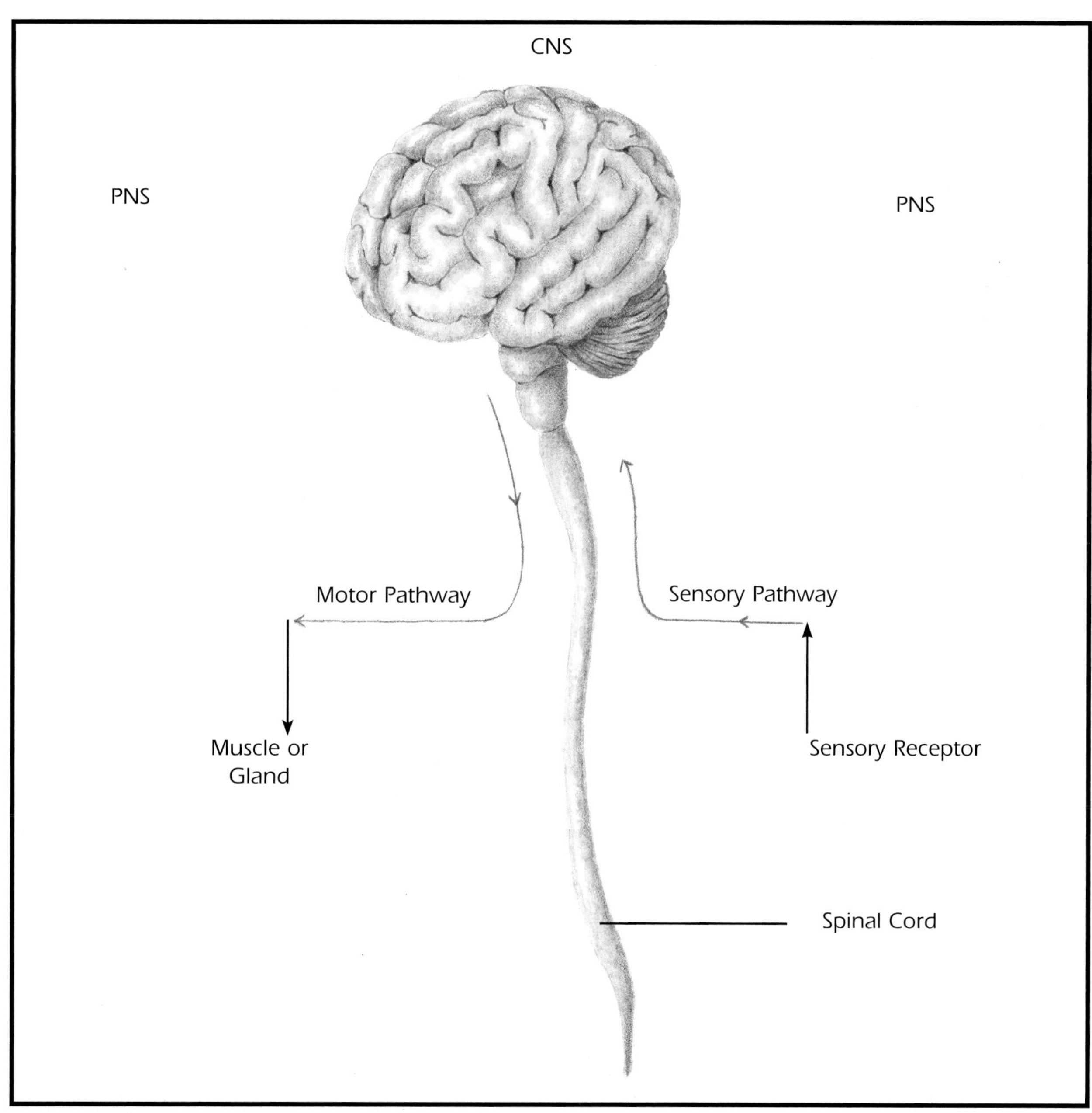
CNS
PNS
PNS
Motor Pathway
Sensory Pathway
Muscle or Gland
Sensory Receptor
Spinal Cord

SECTION 3

Gross Cerebral Structures

Gross Cerebral Structures

Gyri (s., Gyrus)

- Gyri are the *wrinkles*, or folds, on the surface of the cerebral hemispheres.

Sulci (s., Sulcus)

- Sulci are the *valleys*, or crevices, between the gyri.

Convolutions

- Convolutions are the *collective name* for the *gyri* and *sulci*. They are the raised and depressed surfaces of the brain.
- Because brain growth is confined by the skull, the brain folds in on itself as it grows.
- Theorists surmise that the more gyri and sulci one has, the larger one's brain surface is and the more brain capacity one has for brain functions.
- Human brain convolutions are unique—like a fingerprint. However, there are certain sulci and gyri that are common in all human brains.

Fissure

- A fissure is a *deep groove* in the surface of the brain.
- A fissure is deeper than a sulcus; one can stick one's fingers into a fissure. A sulcus is shallow.

Medial Longitudinal Fissure

- The medial longitudinal fissure *separates* the *right* and *left cerebral hemispheres*.
- This fissure runs along the *midsagittal plane*.

Central Sulcus (also called the Sulcus of Rolando)

- The central sulcus *separates* the *frontal* and *parietal* lobes.
- It also *separates* the *primary motor cortex* (M1) from the *primary somatosensory cortex* (SS1).

Precentral Gyrus

- The precentral gyrus is the *primary motor cortex* (M1). This area handles *voluntary motor movement*.
- It is located just anterior to the central sulcus.

Postcentral Gyrus

- The postcentral gyrus is the *primary somatosensory cortex* (SS1). It is located just posterior to the central sulcus.
- This is the part of the brain that mediates the *detection of physical sensation*.

Lateral Fissure (also called Fissure of Sylvius)

- This fissure *separates* the *temporal* lobe from the *frontal* lobe.

CEREBRAL LOBES

- Each hemisphere has four separate lobes.

Frontal Lobes

- The borders of the frontal lobes are the lateral fissure and the central sulcus.
- The frontal lobes mediate *cognition* (intelligence, problem-solving, and short-term memory), *expressive language*, *motor planning*, *mathematical calculations*, and *working memory*.
- The prefrontal lobe mediates *executive functions* (organization, planning, sequencing, and motivation), *self-insight*, and *regulation of emotions*.
- The frontal lobes develop most after birth. Development is not thought to be complete until late adolescence or early adulthood.

Parietal Lobes

- Parietal lobes sit just posterior to the frontal lobes.
- The central sulcus divides the parietal lobes from the frontal lobes.
- The posterior border is the parieto-occipital sulcus and can be seen on a midsagittal cross-section.
- The inferior border is the temporal lobe and the lateral fissure.
- Their functions is *sensory detection, perception, and interpretation.*

Temporal Lobes

- Temporal lobes are the most inferior or caudal lobes.
- They have a poorly defined posterior border—the anterior occipital lobe.
- Their function is *audition (hearing)*, *comprehension of language*, and *long-term memory*.

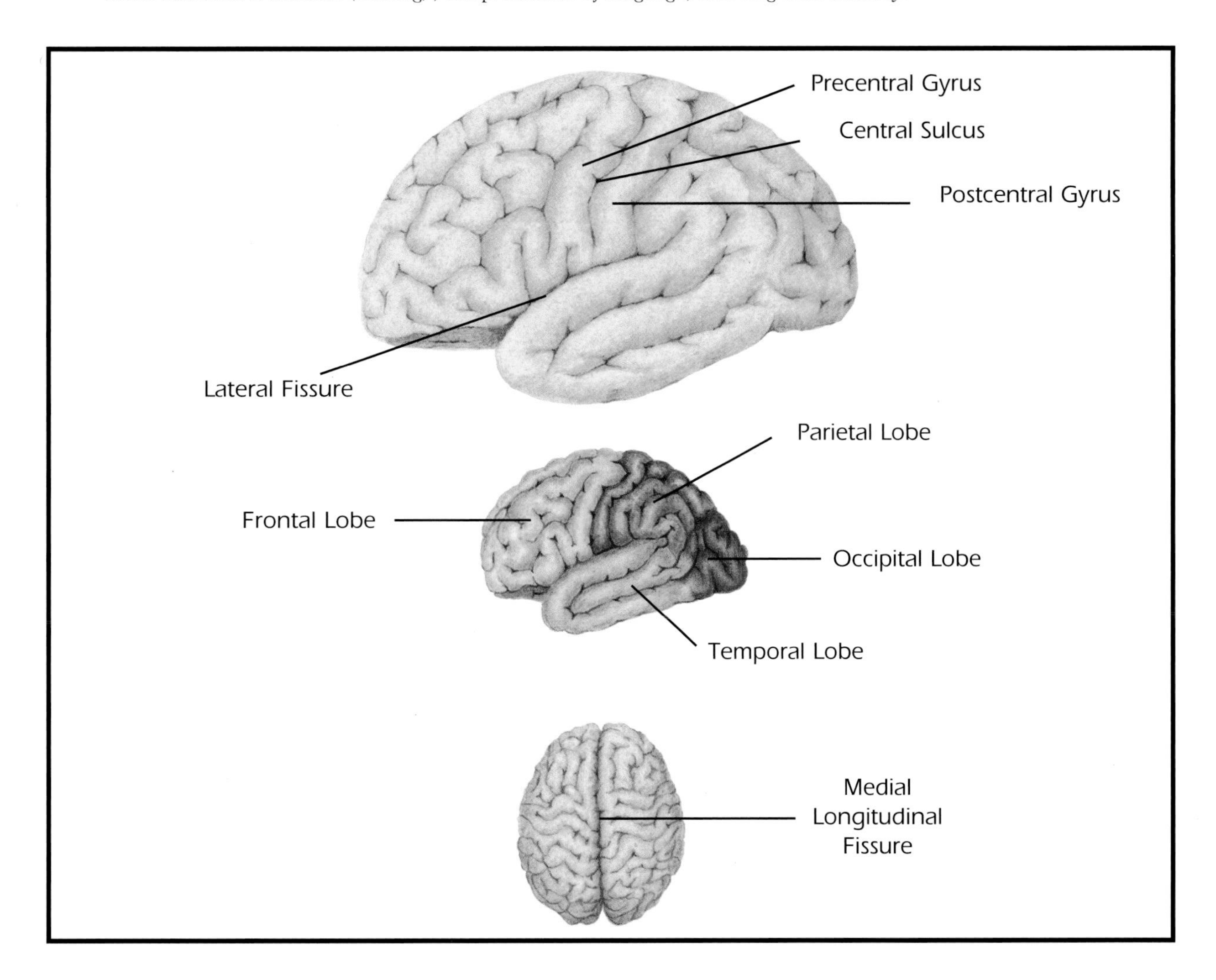

Occipital Lobes

- Occipital lobes are the most posterior lobes.
- They are responsible for the *interpretation of visual stimuli* from the optic pathways.

Insula

- The insula is a portion of the cerebral cortex that lies deep in the lateral fissure.
- It is covered from view by the frontal, parietal, and temporal lobes.
- Some sources consider the insula as a fifth lobe.
- It has a role in the perceptual processing of gustatory information.
- It has a possible role in the interpretation of music.

Right vs Left Hemispheres

Right Hemisphere

- The right hemisphere is largely responsible for the *interpretation of perceptual and spatial information* (eg, reading maps and creating music or art).
- It is responsible for the interpretation of information that is *abstract* and *creative* (as opposed to concrete and logical).
- It is also responsible for the interpretation of *tonal inflections in language* (as opposed to the concrete meaning of words).
- This hemisphere is responsible for taking the literal interpretation of a story and forming abstract symbolism and metaphors.
- It is responsible for the interpretation of the emotional messages underlying the concrete meaning of words.
- The right hemisphere controls *movement* on the *left* side of the body.
- It receives *sensory* information from the *left* side of the body.

Left Hemisphere

- In people who are right-hand dominant, the left hemisphere is usually dominant.
- The left hemisphere plays a large role in human *language* (the expression and interpretation of written and spoken words).
- People with aphasia often have sustained left hemisphere damage.
- It controls *movement* on the *right* side of the body.
- It receives *sensory* information from the *right* side of the body.

Gray Matter vs White Matter

- The cerebral hemispheres consist of gray and white matter.

Gray Matter

- Areas where gray matter covers part of the central nervous system (CNS) are called the cortex (pl., cortices).
- Humans have a cerebral cortex and a cerebellar cortex.
- Gray matter sits on the surface of the cerebrum and the cerebellum.
- It has a grayish or beige appearance because it consists of *nerve cell bodies* (nuclei).
- The gray matter is *nonmyelinated* brain matter. Myelin is a lipid that insulates a nerve and increases conduction velocity.

Ganglia

- Ganglia are collections of *neural cell bodies* (or nuclei) usually located *outside of the* CNS, or in the peripheral nervous system (PNS).
- Example: Dorsal root ganglia.
- The dorsal root ganglia contain the cell bodies of the sensory spinal nerves.

White Matter

- White matter is located beneath the gray matter, in the internal regions of the cerebrum and cerebellum.
- White matter consists of *myelinated fiber tracts* or neuronal axons.

Commissure

- A commissure is any collection of axons (white matter) that *connect one side* of the *nervous system* to the *other*.
- Example: Corpus callosum and the pyramidal decussation.

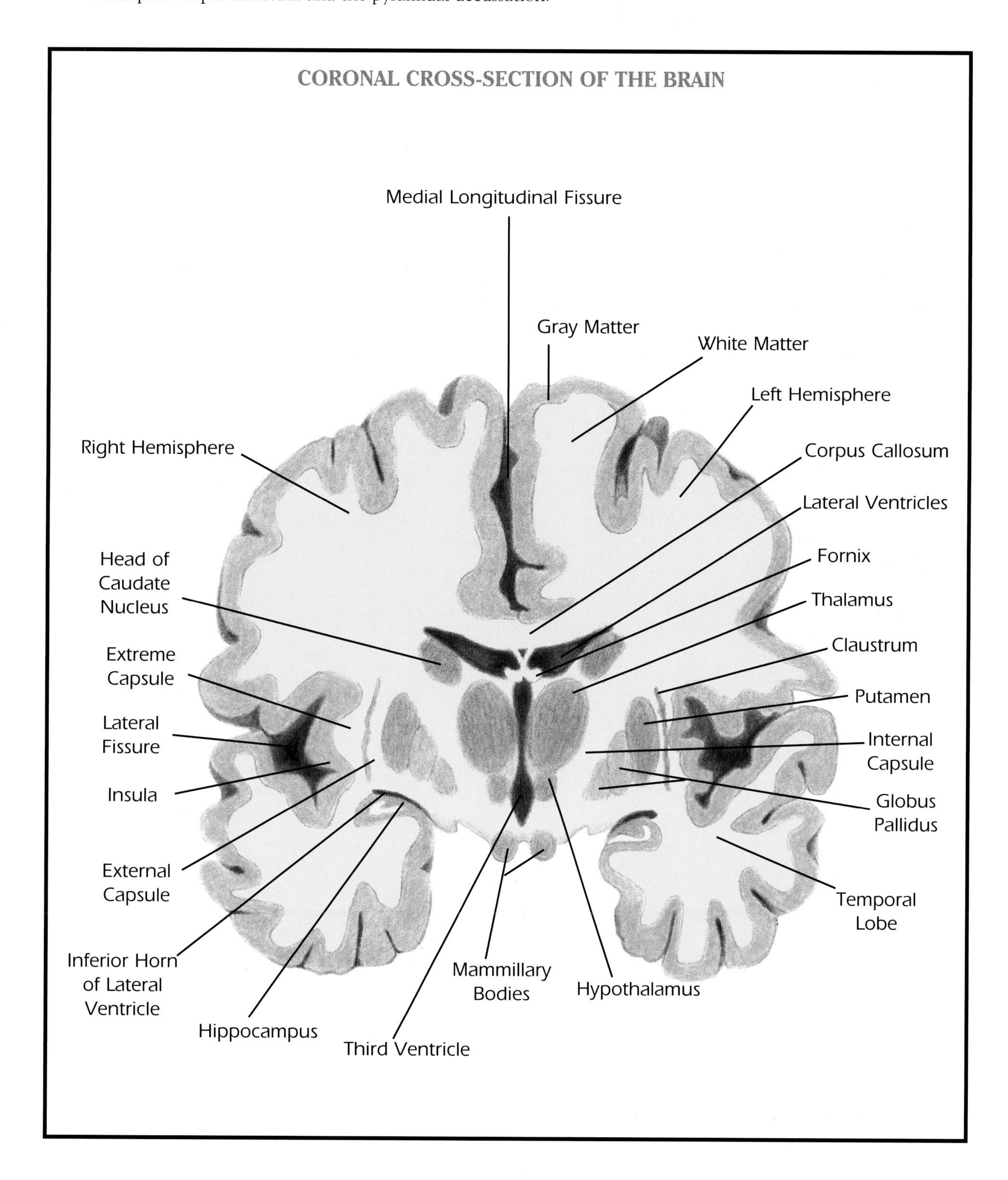

DIENCEPHALON

- In phylogenetic terms, the diencephalon is considered to be an old part of the brain while the cortex is considered to be the newest brain region.
- The diencephalon consists of four structures:
  - Thalamus
  - Hypothalamus
  - Epithalamus
  - Subthalamus

Thalamus

- Thalamus means egg-shaped. There are two thalamic lobes—one in each hemisphere.
- The thalamus contains 26 pairs of nuclei. These primarily *receive sensory data from the sensory systems*. These nuclei then relay the sensory data to specific parts of the cerebral hemispheres.
- All sensory information, except olfaction, travels through the thalamus before it reaches the cortex and is consciously interpreted. The thalamus can be considered as a gateway to the cortex.
- However, the thalamus is not a passive relay of sensory information. Instead, it is a dynamic structure that works collaboratively with the cortex in feedforward, feedback, and loop circuits.
- The thalamus *receives motor information from the cerebral hemispheres* and relays it to the motor receptors, again, in a dynamic process in which the thalamus and cortex work collaboratively.
- The thalamus has a role in *sleep–wake cycles* and *consciousness*.
- It works with the *reticular formation* to alert the brain to important incoming sensory information and to calm the body down. The thalamus acts as a screen for information traveling to the cortex, inhibiting less important information that can be handled at a subcortical level, and alerting the cortex to important information that must be dealt with at a conscious level.
- Important *thalamic nuclei* include the following:
  - Lateral Geniculate Nucleus—responsible for *visual* processing
  - Medial Geniculate Nucleus—responsible for *auditory* processing
  - Ventrolateral Nucleus—responsible for the organization of *motor responses*
  - Ventral Posterolateral Nucleus—responsible for *tactile-sensory* processing
- The previous nuclei have pathways leading to and from the thalamus and traveling to and from specific sensory and motor systems.
- This allows such functions (eg, vision and audition) to be handled at both a conscious and unconscious level. For example, a cortically blind person may not identify the presence of a visual object but may negotiate his or her body movements to accurately avoid contact with the object. Similarly, a person with cortical hearing loss may startle in response to the slam of a door even though the sound of the slammed door was never consciously interpreted.

Hypothalamus

- The hypothalamus is located just anterior and inferior to the thalamus.
- There are two hypothalamic lobes—one in each hemisphere.
- Functions include the following:
  - Regulates the autonomic nervous system (ANS)
  - Releases hormones from the pituitary gland, adrenal glands, and pineal gland
  - Regulates temperature
  - Regulates hunger
  - Regulates sleep–wake cycles (circadian rhythms)
  - Works collaboratively with the limbic system in the expression of emotions

Epithalamus

- The epithalamus is located just posterior to the thalamus and just anterior to the pineal gland.
- This is a very small structure.
- A principal structure of the epithalamus is the *habenula*—a nucleus at the posterior of the epithalamus.

Subthalamus

- The subthalamus is a deep structure.
- It is considered a *thalamic nuclei group* located caudal to the thalamus.
- The subthalamus contains cells that use *dopamine*.
- It is a key structure *connecting feedback* and *feedforward circuits* of the *thalamus* and *basal ganglia*.
- The subthalamus is larger than the epithalamus, and it can be identified in certain coronal cross-sections.

The epithalamus and subthalamus have no distinct borders.

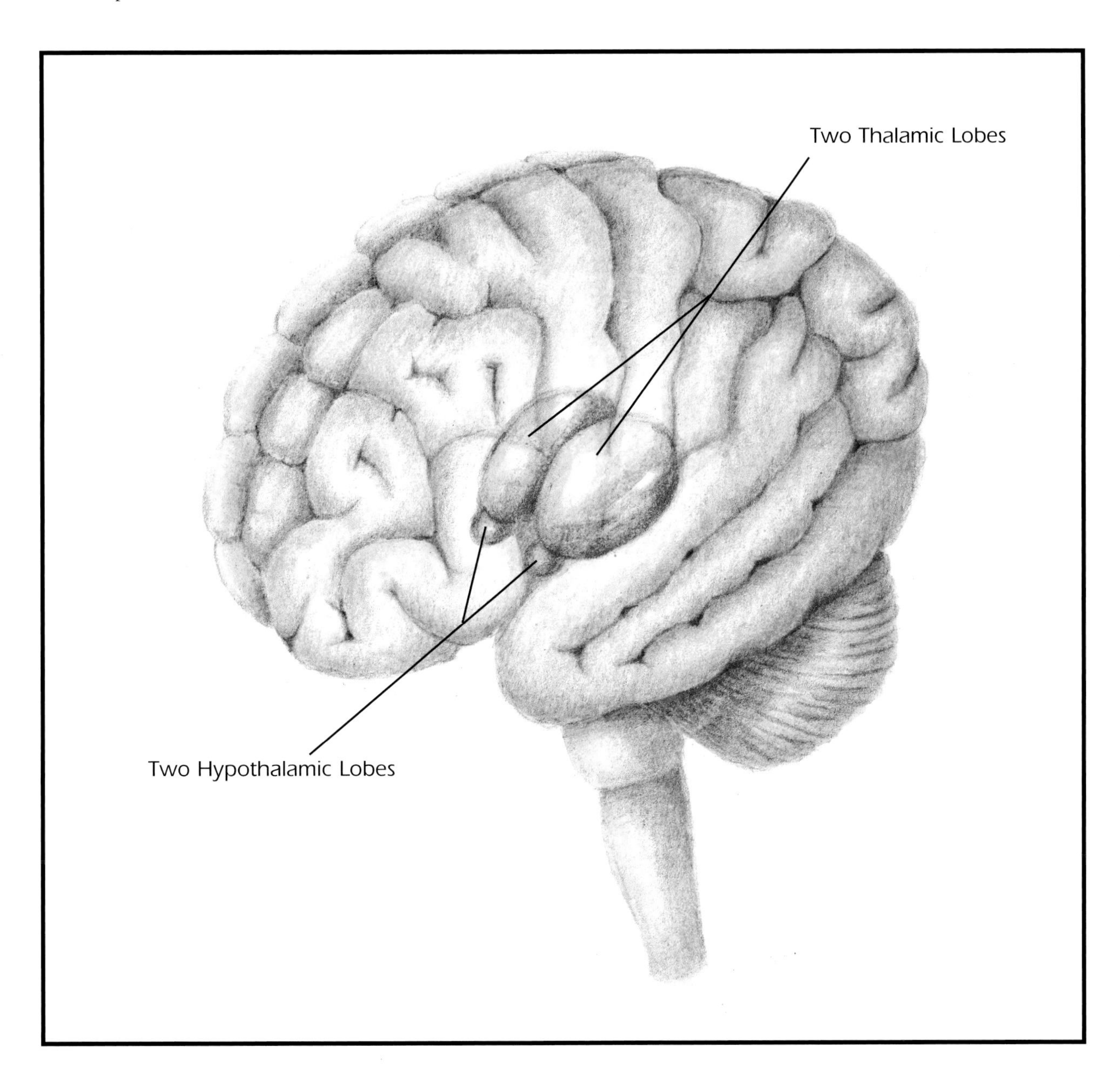

OTHER STRUCTURES OF THE DIENCEPHALON

Pituitary Gland

- There is only one pituitary gland in the body.
- The pituitary gland is an *endocrine gland* that secretes hormones that *regulate growth, reproductive activities*, and *metabolic processes.*
- These hormones include the following:
  - Growth hormone (GH)
  - Prolactin (PRL) (for lactation)
  - Luteinizing hormone (LH) (for reproduction)
  - Follicle stimulating hormone (FSH) (for reproduction)
  - Thyroid stimulating hormone (TSH) (for metabolism)
  - Adrenocorticotrophic hormone (ACTH) (for the regulation of stress)
- The pituitary works collaboratively with the hypothalamus. The synthesis and secretion of the above hormones are controlled by neuropeptides released from the hypothalamus.

Infundibulum

- The infundibulum is the stalk that extends from the hypothalamus and holds the pituitary gland.

Pineal Gland

- The pineal gland is a midline structure located just posterior to the thalamus.
- There is only one pineal gland in the body.
- The pineal gland is innervated by the ANS.
- It has a role in *sexual* and *hormonal functions* and in *sleep–wake cycles.*

Posterior Commissure

- The posterior commissure connects the right and left halves of the diencephalon.
- It is located just above the superior colliculi.
- This structure allows *communication between the hemispheres* if the corpus callosum is lesioned or removed because of pathology.

Anterior Commissure

- The anterial commissure connects the olfactory bulb to the amygdala.
- It is located in the anterior thalamus and passes through the head of the caudate nucleus.
- This structure allows *communication between the hemispheres* if the corpus callosum is lesioned.

Interthalamic Adhesion

- The interthalmic adhesion allows communication between the two thalamic lobes.
- It is located centrally in the thalamus.

Septum Pellucidum

- The septum pellucidum is a sheath-like cover that extends over the medial wall of each lateral ventricle.
- May have a role in the *processing of emotion* along with the limbic system.

STRUCTURES LOCATED NEAR THE DIENCEPHALON (BUT ARE NOT PART OF THE DIENCEPHALON)

Corpus Callosum

- The corpus callosum is the largest commissure in the brain.
- It allows the *right* and *left cerebral hemispheres* to *communicate* with each other.
- This structure arches around the anterior horn of the lateral ventricles.

Optic Chiasm

- A chiasm is a crossing-over point.
- The optic chiasm is a cross-shaped connection located between the optic nerves.
- It is a midline structure specifically located at the base of the brain just superior to the pituitary gland.

Internal Capsule

- The internal capsule is a large fiber bundle that connects the cerebral cortex with the diencephalon.
- All descending *motor messages from the cortex travel through* the *internal capsule* to the thalamus, brainstem, spinal cord, and the skeletal muscles.

MIDSAGITTAL CROSS-SECTION OF THE BRAIN

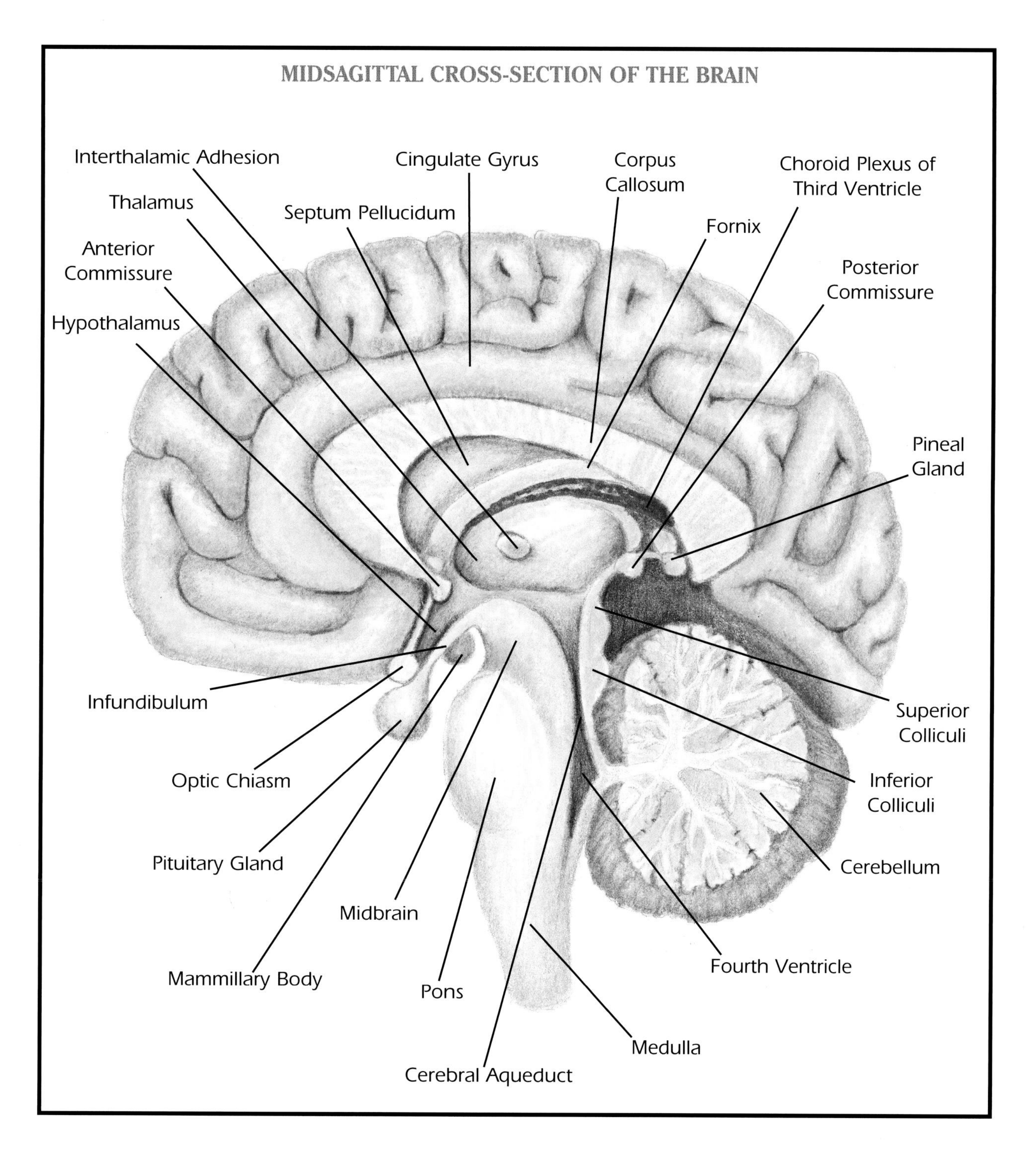

- *Sensory information* from the sensory receptors ascends in the spinal cord, through the brainstem, to the thalamus, through the internal capsule, and to the primary somatosensory cortex (SS1).

Mammillary Bodies

- These structures form two protrusions that sit within the interpeduncular fossa on the anterior side of the midbrain.
- The mammillary bodies are nuclei groups that form attachments with the hypothalamus and fornix and may play a role in the processing of emotion.

Brainstem

- The brainstem is composed of three basic structures:
  - Midbrain
  - Pons
  - Medulla
- The brainstem controls *vegetative functions*:
  - Respiration
  - Cough and gag reflex
  - Pupillary response
  - Swallowing reflex

Midbrain

- The midbrain is the most rostral structure of the brainstem.
- It sits atop of the pons.
- It is just inferior to the thalamus.
- The midbrain has a role in *automatic reflexive behaviors* dealing with *vision* and *audition*.

External Structures of the Midbrain

Cerebral Peduncles

- Cerebral pendunceles are large fiber bundles located on the anterior surface of the midbrain.
- They *carry descending motor tracts* from the cerebrum to the brainstem.

Interpeduncular Fossa

- The interpenduncular fossa is the indentation between the pair of cerebral peduncles.
- The *mammillary bodies* sit within the interpeduncular fossa on the anterior aspect of the midbrain.

Superior and Inferior Colliculi

- The superior and inferior colliculi sit on the posterior surface of the midbrain just beneath the posterior commissure.
- The superior colliculi are located just above the inferior colliculi.
- The *superior colliculi* are a pair of *relay centers for vision*. They communicate directly with the thalamic nuclei (lateral geniculate nuclei) that process visual stimuli. This pathway allows visual information to be processed at an unconscious level.
- The *inferior colliculi* are a pair of *relay centers for audition*. They communicate directly with the thalamic nuclei (medial geniculate nuclei) that process auditory stimuli. This pathway allows auditory information to be processed at an unconscious level.

Internal Structures of the Midbrain (seen on cross-section)

Cerebral Aqueduct (also called Aqueduct of Sylvius)

- The cerebral aqueduct is part of the *ventricular system*. It connects the third and fourth ventricles. Cerebrospinal fluid (CSF) flows through the cerebral aqueduct.

Superior and Inferior Colliculi

- Superior and inferior colliculi can be seen on cross-section.

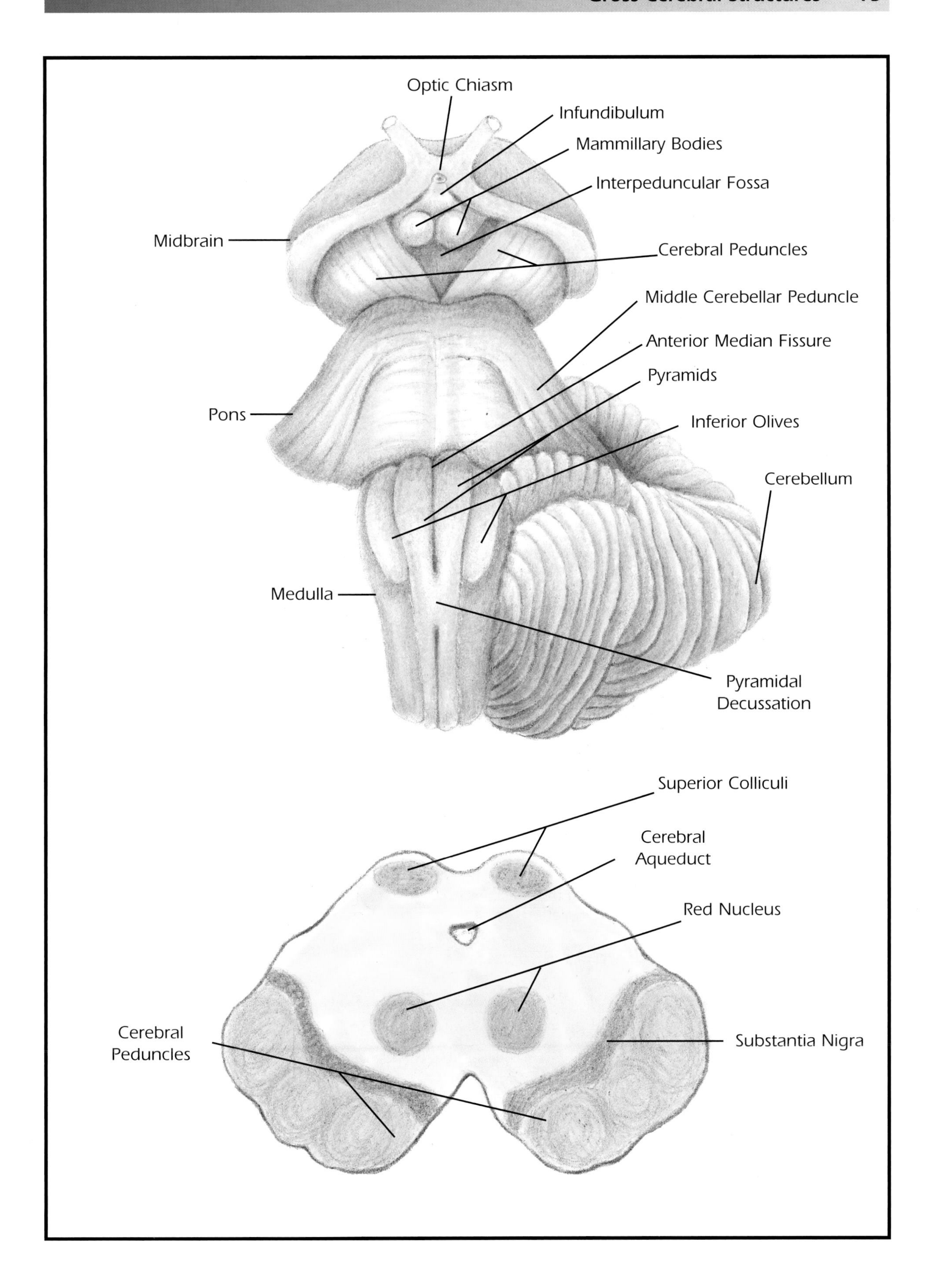
Optic Chiasm
Infundibulum
Mammillary Bodies
Interpeduncular Fossa
Midbrain
Cerebral Peduncles
Middle Cerebellar Peduncle
Anterior Median Fissure
Pyramids
Pons
Inferior Olives
Cerebellum
Medulla
Pyramidal Decussation
Superior Colliculi
Cerebral Aqueduct
Red Nucleus
Cerebral Peduncles
Substantia Nigra

Cerebral Peduncles

- The cerebral peduncles have an inner and outer coat.
- The *outer coat* consists of the *crus cerebri.*
- The *inner coat* consists of the *red nucleus* and the *substantia nigra.*

Tegmentum

- The *substantia nigra* and the *red nucleus* are collectively called the tegmentum.

Tectum

- The *superior* and *inferior colliculi* are collectively called the tectum.

PONS

- The pons is located just caudal to the midbrain and rostral to the medulla.
- The pons is a relay system among the spinal cord, cerebellum, and the cerebrum.
- It largely *mediates motor information on an unconscious level*—for example, shifting weight to maintain balance and making fine motor adjustments in one's muscles to perform precise coordinated limb movement.

External Structures of the Pons

Cerebellar Peduncles

- Cerebellar peduncles largely carry *sensory information* from the pons to the cerebellum about the *body's position in space.*
- Each cerebellar peduncle is paired—one on each side.
- There are three cerebellar peduncles on each side:
  - *Middle Cerebellar Peduncle*
    - This peduncle carries sensory information about the body's position in space to the cerebellum.
  - *Inferior Cerebellar Peduncle*
    - This peduncle is located in the pons and medulla.
    - It carries sensory information about the body's position in space from the pons/medulla to the cerebellum.
    - The cerebellum then analyzes this sensory information and makes decisions about how to readjust the body for precision movement and balance.
    - The cerebellum then sends its decision to the thalamus via the superior cerebellar peduncles. The thalamus sends this motor information back down through the brainstem to the spinal cord and to the skeletal muscles.
  - *Superior Cerebellar Peduncle*
    - This peduncle is located in the pons.
    - It carries sensory information from the pons to the cerebellum.
    - It also carries sensorimotor information from the cerebellum to the thalamus.

Internal Structures of the Pons

Fourth Ventricle

- The fourth ventricle is located in the posterior pons.
- It is a *continuation* of the *cerebral aqueduct.*
- It *leads into the central canal* of the spinal cord.

Corticospinal Tracts

- These are *descending motor tracts* from the M1.

Middle and Superior Cerebellar Peduncles

- These can be seen in a cross-section of the pons.
- They carry *sensory information* from the pons to the cerebellum.

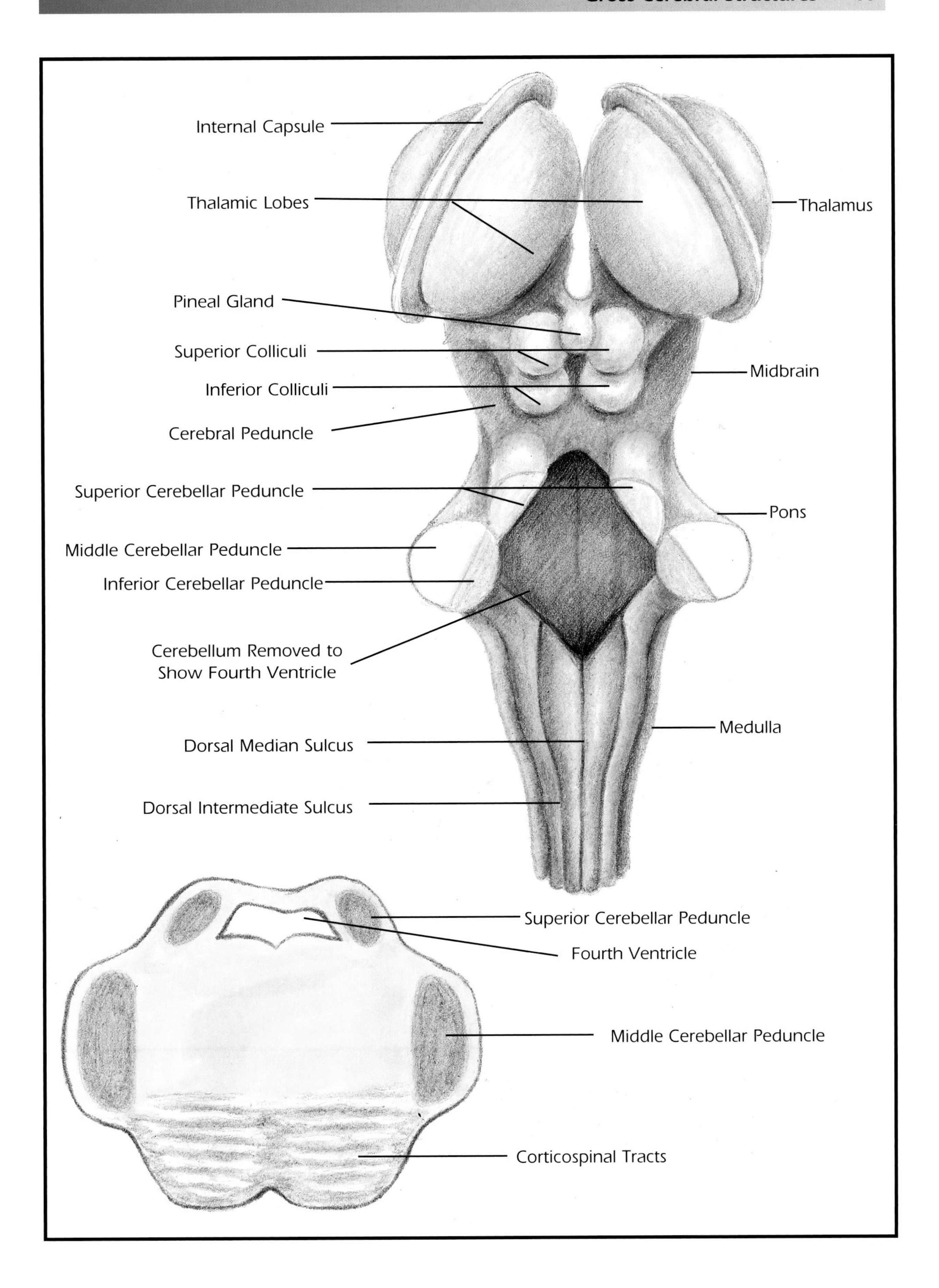
Internal Capsule
Thalamic Lobes
Thalamus
Pineal Gland
Superior Colliculi
Midbrain
Inferior Colliculi
Cerebral Peduncle
Superior Cerebellar Peduncle
Pons
Middle Cerebellar Peduncle
Inferior Cerebellar Peduncle
Cerebellum Removed to
Show Fourth Ventricle
Medulla
Dorsal Median Sulcus
Dorsal Intermediate Sulcus
Superior Cerebellar Peduncle
Fourth Ventricle
Middle Cerebellar Peduncle
Corticospinal Tracts

MEDULLA

- The medulla is just caudal to the pons.
- Because the medulla is long, it is divided into the rostral and caudal medulla.
- The medulla *carries descending motor messages* from the cerebrum to the spinal cord.
- It also *carries ascending sensory messages* from the spinal cord to the cerebrum.

External Structures of the Medulla—Anterior Side

Anterior Median Fissure

- The anterior median fissure divides the medulla into equal right and left halves.
- This fissure continues all the way down the spinal cord.

Pyramids

- The pyramids are two large structures that are divided by the anterior median fissure.
- They are *motor tracts*, or fiber bundles, that carry descending motor information from the cortex to the spinal cord.

Pyramidal Decussation

- This is the crossing-over point where *motor fibers from the left cortex cross to the right* side of the spinal cord. *Motor fibers from the right side of the cortex cross to the left* side of the spinal cord.
- This is why the right cerebral hemisphere controls the left side of the body and the left cerebral hemisphere controls the right side of the body.

Inferior Olives

- The olives are located just lateral to each pyramid.
- They are relay nuclei that *carry ascending sensory information* to the cerebellum. This sensory data pertain to the body's position in space.

External Structures of the Medulla—Posterior Side

Dorsal Median Sulcus

- The dorsal median sulcus is the midline that divides the posterior medulla into equal left and right sides.
- It is not as well-defined as the anterior median fissure.

Dorsal Intermediate Sulcus

- These sulci are located just lateral to the dorsal median sulcus.

Internal Structures of the Medulla (seen in cross-section)

Rostral Medulla

- Caudal end of the fourth ventricle
- Pyramids
- Inferior olivary nuclei
- Inferior cerebellar peduncles

Caudal Medulla

- Central canal (this is where the end of the fourth ventricle meets the spinal cord)
- Pyramids
- Fasciculus gracilis and cuneatus (these are ascending sensory tracts)
- Nucleus gracilis and cuneatus (these are where the nuclei or cell bodies of the fasciculus gracilis and cuneatus are located)

RETICULAR FORMATION

- The reticular formation is diffusely located in the brainstem.
- It consists of two systems:
  - *Reticular Activating System (RAS)*
  - *Reticular Inhibiting System (RIS)*

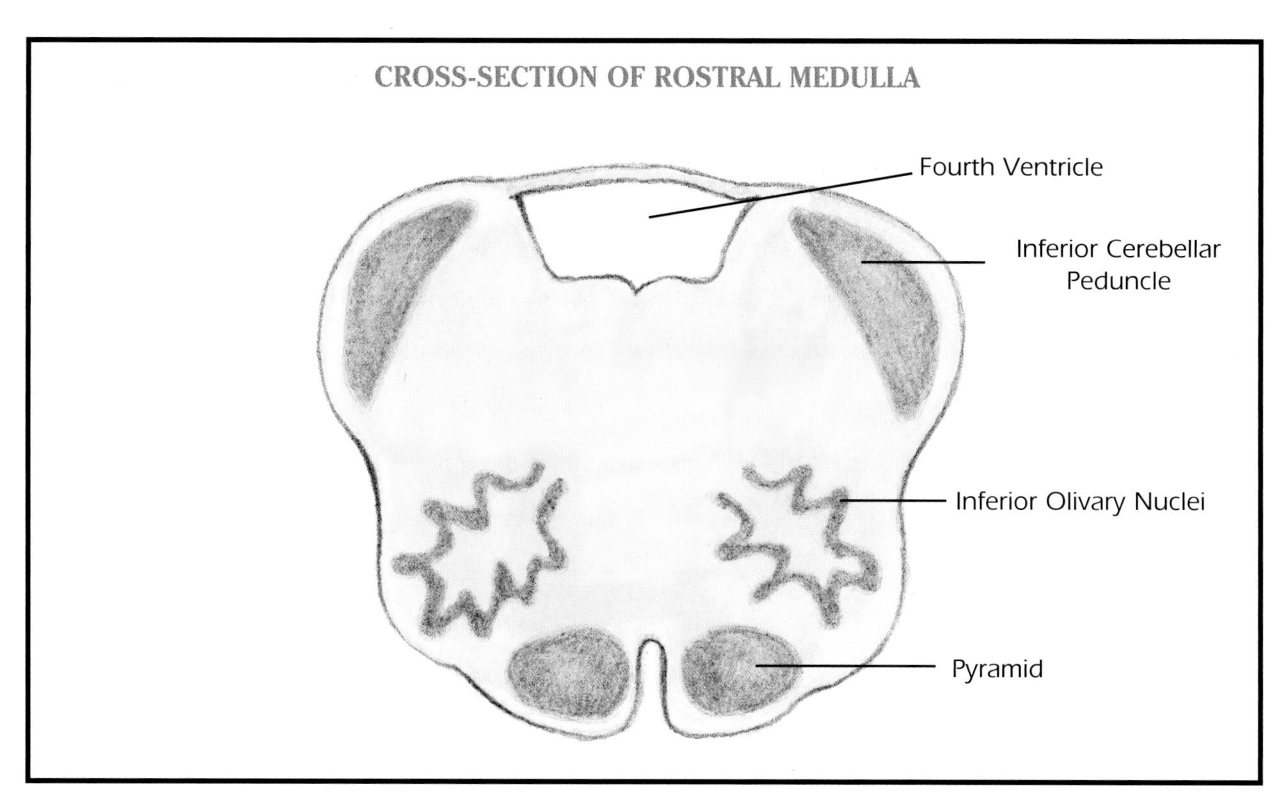

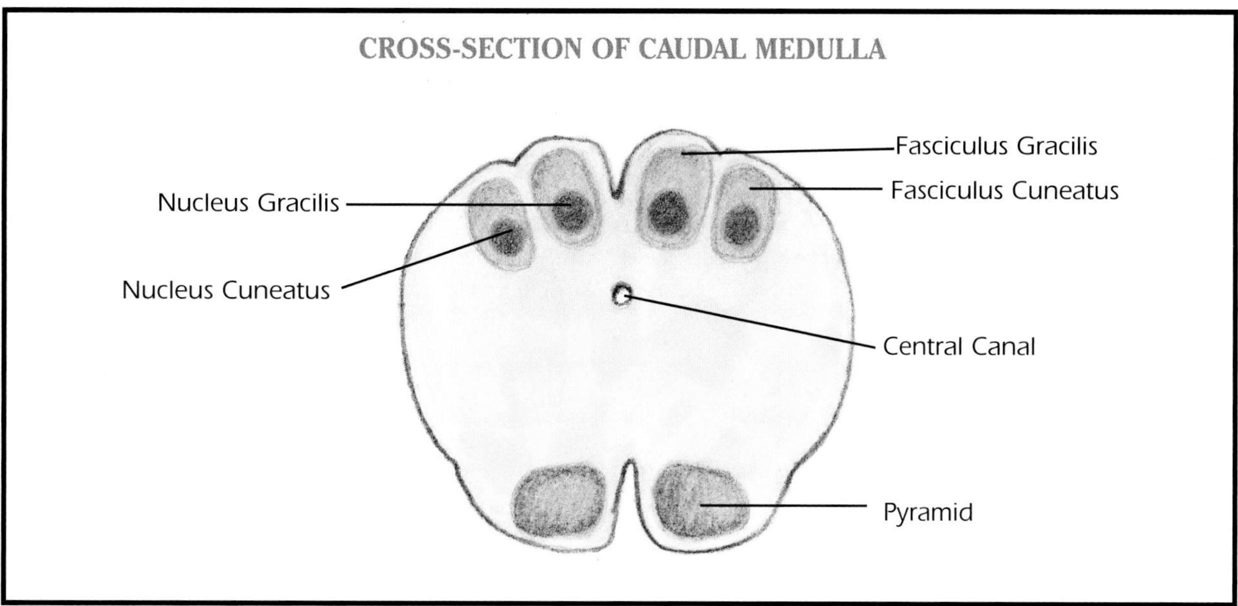

Reticular Activating System

- The RAS is the brainstem center that is involved in states of *wakefulness* and sets the general level of activation of the brain.
- It plays a role in alerting the cortex to attend to important sensory stimuli.
- The RAS is located in the *rostral midbrain.*

Reticular Inhibiting System

- The RIS is the brainstem center that is involved in states of *unconsciousness* such as sleep, stupor, or coma.
- It extends from the *caudal midbrain* to the *caudal medulla.*

Basal Ganglia

- The basal ganglia form an unconscious motor system that operates on a subcortical level.
- They specifically mediate *stereotypic or automatic motor patterns* such as those involved in walking, riding a bike, and writing.
- These are activities that are initially learned by using the cortex to think about how to perform such motor movements.
- Once learned, such motor patterns are stored subcortically. They become integrated by the basal ganglia.
- Recent research suggests that the basal ganglia also appear to work collaboratively in certain *cognitive* and *affective processes that require timing*—such as the ability to appropriately time verbal contributions in group situations and the ability to delay impulses. Some theorists suggest that pathology of the basal ganglia may be involved in attention deficit hyperactivity disorder (ADHD).
- The basal ganglia consists of three primary structures:
  - Caudate Nucleus
  - Putamen
  - Globus Pallidus
- Two other structures that are considered to be part of the basal nuclei are the *subthalamic nucleus* of the diencephalon and the *substantia nigra* of the midbrain.
- The basal ganglia form clusters of neurons within the cerebrum. They are the only ganglia located directly in the CNS.

Caudate Nucleus

- The caudate is an arch-shaped structure that follows the arc of the fornix and lateral ventricles.
- The caudate has a head, body, and tail. The tail attaches to the amygdala of the limbic system.
- It is involved in the planning and execution of a particular automatic movement pattern.
- It is also involved in the evaluation of that movement's appropriateness.
- The caudate has strong connections with the frontal lobe and interacts with it in *motor planning*.
- The caudate also plays a role in the *inhibitory control of movement*. It acts like a brake on certain motor activities.
- When the brake is not working—when the caudate is damaged—extraneous, purposeless movements appear. Examples include tics and tardive dyskinesias (tongue protrusions, facial grimacing, and lip smacking).
- Tardive means that the disorder occurred after chronic use of certain drugs that affect the caudate nucleus.

Putamen and Globus Pallidus

- The putamen and globus pallidus are located just lateral to the internal capsule.
- These are *excitatory structures*. If the caudate nucleus acts like a brake to inhibit movement, the putamen and globus pallidus serve as an activating mechanism.
- The basal ganglia's ability to achieve balance between inhibitory and excitatory movement is a key component in the precision and timed execution of specific behaviors and actions.

Subthalamic Nucleus

- The subthalamic nucleus is a *thalamic nuclei group* located caudal to the thalamus.
- It contains cells that use *dopamine*.
- It is a key structure *connecting feedback* and *feedforward circuits* of the *thalamus* and *basal ganglia*.

Substantia Nigra

- The substantia nigra is located in the *midbrain*.
- The red nucleus and the substantia nigra of the midbrain form the inner coat of the cerebral peduncles.
- The substantia nigra produces *dopamine*—a neurotransmitter that functions in movement and mood regulation.
- The axons of the substantia nigra form the *nigrostriatal pathway*. This pathway is referred to as a *dopaminergic* pathway supplying dopamine to the striatum.

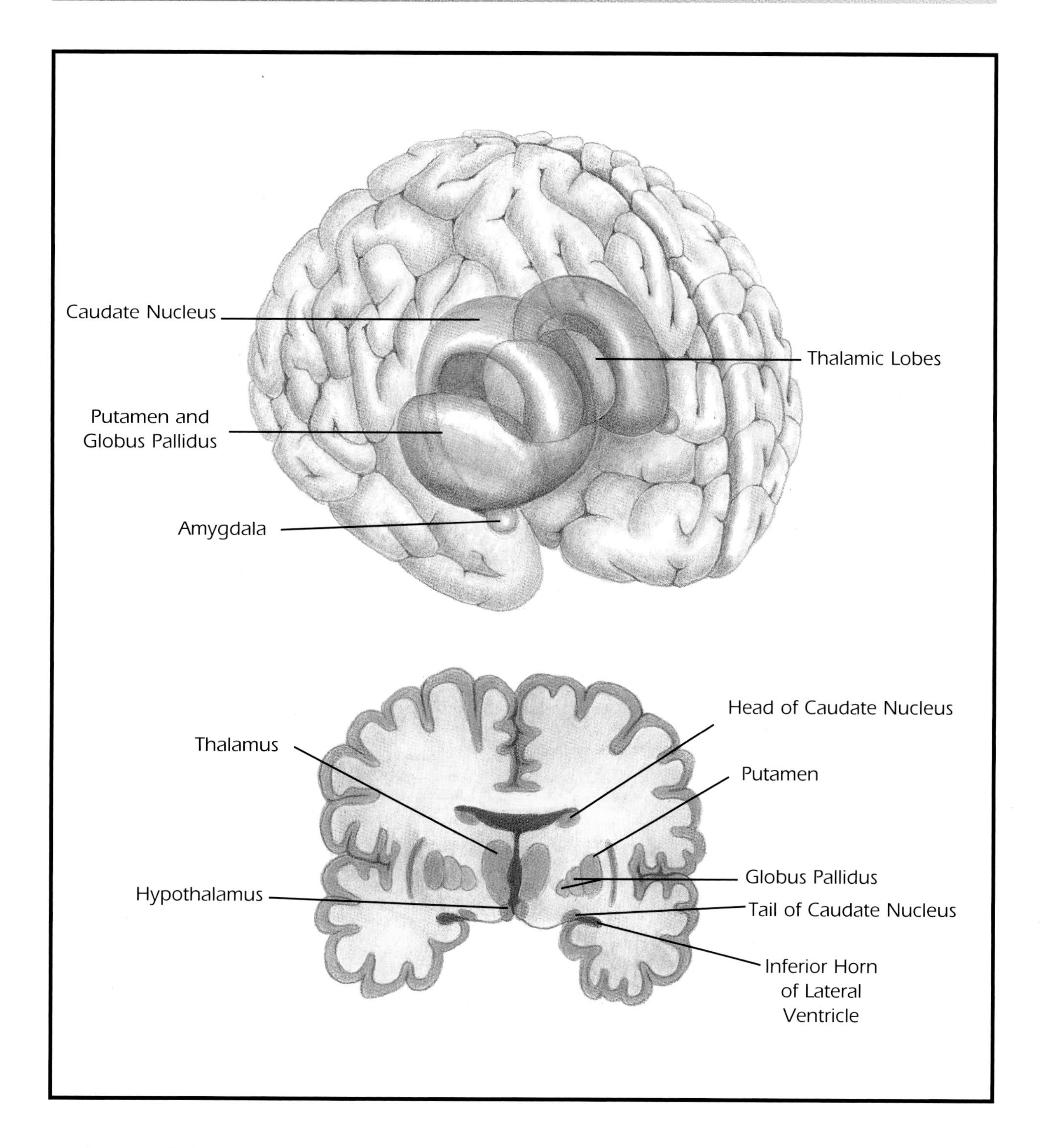

Pathology of the Basal Ganglia

- The basal ganglia have been implicated in disorders involving the inability to appropriately time one's social expression and participation. For example, some researchers have suggested that *ADHD*—which involves difficulty in timing one's verbalizations and movements, being able to delay urges and actions, and responding impulsively—may be related to pathology of the basal ganglia.
- Deterioration of the basal ganglia has been identified as a key component in *Parkinson's disease*. When *dopamine neurons degenerate*, automated movements that are controlled by the basal ganglia (such as walking or reacting to a sudden fall) become inhibited.

Extrapyramidal System

- The basal ganglia are considered to be an extrapyramidal system—or a *motor system* that does *not use the pyramids* to send motor messages to the skeletal muscles.

Corpus Striatum

- Corpus striatum is a collective name for the caudate, putamen, and globus pallidus.

Neostriatum

- Neostriatum is a collective name for the caudate and putamen.

Lenticular Nucleus

- Lenticular nucleus is a collective name for the globus pallidus and putamen.

Paleostriatum

- Paleostriatum is another name for the globus pallidus—because of its striped appearance.

Claustrum

- The claustrum is not considered to be part of the basal ganglia although it is a group of nuclei located just lateral to the extreme capsule and just medial to the insula.
- Its function is not well understood.

CEREBELLUM

- Cerebellum means "little brain," and in many ways, it is like a brain unto itself.
- The cerebellum has two hemispheres that are connected by a vermis—much like the corpus callosum connects the cerebral hemispheres.
- It has a cerebellar cortex with an outer coat of gray matter and an inner core of white matter.
- The cerebellum has three lobes.
- The traditional view of the cerebellum holds that it is a sensorimotor system that oversees *proprioception* (or the unconscious awareness of the body's position in space). In this view, the cerebellum is a sensory and motor system that receives sensory information from joint and muscle receptors concerning the body's position. The cerebellum uses this sensory information to make decisions about how to adjust the body for the coordinated, precision control of movement and balance. These decisions are made on an unconscious level—the messages never reach the cortex for conscious awareness.
- More recent research suggests that the cerebellum has a role in a wider range of functions including *attention shifting, practice-related learning, spatial organization*, and *memory*. The cerebellum appears to work collaboratively with cortically based cognitive functions to predict and prepare functional responses to environmental demands.
- An additional function of the cerebellum is the *regulation of speech*. The cerebellum is largely responsible for the timing and fluidity of speech.

External Structures of the Cerebellum

- The cerebellum has *two hemispheres*; each hemisphere has *three lobes*.

Flocculonodular Lobe

- The flocculonodular lobe is also called the *archicerebellum* (ie, ancient brain) because it used to be considered the oldest part of the cerebellum (in phylogenetic terms).
- It is believed to have developed in organisms without limbs.
- It plays a role in *trunk control, postural reflexes*, and *balance*.

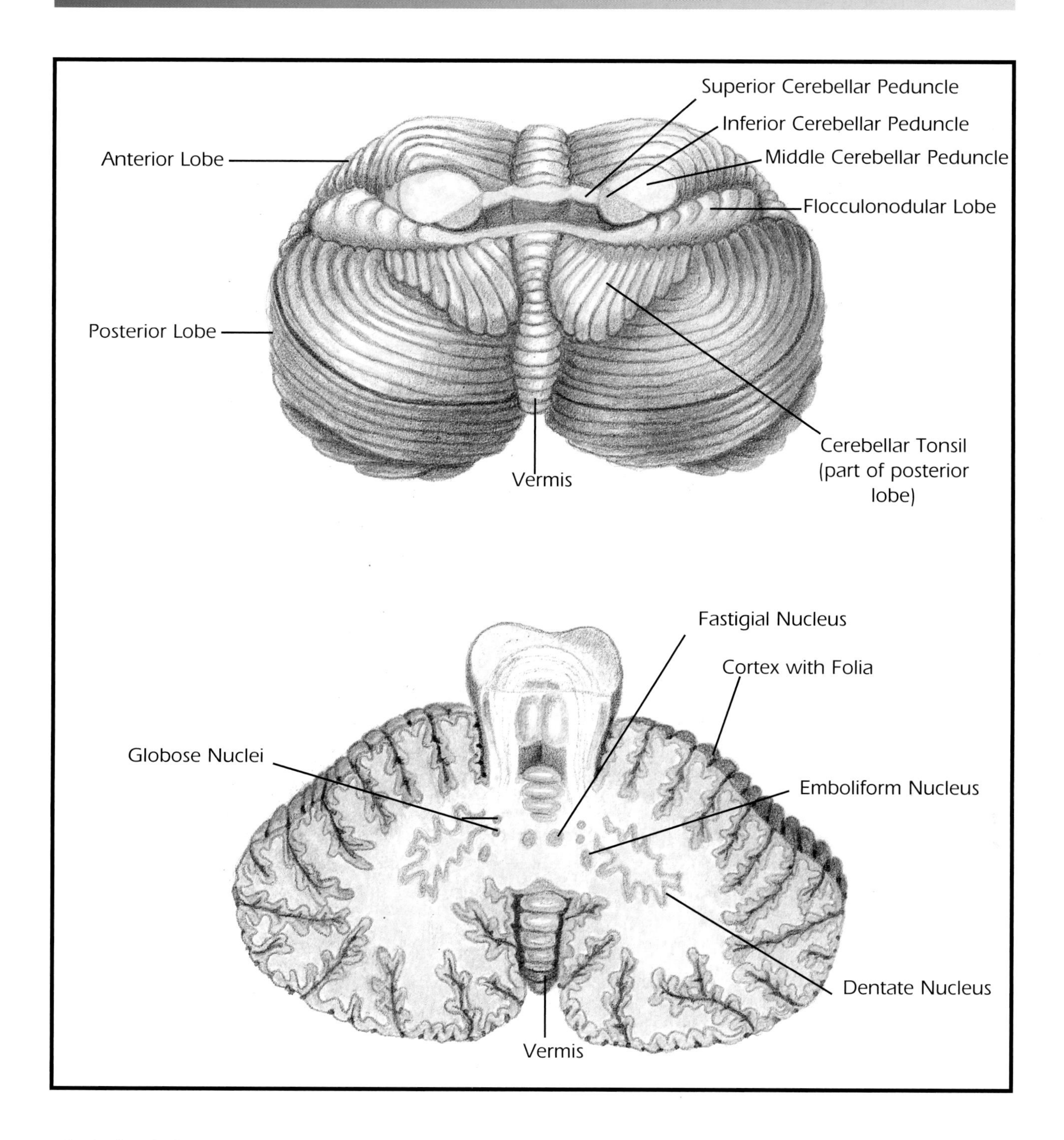

Anterior Lobe

- The anterior lobe is also called the *paleocerebellum* (ie, old brain).
- It is often large in size in organisms with limbs.
- It functions in *extremity control*, *postural adjustments*, and *stereotypic movement patterns*.

Posterior Lobe

- The posterior lobe is also called the *neocerebellum* (ie, new brain).
- It is commonly large in size in organisms having a cortex.
- It plays a role in *motor planning* (praxis) and the *precise timing* and *coordination of multiple muscle groups*.
- Some researchers have found evidence that the *coordination of cognitive functions* may take place in this lobe.

Lateral vs Intermediate Zones

- Functionally, the cerebellum can also be organized into *lateral* and *intermediate zones*.
- The *lateral zone* is believed to have a large role in *motor planning*.
- The *intermediate zone* is believed to function in *posture and trunk and limb movements*.

Three Primary Fissures

- *Primary fissure*—separates the anterior and posterior lobes
- *Posterolateral fissure*—divides the flocculonodular and posterior lobes
- *Horizontal fissure*—divides the posterior lobe in half

Cerebellar Peduncles

- The peduncles are composed of axons or fibers that travel between the cerebellum and the brainstem (superior and middle cerebellar peduncles travel through the pons; inferior cerebellar peduncle travels through the medulla).
- They are located in the anterior lobe of the cerebellum.

Vermis

- The vermis is a midline structure that has a role in the integration of information used by the right and left cerebellar hemispheres.
- Some have also suggested that the vermis has a role in *emotion* and the *timing* of *appropriate, affective responses*.

Internal Structures of the Cerebellum

Cortex with Folia

- The outer covering of the cerebellum is lined with gray matter that has many folds, called *folia*.

White and Gray Matter

- The interior of the cerebellum consists of a central mass of white matter surrounded by gray matter.
- There are also some smaller masses of gray matter within the central white core. These are the cerebellar nuclei.

Four Pairs of Nuclei

- There are four pairs of nuclei—one in each hemisphere:
  - Dentate nuclei (looks similar to the inferior olivary nuclei of the medulla)
  - Globose nuclei
  - Emboliform nuclei
  - Fastigial nuclei

Vermis

- The vermis can also be seen in the interior of the cerebellum.
- As noted above, evidence is growing that the vermis has a role in *emotional processing* and the *timing of appropriate affective responses*.

Pathology of the Cerebellum

- The most evident pathologies of the cerebellum take the form of *motor incoordination*, *decreased proprioception*, *ataxia*, and *dysarthric speech*.
- A growing body of research, however, suggests that damage to the cerebellum may also result in *impaired cognitive functions* and a *decreased ability to shift attention and respond appropriately to social situations*.
- Several studies have shown that the most consistent site of neural abnormality in people with *autism* is the cerebellum.
- Autopsy and neuroimaging reports have also indicated the presence of cerebellar abnormality in a variety of conditions including *ADHD*, *unipolar depression*, *bipolar disorder*, and *schizophrenia*.

Limbic System

- Phylogenetically, the limbic system is considered to be a very old part of the brain.
- It is located deep within the core of the brain.
- The limbic system appears to be the *source* of our *raw emotions* before they are modulated by the frontal lobes.
- The limbic system is also a storehouse for *long-term memories*—particularly memories that have a strong emotional component.

Cingulate Gyrus

- Cingulate gyrus is the most medial and deepest gyrus in the frontal and parietal lobes.
- It sits right above the corpus callosum.
- The cingulate has vast connections to the other structures within the brain's emotional limbic system.
- It plays a role in *decision-making* regarding which actions to take in response to sensory data.

Parahippocampal Gyrus

- Parahippocampal gyrus is the most medial and deepest gyrus in the temporal lobes.
- It folds back on itself at its anterior end to become the *uncus*.
- It relays information between the hippocampus and other cerebral areas—particularly the frontal lobes.
- It functions when we *compare a present event to an event stored in long-term memory* in order to decide how to handle a present situation.

Uncus

- The uncus is the bulb-like, anterior end of the parahippocampal gyrus.

Fornix

- The fornix bodies are a pair of arch-shaped fibers that begin in the uncus and wrap around to the mammillary bodies.
- The fornix is a relay system for messages generated by the limbic system.

Amygdala

- The amygdala is an almond-shaped group of nuclei located in each anterior temporal lobe.
- It has neural connections to the prefrontal cortex, occipitofrontal cortex, caudate nucleus, hippocampus, ventral tegmental area, nucleus accumbens, hypothalamus, thalamus, and anterior cingulate.
- It plays a primary role in the mediation of *fear*, *anger*, and *anxiety*.
- It also plays a role in the *perception of social cues* and the generation of *feelings of empathy*.
- The amygdala has an additional role in the memory formation of emotionally arousing events.

- The size of the amygdala has been found to be significantly smaller in some people with *autism* who have difficulty perceiving and interpreting social cues. A significant number of children with autism also have abnormally densely packed neurons in the amygdala, suggesting a deficiency of dendrites. The neuronal branches connecting neurons appear to be diminished.
- Patients with *amygdala damage or pathology* may not be able to interpret emotionally laden social cues. For example, they may not recognize the expression of fear on another's face or interpret someone's verbal tone as indicative of irritation.
- Researchers have found that in the normal human brain, a significantly greater number of neural pathways run from the amygdala to the cortex than from the cortex to the amygdala. This suggests that the amygdala may, in part, be responsible for the generation of anxiety and worry. It alerts the cortex that a condition of crisis exists.
- Because far fewer pathways run from the cortex to the amygdala, it is likely that the amygdala is better at generating anxiety than the cortex is at calming one's worries.
- The amygdala is also a critical component of the *brain reward system* (see Section 29)—a group of neuroanatomical structures, chemicals, and pathways that are responsible for the processes of addiction and relapse as well as the experience of pleasure and aversion.
- *Pathology* of the amygdala may be linked to *social phobia*, *autism*, and *addiction*.

Olfactory Bulb and Tract

- The olfactory bulb and tract is also known as cranial nerve 1.
- The olfactory tract connects to limbic system structures—it travels directly to the hippocampus.
- This connection accounts for the deep association between specific odors and long-term memories that hold emotional significance.

Hippocampus

- The hippocampus is located within the parahippocampal gyrus.
- In a coronal section, the hippocampus looks like a sea horse (in Latin, hippocampus means sea horse).
- It is one of the major storehouses in the brain for *long-term memory*—particularly for memories that are traumatic or emotionally laden.
- In psychoanalytic terms, the hippocampus has been proposed as the storehouse of *repressed memories* that are gated by the processes of the amygdala.
- Research has shown that *neurogenesis*—or the cell birth and maturation of neurons—occurs in the hippocampus throughout life. This may account for the clinical finding that long-term memory is often spared in brain trauma, while short-term memory commonly becomes impaired. Long-term memory is believed to be a function of the hippocampus, while short-term memory occurs largely in the frontal lobes. Neurogenesis in the hippocampus may help to maintain long-term memory despite brain damage that impairs other cognitive functions.
- There has also been evidence that patients with depression or post-traumatic stress disorder have reduced hippocampal volume and cell proliferation—in other words, a reduction in hippocampal neurogenesis. Animal models have shown that chronic stress reduces cell proliferation in the hippocampus. Researchers have found that several classes of *antidepressant drugs* actually enhance hippocampal neurogenesis—a finding that may, in part, account for antidepressant drug effectiveness.

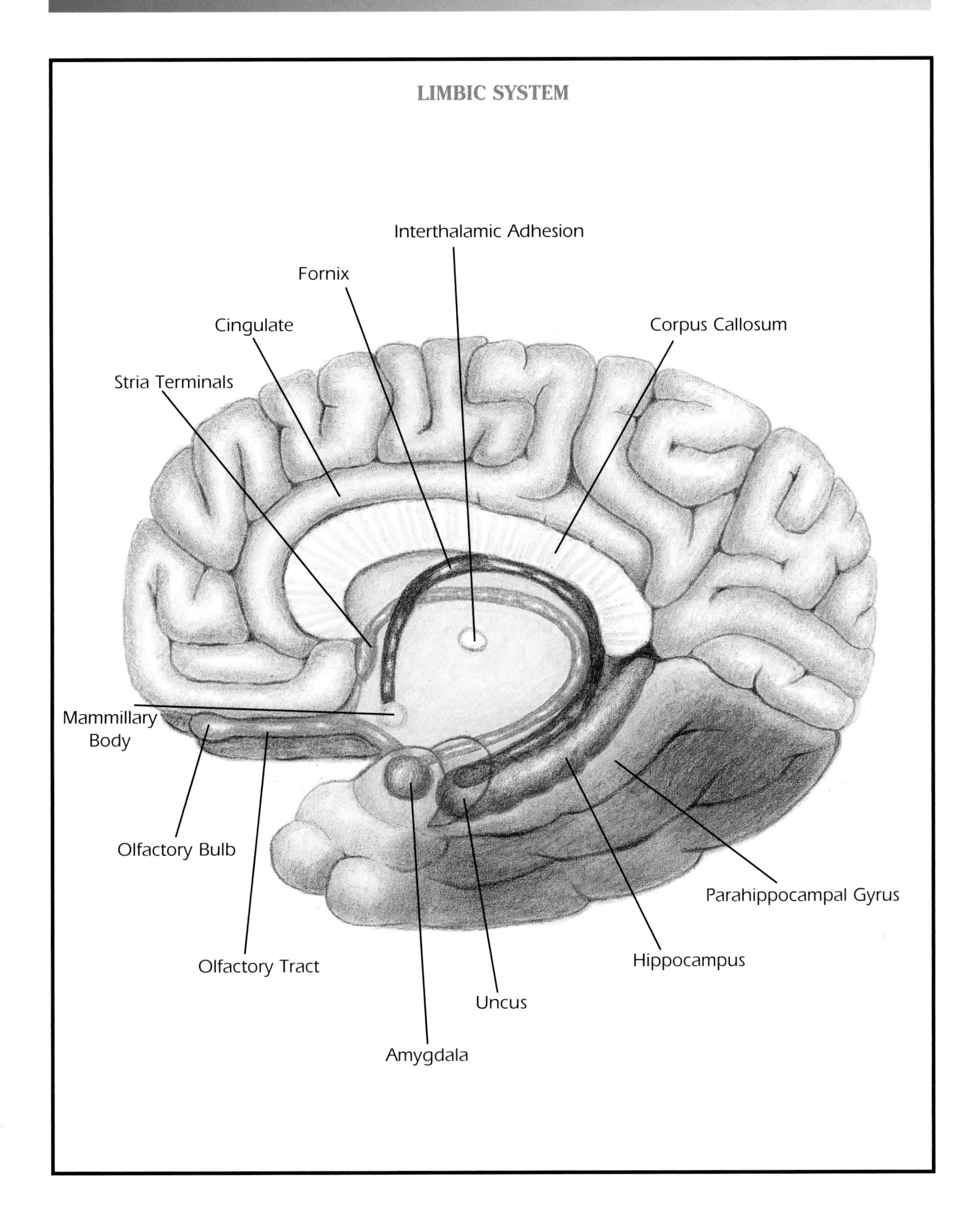
LIMBIC SYSTEM
Interthalamic Adhesion
Fornix
Cingulate
Corpus Callosum
Stria Terminals
Mammillary Body
Olfactory Bulb
Parahippocampal Gyrus
Olfactory Tract
Hippocampus
Uncus
Amygdala

SECTION 4

Ventricular System

Ventricular System

- In embryonic development, the brain begins as a *flat plate* that *fuses* into a *tube.*
- The *space within the tube* becomes the *ventricular system.*
- The *walls of the tube* become the brain; the *ventricles* are the *hollow spaces* in the brain that contain cerebrospinal fluid (CSF).

There Are Four Ventricles in the Brain

- One pair of lateral ventricles (one in each hemisphere)
- One third ventricle
- One fourth ventricle

Two Lateral Ventricles

- There is *one* lateral ventricle in each hemisphere.
- The lateral ventricles are divided by the *septum pellucidum*—a thin partition covering the medial wall of each lateral ventricle.
- Each lateral ventricle has three horns:
    - *Anterior horn*—projects into the frontal lobe
    - *Inferior horn*—projects into the temporal lobe
    - *Posterior horn*—projects into the occipital lobe

One Third Ventricle

- The third ventricle is surrounded by the diencephalon.
- The *thalamic lobes* form the *walls* of the third ventricle.
- The *hypothalamic lobes* form the *floor* of the third ventricle.

One Fourth Ventricle

- The fourth ventricle is located among the pons, rostral medulla, and the cerebellum.

Choroid Plexus

- The choroid plexus are the *vascular structures* in the brain that protrude into the ventricles and *produce* the CSF.
- All of the ventricles contain choroid plexus, but the lateral ventricles contain the most.

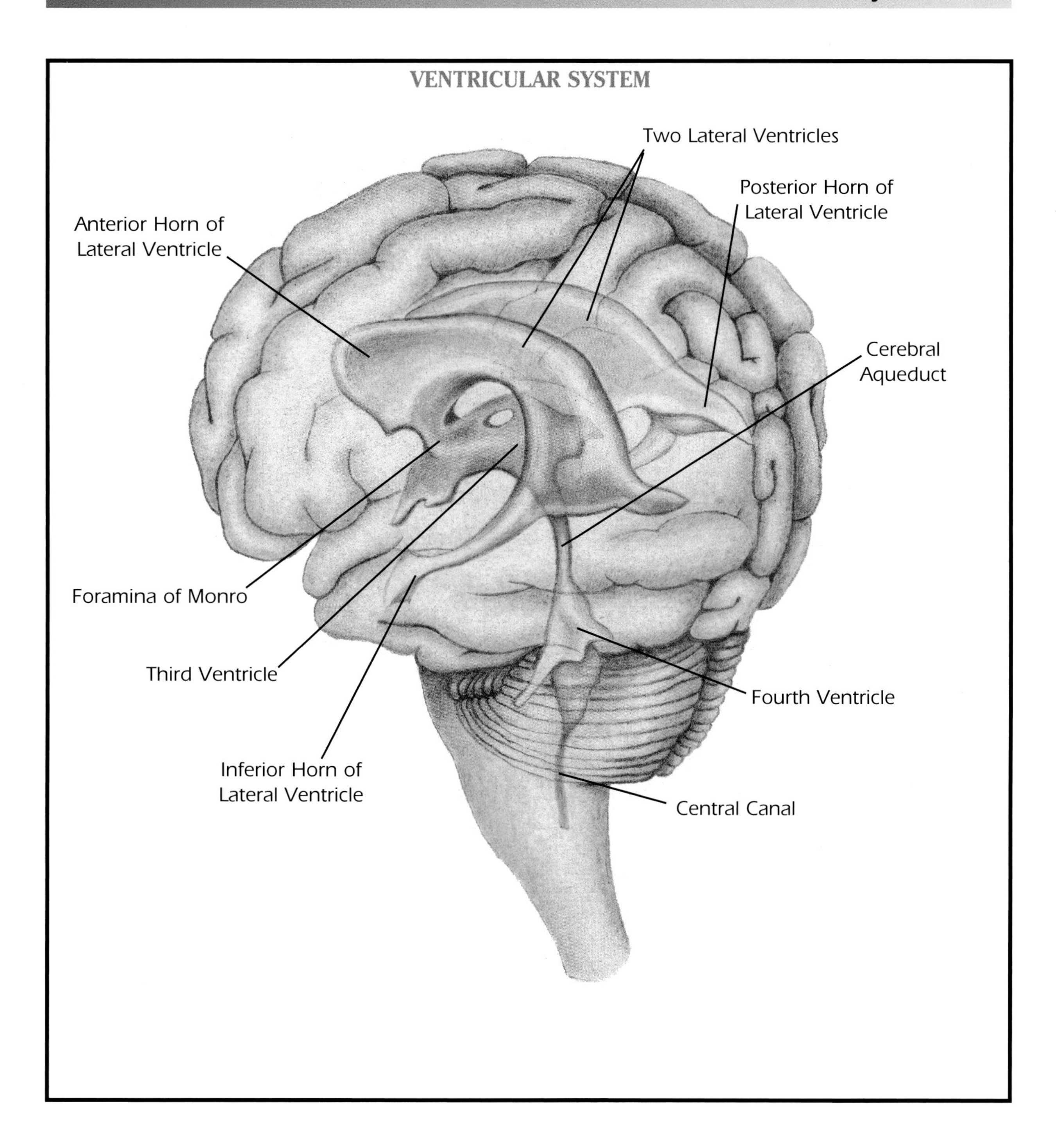
VENTRICULAR SYSTEM
Two Lateral Ventricles
Posterior Horn of Lateral Ventricle
Anterior Horn of Lateral Ventricle
Cerebral Aqueduct
Foramina of Monro
Third Ventricle
Fourth Ventricle
Inferior Horn of Lateral Ventricle
Central Canal

Cerebral Aqueduct (also called Aqueduct of Sylvius)

- The cerebral aqueduct is a narrow channel that descends through the midbrain.
- It *connects* the *third* and *fourth ventricles.*
- It is a common site of blockage.

Central Canal

- The central canal begins in the caudal medulla and descends all the way down the spinal cord.
- It contains CSF.
- The central canal *connects* the *ventricular system* with the *spinal cord.*

Foramina of Monro

- There are *two* foramina of Monro—one in each hemisphere.
- They are small channels that *connect* the *lateral ventricles* with the *third ventricle.*

Foramen of Magendie (also called Median Aperture)

- There is only *one* foramen of Magendie. This is an *opening* in the *fourth ventricle* (in the rostral medulla).
- The foramen of Magendie opens to the *subarachnoid space* below the cerebellum. This space is located above the brain and beneath the skull.
- The *subarachnoid space* is the space between the *arachnoid membrane* and the *pia matter.*
- This is also a potential site of CSF blockage.

Foramina of Luschka (also called Lateral Apertures)

- There are *two* foramina of Luschka. These are *openings* in the *fourth ventricle* (in the pons).
- These open to the *subarachnoid space.*
- The formina are potential sites of blockage.

CEREBROSPINAL FLUID

- The CSF is a clear, colorless fluid that bathes and nourishes the brain and spinal cord.

Arachnoid Villi

- The CSF is *reabsorbed* in the *arachnoid villi* and returns to *blood circulation* through the venous sinuses.
- The arachnoid villi are *projections* of the *arachnoid matter* into the *dura matter.*

Cerebrospinal Fluid Pressure

- CSF maintains a *constant circulatory pressure*—unless a problem occurs.
- The formation of CSF is independent of the pressure.
- This is important with regard to *hydrocephalus.*
- Even if the CSF pressure increases, the CSF continues to be produced.
- There is *no* neurologic mechanism that detects too much CSF.

Composition of Cerebrospinal Fluid

- The *composition* of CSF is used for *diagnostic purposes* to identify disease processes.
- Physicians examine the rate of pressure and the composition.
- Example: *Spinal tap* (or a lumbar puncture) is a procedure in which the spinal cavity is punctured with a needle to extract CSF for diagnostic purposes.

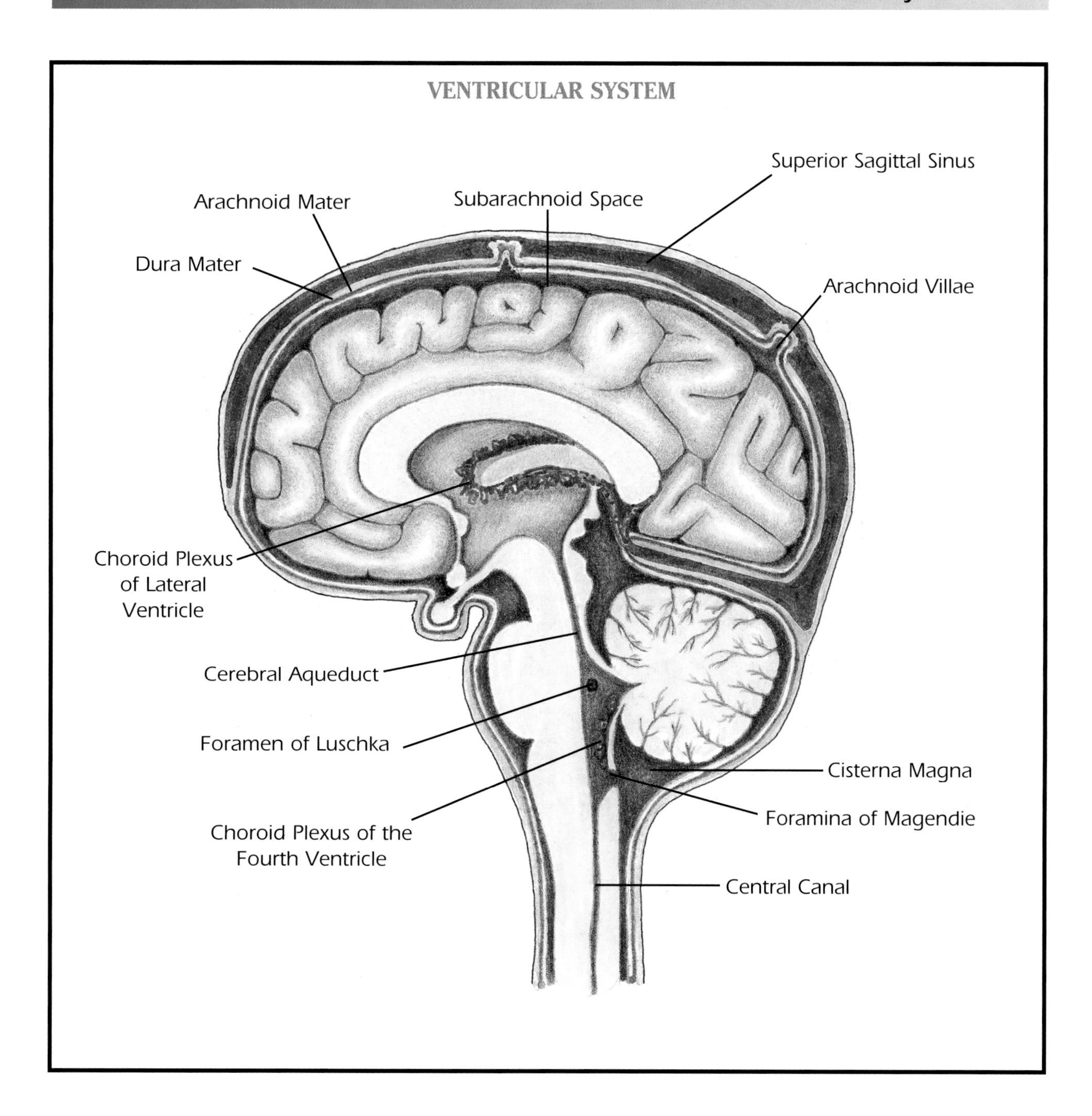
VENTRICULAR SYSTEM
Superior Sagittal Sinus
Arachnoid Mater
Subarachnoid Space
Dura Mater
Arachnoid Villae
Choroid Plexus of Lateral Ventricle
Cerebral Aqueduct
Foramen of Luschka
Cisterna Magna
Foramina of Magendie
Choroid Plexus of the Fourth Ventricle
Central Canal

Flow of Cerebrospinal Fluid

- The CSF is produced in the choroid plexus; travels through the ventricles, subarachnoid space, and spinal cord; and then returns to the circulatory system.

Function of the Cerebrospinal Fluid

- A function of CSF is protection of the brain—the fluid acts like a shock absorber.
- Another function is the exchange of nutrients and waste—CSF plays a role in the transfer of substances between the blood and the nervous tissue.
- The CSF helps in diagnosis—the CSF is examined for its rate of pressure and its fluid composition.
- It also transports hormones—the CSF has a role in the transport of some hormones throughout the CNS.

HYDROCEPHALUS

- Hydrocephalus is a *buildup of pressure* and *fluid* that results in *compression* of *neural tissue* and *enlargement* of the *ventricles*.
- It can occur in both infants and adults.

Noncommunicating Hydrocephalus

- Noncommunicating hydrocephalus occurs when *blockage* in the ventricular system prevents the CSF from reaching the arachnoid villi for reabsorption.

Communication Hydrocephalus

- Results from *impaired reabsorption* of CSF that does not occur from blockage.

Congenital Hydrocephalus in Infants

Etiology

- *Blockage* (particularly in the foramina of Luschka and Magendie)
- *Excessive production of CSF for unknown reasons*
- *Meningitis*—causing adhesions and resultant blockages in the subarachnoid space
- *Tumors of the choroid plexus*—causing excessive CSF production
- *Hemorrhage or inflammation*—the ependyma (or the lining of the ventricles) is especially sensitive to viral infections during embryonic development.

Common Sites of Blockage

- Cerebral aqueduct
- Foramen of Luschka
- Foramen of Magendie

Pathological Effects

- Infant's skull expands to accommodate the increased fluid. Cranial sutures separate.
- Head expansion and bulging of the fontanels
- Compression of neural tissue
- Because the skull can expand, increased intracranial pressure is usually not present; intelligence is often spared.

Treatment

- If the hydrocephalus was caused by a blockage, then treatment requires a *shunt*, or tube, that *bypasses* the *blockage*.
- If there is excessive production of CSF, treatment requires a *shunt* usually placed from the *fourth ventricle* to the *abdomen* to drain the excess CSF.
- If diagnosed, hydrocephalus can be successfully treated in utero.

Adult Onset Hydrocephalus

Etiology

- *Tumors*
- *Meningitis*
- *Hemorrhage and inflammation*
- *Unknown causes*

Pathological Effects

- Enlarged ventricles and rapid atrophy of neural tissue—there is no place for the fluid to go.
- Increased intracranial pressure
- Headache and vomiting
- Cognitive deterioration
- This condition is life-threatening unless treatment occurs quickly.

Treatment

- Treatment for an adult usually involves *surgical shunt placement* (to the abdomen) to drain the excess fluid.
- In *noncommunicating hydrocephalus*, surgical shunt placement is attempted to bypass the blockage.
- In *communicating hydrocephalus*, attempts to clear the arachnoid villi of exudate are made first. If this is unsuccessful, surgical shunting is often indicated.

Normal Pressure Hydrocephalus

- A form of hydrocephalus that develops in adulthood—usually in the fifth, sixth, or seventh decade.
- This occurs because the arachnoid villi cannot absorb the CSF. The CSF pressure remains normal.
- It is characterized by the following:
  - Unsteady gait
  - Progressive dementia
  - Urinary incontinence
- Treatment usually involves surgical shunt placement.

SECTION 5

The Cranium

Skull

- The skull is the bony framework of the head.
- It supports, anchors, and protects the brain.
- It is composed of *14 bones of the face*, *28 adult teeth*, and 8 *cranial bones*.

Cranium

- The cranium is the portion of the skull that *encloses the brain*.
- It consists of *eight separate fused bones*: frontal bone, occipital bone, sphenoid bone, ethmoid bone, two temporal bones, and two parietal bones.

Suture Lines

- The suture lines are junctions between the skull bones. These are areas where the bones have fused.
- The cranial sutures begin to fuse at 2 months and are complete at 18 months.
- There are three suture lines: *coronal*, *sagittal*, and *lambdoid*.

Coronal Suture

- Runs along the *coronal plane*
- Connects the *frontal bone* with the *parietal bones*

Sagittal Suture

- Runs along the *midsagittal plane*
- Connects the *two parietal bones*

Lambdoid Suture

- Connects the *two parietal bones* to the *occipital bone*

Fontanels

- The fontanels are *non-ossified spaces*, or soft spots, located between the cranial bones of a fetus and newborn.
- These allow the skull to expand to accommodate the growing brain.
    - Anterior fontanel
    - Posterior fontanel
    - Sphenoid fontanel
    - Mastoid fontanel

Floor of the Cranial Cavity

- The *undersurface of the brain* sits within the floor of the cranial cavity.
- The cranial cavity holds the anterior–inferior aspect of the frontal lobes, the inferior aspect of the temporal lobes, and the inferior aspect of the cerebellum.

Fossa

- The undersurface of the brain sits in three cranial sections or fossae:
    - Anterior cranial fossa—primarily supports the *frontal lobes*.
    - Middle cranial fossa—supports the *anterior–inferior temporal lobes* and the *diencephalon*.
    - Posterior cranial fossa—supports the *cerebellum*.

Sharp Edges of the Fossa

- The fossae are problematic in *brain injuries* caused from motor vehicle accidents (MVA).
- Sharp edges of each fossa can *shear brain tissue* and *vessels* in MVAs.

Foramina

- The foramina are *openings* in the skull for the passage of *blood vessels* and *nerves*.

Foramen Magnum

- The foramen magnum is the *largest foramen* in the skull.
- It is the opening through which the *brainstem connects with the spinal cord*.
- The foramen magnum is located in the *occipital bone*.

Function of the Cranial Bones

- The cranial bones provide protection for the brain.

Skull Fractures

- One of the weakest cranial sutures is at the *pterion*—a site that joins the frontal, parietal, temporal, and sphenoid bones.
- The pterion *often fractures upon strong impact* causing penetration of bone fragments to enter the brain; cerebral arteries are easily ruptured as a result.
- Frontal and parietal bone fractures are often seen as a result of MVAs.

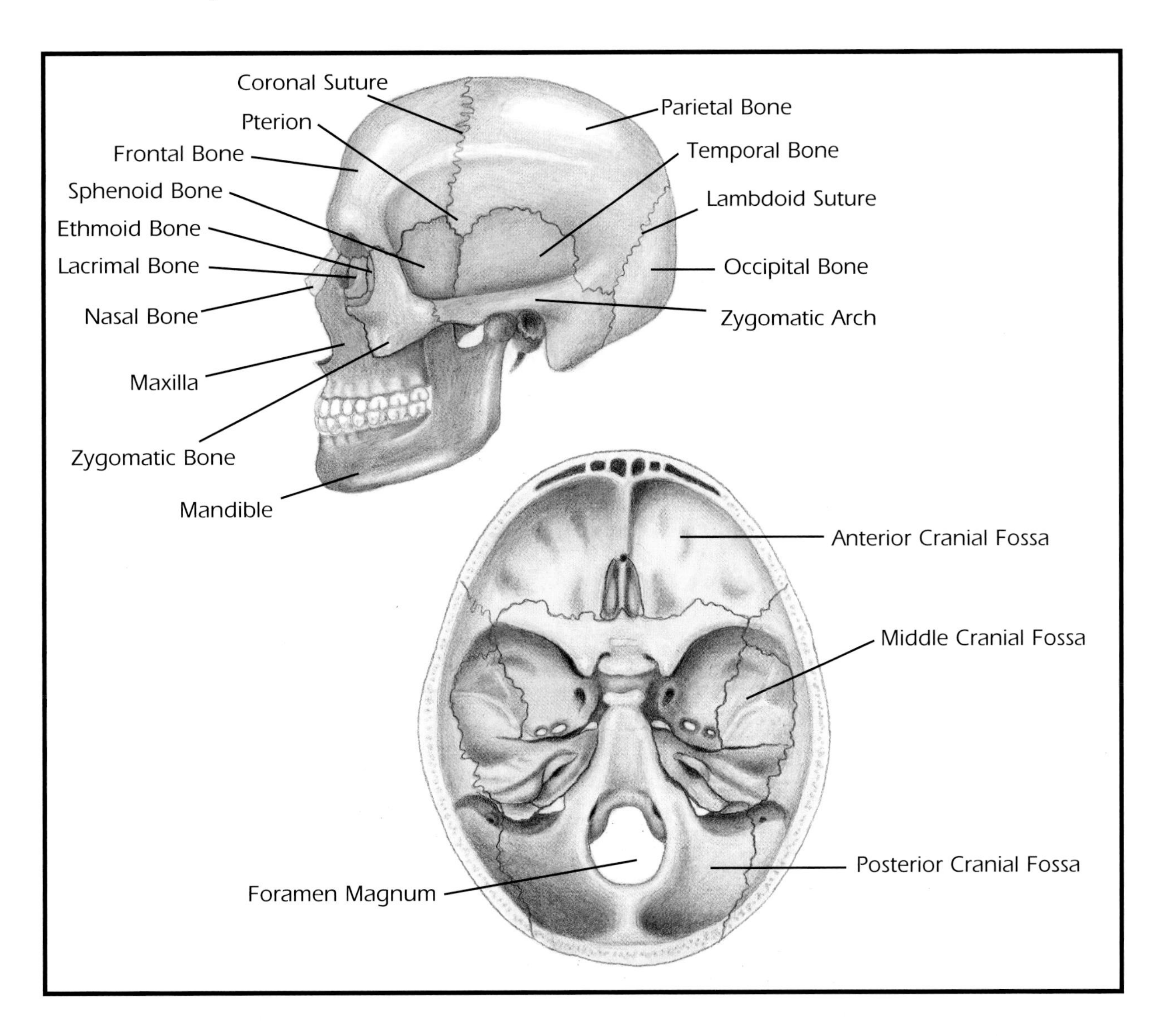

SECTION 6

The Meninges

Location

- The *meninges* are located between the *skull* and *brain*, and they *cover* the *spinal cord*. They form a seal around the central nervous system (CNS).
- There are three layers of meninges: *dura mater*, *arachnoid mater*, and *pia mater*.
- Following are the layers of structures as they are positioned between the skull and brain:
  - Skull
  - Epidural Space
  - Dura Mater
  - Subdural Space
  - Arachnoid Mater
  - Subarachnoid Space
  - Pia Mater
  - Brain

Dura Mater

- The dura is the *outermost meningeal layer*.
- It is a very *tough* and *thick membrane* that is *attached* to the *inner surface* of the *cranium*.
- The dura has two projections that extend into the brain:
  - **Falx Cerebri**—extends into the *medial longitudinal fissure*
  - **Tentorium**—the horizontal shelf of dura that sits between the *occipital lobe* and the *cerebellum*
- The falx cerebri and tentorium decrease linear and rotary forces on the brain.

Dural Sinuses

- The dural sinuses *function as large veins*.
- They are located above the frontal and parietal lobes.
- The sinuses function as a circulatory system.

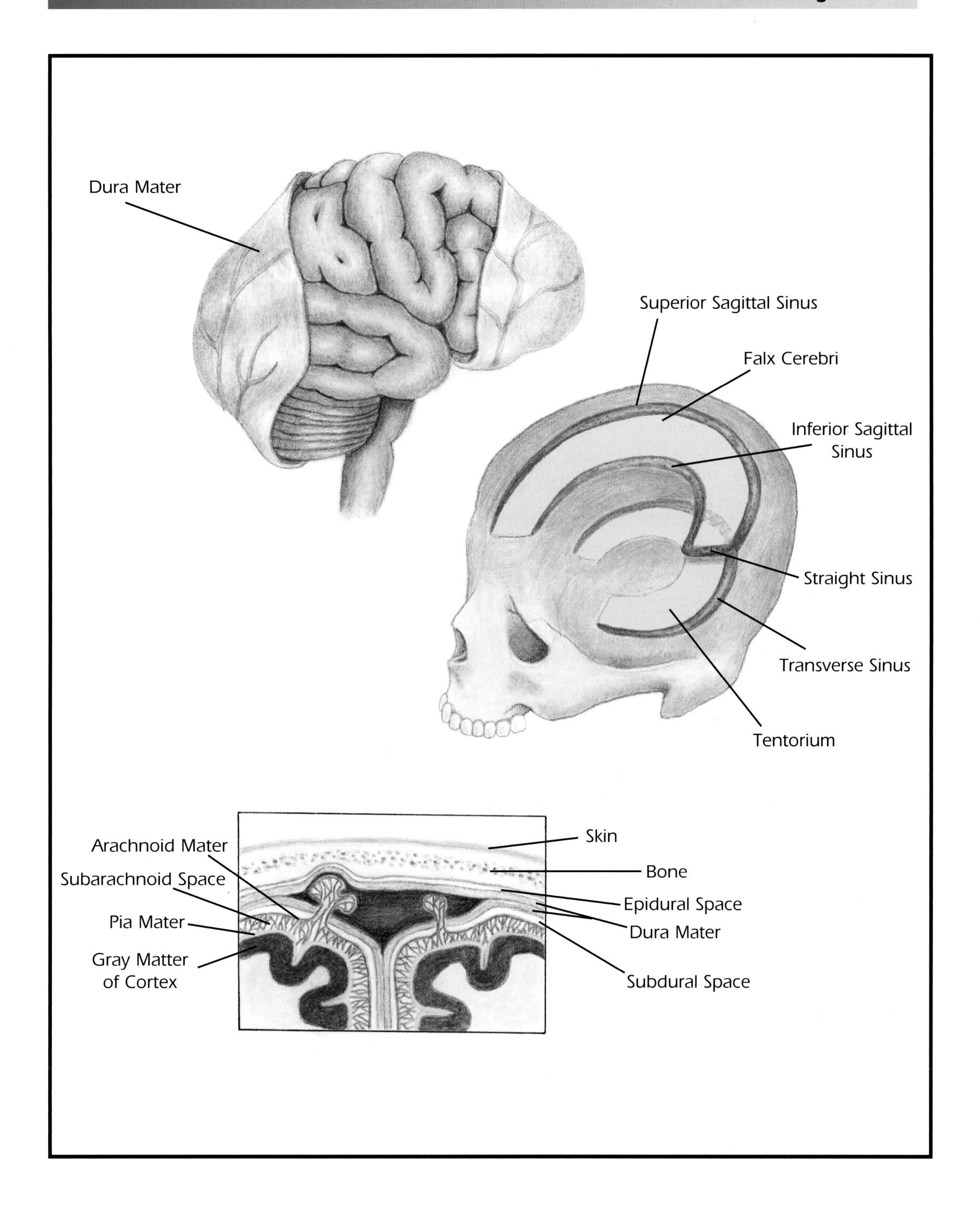
Dura Mater
Superior Sagittal Sinus
Falx Cerebri
Inferior Sagittal Sinus
Straight Sinus
Transverse Sinus
Tentorium
Arachnoid Mater
Subarachnoid Space
Pia Mater
Gray Matter of Cortex
Skin
Bone
Epidural Space
Dura Mater
Subdural Space

- Cerebral veins empty into the sinuses.
- The cerebral veins also receive cerebrospinal fluid (CSF) from the subarachnoid space via the arachnoid villi.
- These fluids are then returned to their general circulatory systems.

Blood Supply of the Dura

- The blood supply of the dura comes from the *middle meningeal artery*.
- This artery often ruptures during head injury resulting in hemorrhages in the subdural space—or *subdural hematomas*.
- Subdural hematomas can also result from a *cerebrovascular accident*.
- Hemorrhages in the subdural space cause *increased cranial pressure* that *compresses neural tissue*. This is fatal if not treated promptly.

Dural Neuronal Innervation

- The dura has *neuronal innervation*—it is innervated by the nervous system and can experience physical sensation.
- The *brain is unable to detect pain*, having no pain receptors of its own.
- Some *headaches* are caused by the *constriction of meningeal membranes*.

Arachnoid Mater

- The arachnoid mater is the *middle meningeal layer*.
- It is located just below the *subdural space*.
- The arachnoid looks like a spider web (arachnoid means spider).
- It protects the brain and *acts as a seal around the* CNS.

Subarachnoid Space

- Beneath the arachnoid is the *subarachnoid space*, and it *holds the* CSF.

Cisterns

- The cisterns are openings or large spaces in the subarachnoid space.

Cisterna Magna (also called the Cerebellar Medullary Cistern)

- This is the *largest subarachnoid cistern.*
- It is located between the cerebellum and the medulla.
- It is often used as a *shunt placement.*

Pia Mater

- The pia is the *deepest meningeal layer.*
- It is located right on the gyri and sulci of the brain and on the spinal cord.

Blood-Brain Barrier

- The blood-brain barrier consists of the *meninges*, the *protective glial cells*, and the *capillary beds* of the brain.
- It is responsible for the *exchange of nutrients* between the CNS and the vascular system.
- The barrier also acts as a wall that *controls which molecules* in the bloodstream will be able to *enter the CNS.*
- Oxygen, sugars, and amino acids are allowed entrance while most other compounds are not.
- This protective mechanism ensures that the brain will not be exposed to toxins that could impair its function. However, because of the efficiency of the blood-brain barrier, approximately 98% of therapeutic drugs cannot access the brain. Pharmaceuticals cannot presently reach brain regions that are implicated in such diseases as meningitis, rabies, tumors, Alzheimer's, and multiple sclerosis.

SECTION 7

Spinal Cord Anatomy

SPINAL CORD ANATOMY

Boundaries of the Spinal Cord

- Boundaries of the spinal cord (SC) extend from the *foramen magnum* to the *conus medullaris*.
- The *conus medullaris* is the *end* of the SC at the *L1 – L2 vertebral area.*
- The SC then becomes the *cauda equina* (means "horse's tail").
- The cauda equina are *spinal nerves* that have *not yet exited* the *vertebral column*.

Enlargements of the Spinal Cord

- The SC has an hourglass shape. *Enlargements* occur in the *cervical* and *lumbar* sections.
- The cervical enlargement is due to the *brachial plexus*—a network of spinal nerves from C5 – T1 that extend from the cervical vertebrae to the upper extremities.
  - When the spinal nerves of the brachial plexus enter the SC (and synapse with SC tracts), they account for the large area of white matter in the cervical SC.
- The lumbar enlargement is due to the *lumbar plexus*—a network of spinal nerves from L1 – S3 that extend from the lumbar vertebrae to the lower extremities.
  - When the spinal nerves of the lumbar plexus enter the SC (and synapse with SC tracts), they account for the large area of white matter in the lumbar SC.

Anterior Median Fissure

- The anterior median fissure continues from the anterior aspect of the medulla to the end of the SC.

Dorsal Median Sulcus

- The dorsal median sulcus continues from the posterior aspect of the medulla to the end of the SC.

Dorsal Intermediate Sulcus

- The dorsal intermediate sulcus continues from the posterior aspect of the medulla and extends only throughout the thoracic levels of the SC.
- This sulcus separates two ascending sensory pathways—the fasciculus gracilis and cuneatus of the dorsal columns.

Central Canal

- The central canal contains cerebrospinal fluid (CSF).

SPINAL CORD ANATOMY: PNS VS CNS

Spinal Nerves

- The spinal nerves are located in the peripheral nervous system (PNS).
- The spinal nerves consist of (a) ascending sensory pathways and (b) descending motor pathways.
- *Ascending sensory spinal nerves* extend from a sensory receptor to the dorsal rootlets.
- *Descending motor spinal nerves* extend from the ventral horn of the SC to skeletal muscles.
- There are 31 pairs of spinal nerves: 8 cervical, 12 thoracic, 5 lumbar, 5 sacral, and 1 coccygeal.

Dorsal Root Ganglion

- Dorsal root ganglion contains the *cell bodies* of *sensory nerves* that are part of the somatic PNS.
- Each sensory nerve has its own *dorsal root ganglion*.
- The dorsal root emerges from the dorsal ganglia.

Dorsal Root and Rootlets

- The dorsal roots are ascending spinal nerves that *carry sensory data* from the sensory receptors (in the PNS) to the dorsal horn of the SC.
- Dorsal roots are *axon bundles* that emerge from a spinal nerve.
- The dorsal root leads into the dorsal rootlets—which are thin string-like axons that emerge from the dorsal root and synapse in the dorsal horn of the SC.
- The dorsal root and rootlets are considered to be *within the PNS*.

Dorsal Horn

- The dorsal horn is considered to be *part of the central nervous system (CNS)*.
- The dorsal horn contains the *cell bodies* of many of the *sensory SC tracts*.
- In the dorsal horn, the dorsal rootlets (of the PNS) may synapse on *interneurons*. These interneurons then synapse with SC tracts, or the rootlets may synapse directly on the cell bodies of the SC tracts.
- When the spinal nerves have synapsed with a SC tract, the SC tract ascends through the SC and brainstem and travels to the cortex.

Ventral Horn, Root, and Rootlets

- *Descending motor SC tracts* travel from the cerebrum down through the brainstem and SC.
- Motor SC tracts synapse with *interneurons* in the *ventral horn*.
- These interneurons then synapse with *motor spinal nerves* and exit the ventral horn through the ventral rootlets.
- The *ventral rootlets* merge into the ventral roots and extend to skeletal muscles.
- The ventral horn, rootlets, and root are all considered to be *within the PNS*.

SPINAL NERVES AND THE DERMATOMES

Dermatome Distribution

- A dermatome is a *skin segment* that receives its *innervation* from a *specific spinal nerve*.

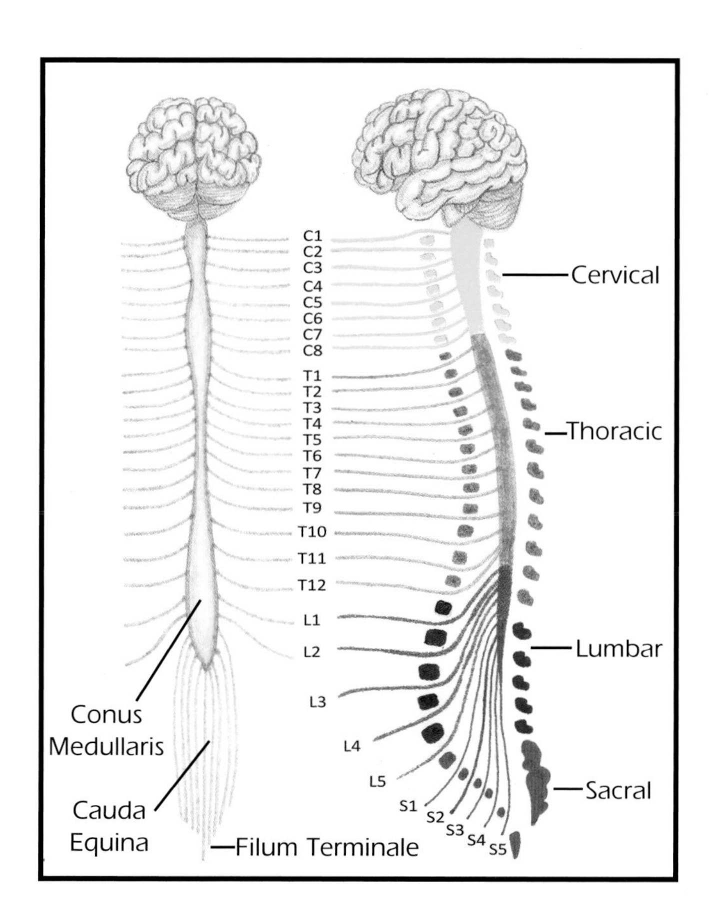

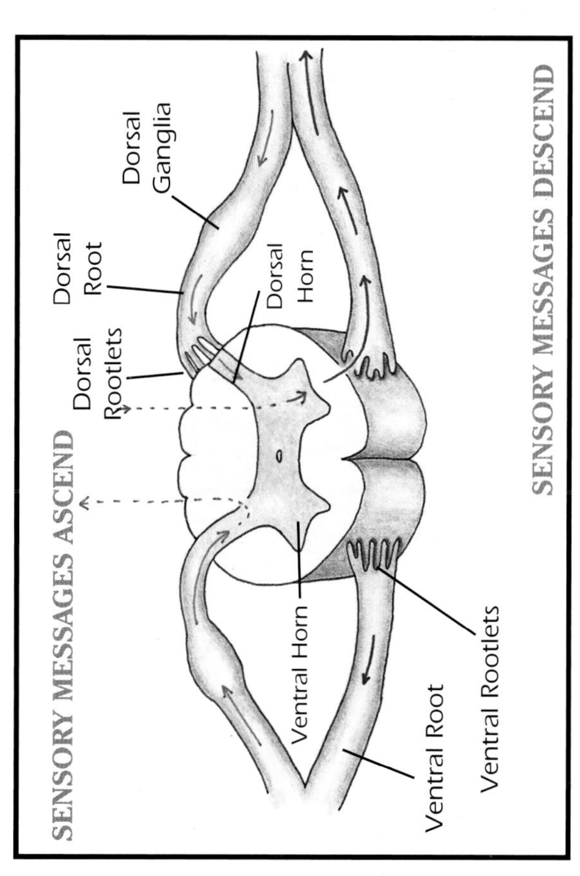

Referred Pain

- Referred pain occurs when a *specific body region* shares its *spinal nerve innervation* with a *separate dermatomal skin segment.*
- The pain experienced by the body part is misinterpreted by the cortex as pain coming from a separate dermatomal skin segment.
- Example: *Referred pain in heart attack.*
  - The spinal nerves that innervate the *heart* share *interneurons* that innervate the *left arm* (T1).
  - When the heart experiences pain, the *cortex misinterprets the origin of the pain* as coming from the *medial aspect* of the *left arm.*
  - This cortical misinterpretation occurs because the cortex does not have prior experience interpreting pain from the heart. The cortex relies on past experience when interpreting pain from a visceral source. Because it is uncommon for pain sensations to originate in the viscera, the cortex initially interprets the pain as coming from the left arm (T1 dermatome).
  - As the pain increases, the cortex is able to correctly identify the source of the pain as the heart.

Clinical Use of the Dermatomal Distribution

- When a therapist performs a sensory evaluation and determines that a specific body region does not register sensation, the therapist is then able to identify the lesion level.
- For example, if a patient cannot perceive sensation on the dorsal forearm, the therapist is able to determine that there is some impairment at C6 level.

Transcutaneous Electrical Nerve Stimulation Unit

- The use of transcutaneous electrical nerve stimulation (TENS) is based on the dermatomal distribution.
- The therapist places the TENS unit on the identified dermatome region to stimulate nerve regeneration or to reduce pain in a peripheral nerve injury.

Spinal Nerves and the Vertebral Column

Relationship of the Spinal Cord to the Vertebral Column

Ontogenetic Development

- The SC is the same length as the vertebral column in utero.
- However, the vertebral column continues to grow after birth. The SC does *not* continue to grow.
- The *adult SC ends* at the *L1 – L2 vertebral region.*

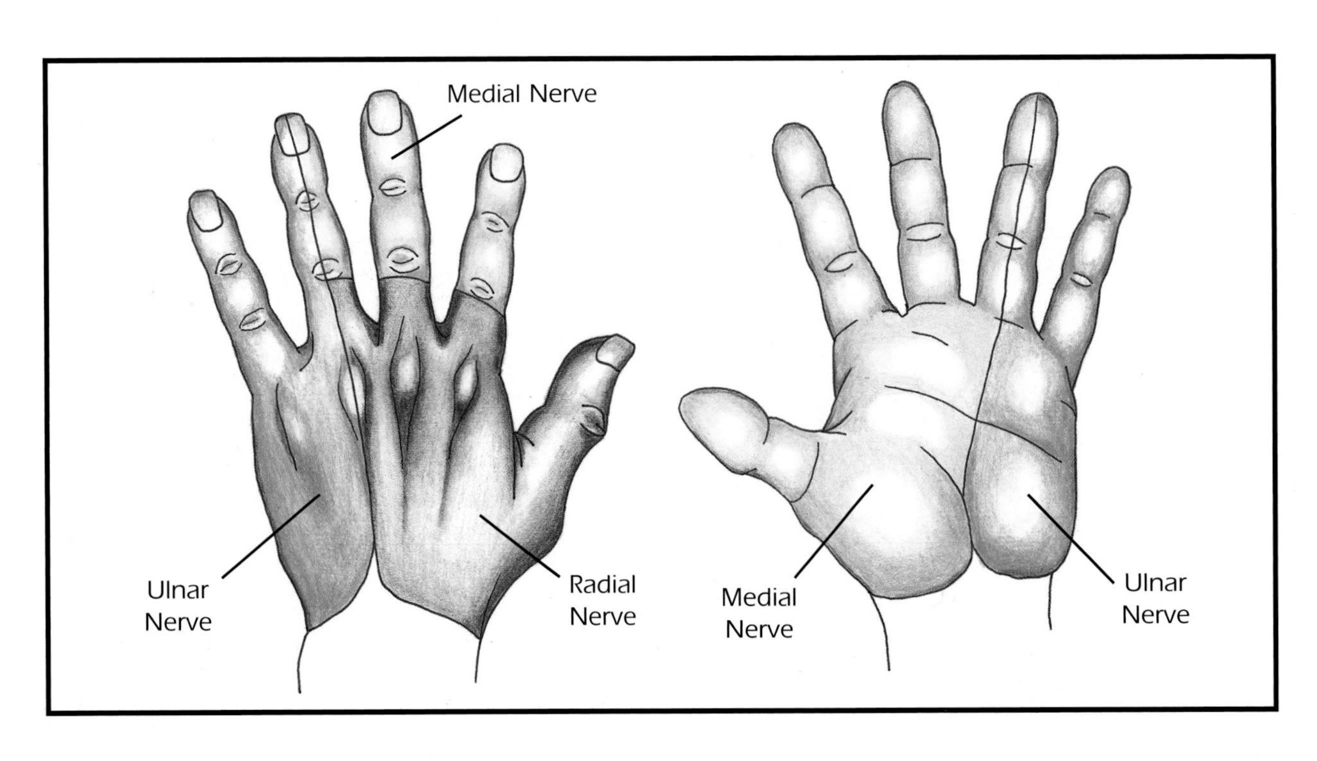

- The remainder of the spinal nerves (the cauda equina) must descend through the vertebral column to exit their intervertebral foramina.

There Is One More Pair of Spinal Nerves Than There Are Vertebrae

- From C1 – C7 the spinal nerves exit *above* their corresponding vertebrae.
- There is a *pair of C8 spinal nerves* but *no C8 vertebra.*
- This means that the C8 *spinal nerve* must exit *below C7 vertebra* and *above T1 vertebra.*
- *T1 spinal nerve* exits *below T1 vertebra* and *above T2 vertebra.*
- From C8 down, the spinal nerves exit below their corresponding vertebrae.

Intervertebral Discs

- *Nucleus pulposus* is a soft, pulpy, highly elastic tissue in the center of the intervertebral disc.
- *Annulus fibrosus* is the more fibrous outer covering of the disc.

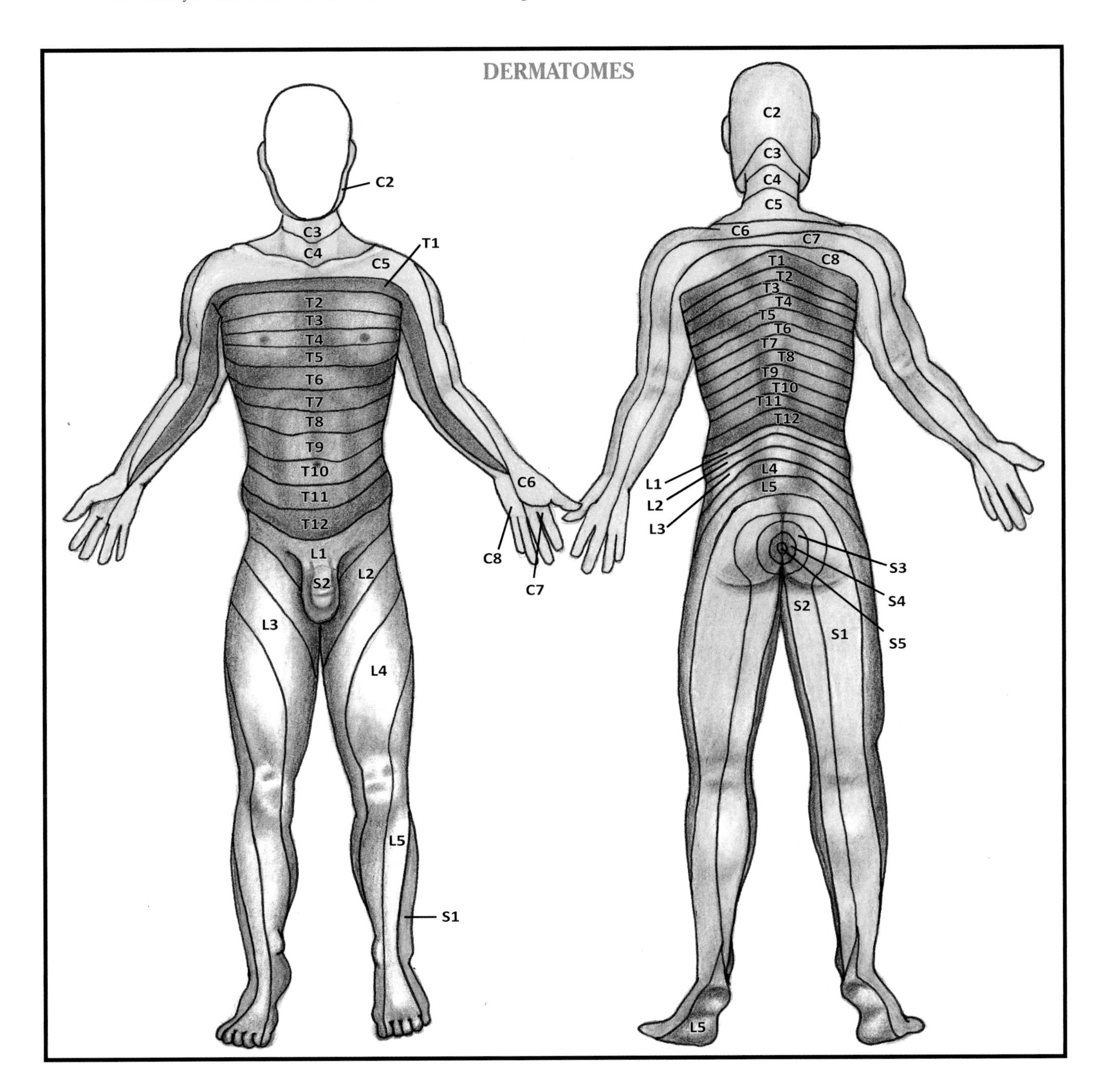

Ruptured Disc

- The nucleus pulposus is the part of the disc most likely to rupture.
- When it ruptures, it travels to the place of least resistance—the intervertebral foramina.
- This results in a *pinched spinal nerve* because the nerves exit the vertebral column through the intervertebral foramina.
- The disc has *herniated.*

Cervical Rupture

- *Cervical* nerves exit through the foramina *above* their corresponding vertebra.
- When a cervical disc has ruptured, the nerve *above the rupture* will be impinged.
- Example: A ruptured C5 disc will impinge the C5 spinal nerve.

Lumbar Rupture

- The *lumbar* and *sacral* nerves exit through the foramina *below* their corresponding vertebrae.
- Because the cauda equina forms the lumbar plexus, a ruptured lumbar or sacral disc will often *impinge several spinal nerves.*
- An example is *sciatica*—pain that radiates down the leg due to a ruptured disc in the lumbar or sacral regions. A ruptured disc in the lumbosacral regions will impinge several spinal nerves that innervate the lower extremities.

Cross-Sections of the Spinal Cord

Cervical Levels

- The cervical sections are *large* and *oval* in appearance.
- They consist of a large amount of *white matter.* The white matter consists of the axons of the sensory and motor tracts.
- The descending motor tracts have not yet exited the SC.
- Most of the ascending sensory tracts have already entered the cord.
- The amount of *gray matter* (nerve cell bodies) is small.

Thoracic Levels

- The thoracic levels are also *oval shaped* but are smaller than the cervical levels.
- The thoracic sections have a *lateral horn* (also called an intermediolateral horn).
- The lateral horn is part of the autonomic nervous system (ANS).
- This is where the *cell bodies* for the *sympathetic nervous system* are located.

Lumbar Levels

- The lumbar sections are more *round* in shape and *large.*
- They have the *largest amount* of *gray matter.*

Sacral Levels

- The sacral sections appear similar to the lumbar sections but are much smaller.
- Most of the *descending motor tracts* have *already exited* the SC.
- Many of the *ascending sensory tracts* have *not yet entered* the SC.

Organization of the Internal Spinal Cord

White Matter

- White matter consists of *myelinated axons.*
- The white matter is divided into three pairs of *funiculi:*
  - Anterior
  - Lateral
  - Dorsal
- The SC tracts are located in the funiculi.

Gray Matter

- Gray matter contains the *cell bodies* of the *sensory SC tracts* (in the dorsal horn) and the *cell bodies* of the *motor spinal nerves* (in the ventral horn).

- The cell bodies for the motor spinal nerves (that innervate the skeletal muscles) are *organized in a precise pattern in the ventral horn*:
  - The cell bodies for the motor spinal nerves that innervate the *proximal muscle groups* are located in the *medial ventral horn*.
  - The cell bodies for the motor spinal nerves that innervate the *distal muscle groups* are located in the *lateral ventral horn*.

Blood Supply of the Spinal Cord

- The blood supply of the SC comes from the *vertebral arteries*.
- The vertebral arteries are two branches that give rise to *one anterior spinal artery* and *two posterior spinal arteries*.

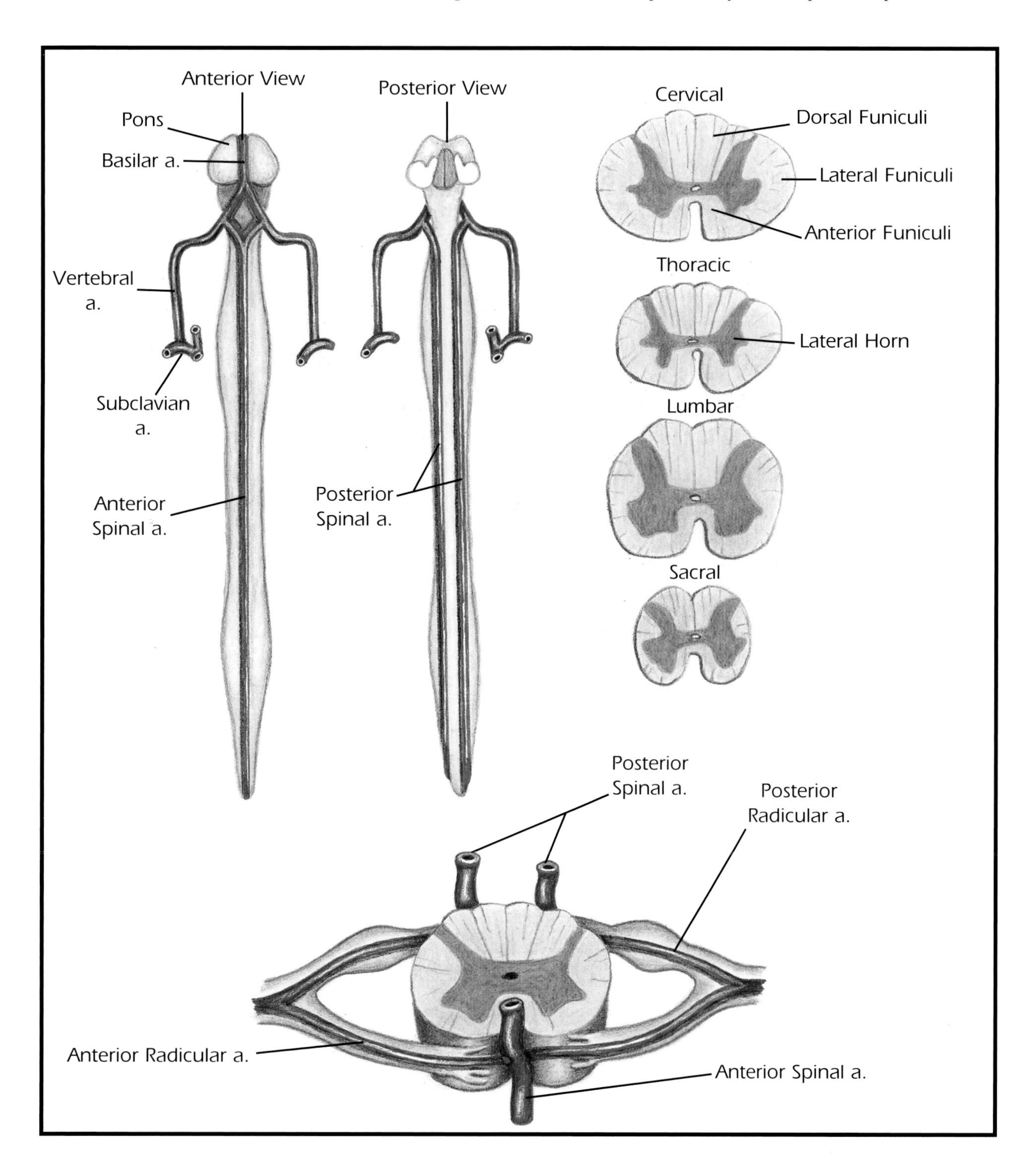

Anterior Spinal Artery (1)

- The vertebral artery traverses the *medulla* and sends off a branch called the *anterior spinal artery*.
- The anterior spinal artery descends down the medulla and the *anterior* aspect of the SC.
- It runs along the *anterior median fissure*.
- It supplies the *anterior* aspect of the SC.

Posterior Spinal Arteries (2)

- The vertebral artery also gives rise to *two posterior spinal arteries*.
- These descend down the *dorsal intermediate sulci* on the *posterior* aspect of the SC.
- They supply the *posterior* aspect of the SC.

Radicular Arteries

- The radicular arteries *encircle the SC* at all levels.
- The radicular arteries meet up with and supply the anterior spinal artery and the *posterior spinal arteries*.

Meninges of the Spinal Cord

- The meninges of the SC are the same as those of the brain.
- Their function is to protect and anchor the SC.
- There are three layers of meninges:
  - *Dura mater*
  - *Arachnoid mater*
  - *Pia mater*

Dura Mater

- The dura is the *most superficial* and *thickest* membrane.

Arachnoid Mater

- The arachnoid mater is the middle meningeal membrane.
- CSF bathes the SC in the subarachnoid space.

Pia Mater

- The pia mater is the *deepest* and *thinnest* membrane.
- It adheres to the SC.
- It sends off two projections: *filum terminale* and *dentate ligaments*.

Filum Terminale (Projection of the Pia)

- The filum terminale is a slender median fibrous thread that *attaches* the *conus medullaris* to the *coccyx*.
- It *anchors* the *end* of the SC to the *vertebral column*.

Dentate Ligaments (also called Denticulate Ligaments; Projections of the Pia)

- The dentate ligaments are a series of *22 triangular bodies* that anchor the SC.

Spinal Reflex Arc

- A spinal reflex arc is a reflex that is mediated at the SC level. There is no cortical involvement—no conscious decision-making.
- Spinal reflexes allow sensory information to be processed and acted upon quickly, without cortical processing.

Pathway of a Spinal Reflex Arc

- A *sensory receptor in the PNS* sends a message along an *ascending sensory spinal nerve*.
- The sensory spinal nerve travels to the *dorsal horn* where it synapses on an *interneuron*.
- The interneuron synapses on a *motor cell body* located in the *ventral horn*.
- The motor cell body in the ventral horn relays the message to a *motor spinal nerve* in the PNS.
- The message travels to a *skeletal muscle group* for action in response to the initial sensory message.

Deep Tendon Reflexes

- A deep tendon reflex is a *reflex arc* in which a muscle contracts when its tendon is percussed.
- Also called *myotatic reflexes*, *monosynaptic reflexes*, and *muscle stretch reflexes*.
- Deep tendon reflexes work on the principle of the *spinal reflex arc*.

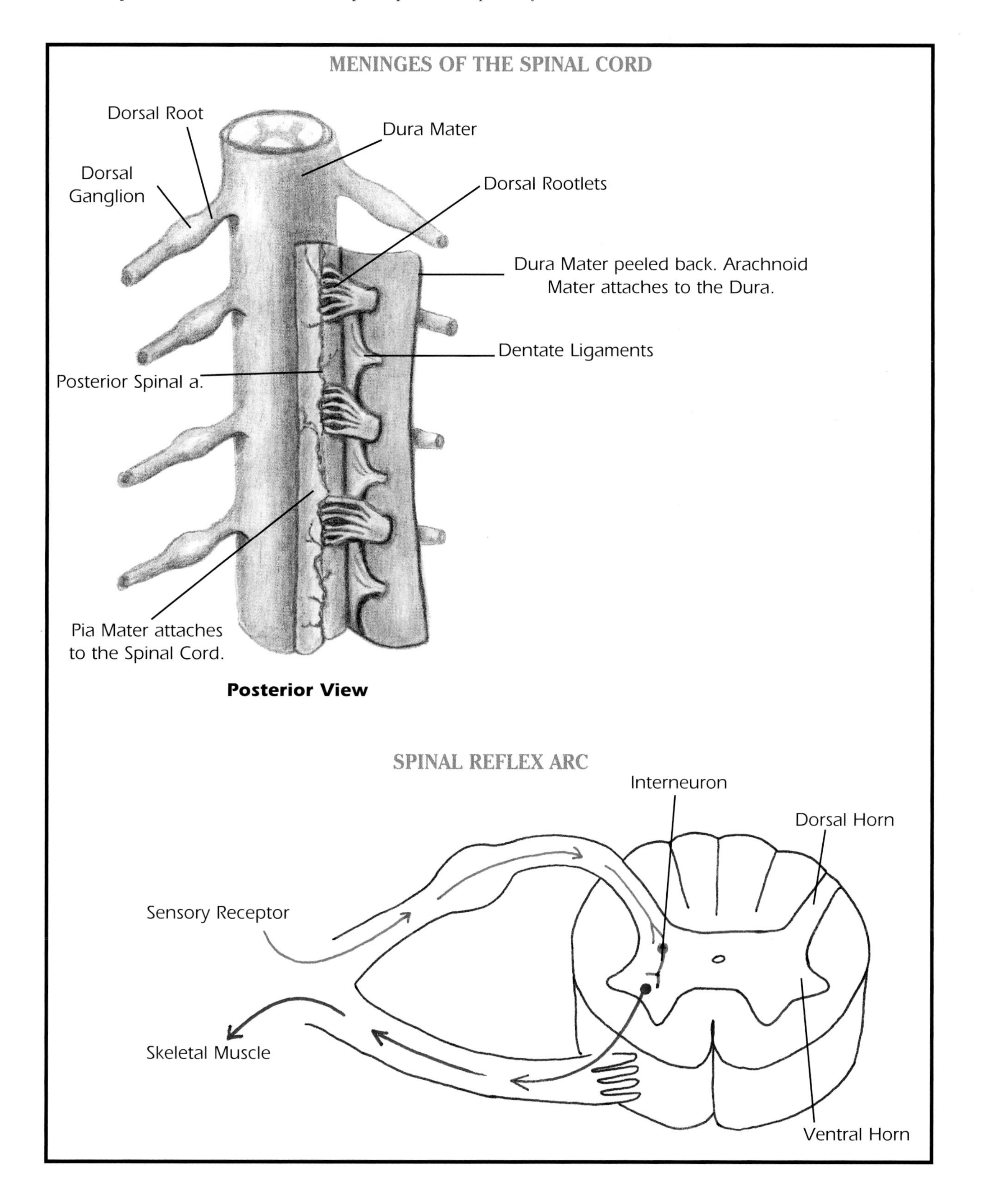

- Common deep tendon reflexes are as follows:
  - *Biceps, brachioradialis, triceps, patella,* and *Achilles tendon.*
- In an *upper motor neuron injury*, deep tendon reflexes become *hyper-reflexive.*
- In a *lower motor neuron injury*, deep tendon reflexes become *hyporeflexive.*

Withdrawal Reflex

- The withdrawal reflex is a spinal reflex that works similarly to the *spinal reflex arc.*
- This reflex is a *protective mechanism* that allows reflexive withdrawal of a body part from physical danger while simultaneously *adjusting posture* to avoid imbalance. Quickly pulling one's hand from a hot stove is an example of a withdrawal reflex. The hand is pulled away from the stove before the cortex consciously perceives pain.
- The withdrawal reflex is *polysynaptic*—in other words, signals (to withdrawal a limb) that enter one side of the SC will synapse with interneurons on the contralateral side of the cord. For example, if the right hand is withdrawn, the left limbs will also adjust to maintain balance.
- A painful stimulus to one side of the body will *activate the flexors* and *inhibit the extensors* on the *ipsilateral* side.
- Simultaneously, the *flexors will be inhibited* and the *extensors activated* on the *contralateral* side of the body.
- This allows for the quick withdrawal of the limb while maintaining balance.

Upper vs Lower Motor Neurons

- Motor neurons carry motor messages from different areas of the nervous system.
- Motor neurons are divided into two categories: upper motor neurons (UMN) and lower motor neurons (LMN).

Upper Motor Neurons

- An UMN carries motor messages from the *primary motor cortex* to the following:
  - The *cranial nerve nuclei* (located in the brainstem).
  - *Interneurons* in the *ventral horn.* An UMN travels up to the ventral horn and synapses with an interneuron that connects to a motor cell body located in the ventral horn.
  - UMNs are considered to be part of the CNS.

Lower Motor Neurons

- A *LMN* carries motor messages from the *motor cell bodies* in the *ventral horn* to the *skeletal muscles* in the periphery.
  - LMNs are considered to be part of the PNS.
  - LMNs include the cranial nerves, spinal nerves, cauda equina, and the ventral horn.

Upper Motor Neuron Lesion

- In an UMN lesion, *spasticity* occurs *below* the *lesion level.*
  - Spasticity occurs because the spinal reflex arcs below the lesion level remain intact.
  - The spinal reflex arcs operate without cortical modification.
  - Thus, increased muscle tone (or spasticity) occurs.
- In an UMN lesion, *flaccidity* occurs *at* the *lesion level.*
  - Flaccidity occurs because the spinal reflex arc at the lesion level is lost.
  - Thus, nothing is innervating the muscles.
  - The muscles (at the lesion level) lose all tone and become flaccid.
  - In reality, this type of lesion involves both UMN and LMN damage. Because the motor cell bodies in the ventral horn are no longer innervated, this type of lesion is also considered a LMN lesion.

Lower Motor Neuron Lesion

- In a LMN lesion, *flaccidity* occurs *at* and *below* the *lesion level.*
  - Flaccidity occurs in all LMN lesions because a LMN does not involve any spinal reflex arcs.
  - Because spinal reflex arcs are not part of LMN lesions, there is nothing that continues to innervate the muscles.

Congenital Anomalies of the Spinal Cord

Spina Bifida

- Spina bifida is a bony defect involving the incomplete closure of the neural tube during fetal development. A section of the SC does not unite in midline. Usually the opening occurs in the low thoracic and lumbar sections of the SC.
- Meningocele
  - A meningocele occurs when the meninges and the CSF protrude through the opening of the vertebral column.
  - This results in compression of the SC and some nerve roots.
  - There is a visible cyst on the infant's back that is filled with CSF and neural tissue.
- Meningomyelocele
  - This is a severe form of spina bifida.
  - It occurs when the meninges, SC, and spinal nerves all protrude.
  - All motor and sensory information below the level of the cyst are lost.

Arnold Chiari Malformation

- Arnold Chiari malformation involves displacement of the cerebellar vermis, brainstem, and fourth ventricle.
- The cerebellar tonsils are often displaced downward into the upper cervical canal.
- Arnold Chiari malformation is often associated with a meningomyelocele.
- Individuals with Arnold Chiari malformation may present with hydrocephalus and cerebellar and lower cranial nerve signs.

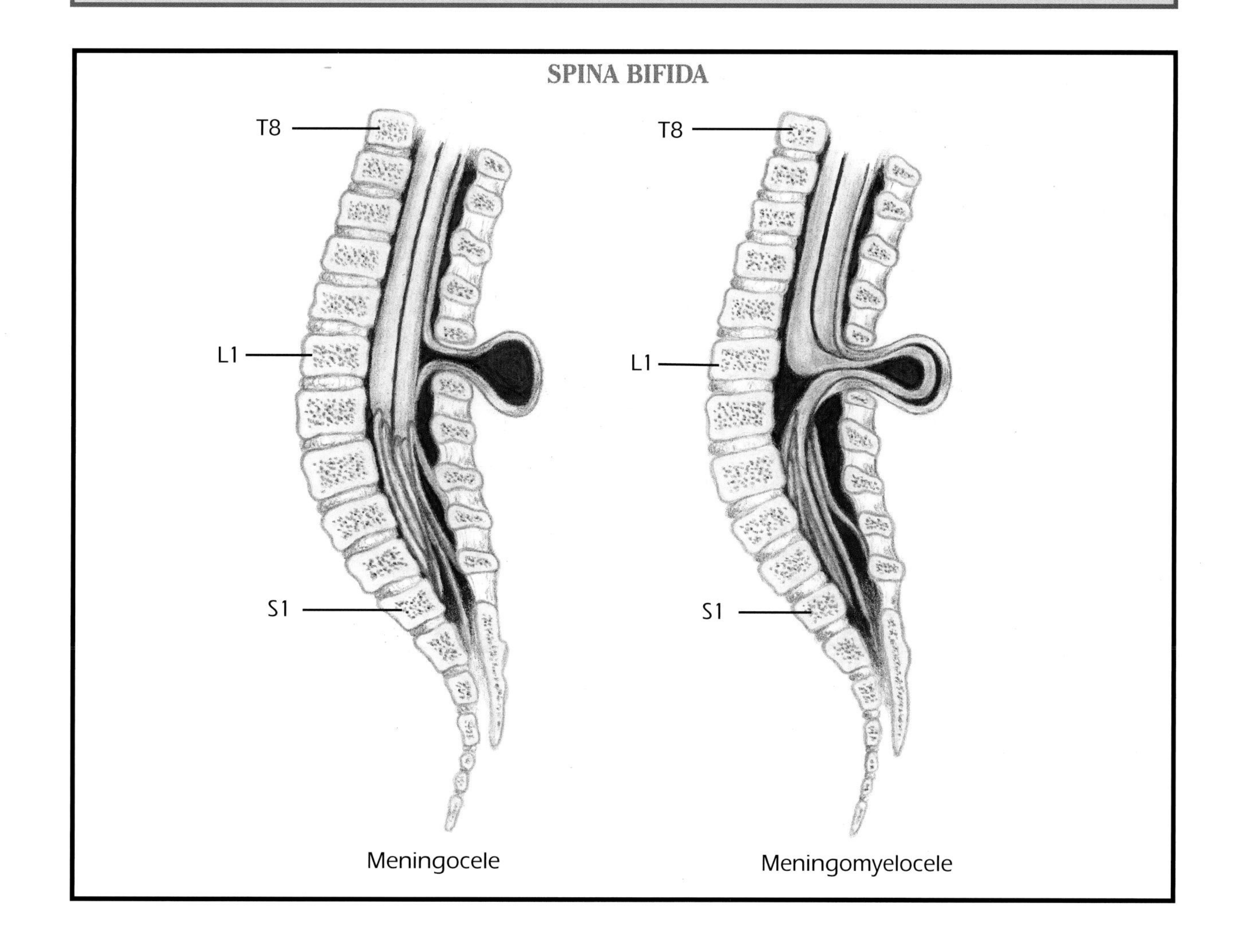

SECTION 8

The Cranial Nerves

THE CRANIAL NERVES

Cranial Nerve Anatomy

- There are *12 pairs* of cranial nerves (CNs).
- The CNs are considered to be part of the *peripheral nervous system (PNS)*.
- Their nuclei (cell bodies) are located in the brainstem.
- CN *nuclei* are considered to be part of the *central nervous system (CNS)*.
- CNs begin exiting the brain at the midbrain level and lead all the way down the medulla.
- CNs use Roman numerals (however, Arabic numbers will be used in this text).

List of Cranial Nerves

1. Olfactory Nerve
2. Optic Nerve
3. Oculomotor Nerve
4. Trochlear Nerve
5. Trigeminal Nerve
6. Abducens Nerve
7. Facial Nerve
8. Vestibulocochlear Nerve
9. Glossopharyngeal Nerve
10. Vagus Nerve
11. Accessory Nerve
12. Hypoglossal Nerve

Function

- CNs carry *sensory* and *motor* information to and from the following receptors of the head, face, and neck:
  - Special sense receptors (for vision, audition, olfaction, gustation, and equilibrium)
  - Somatosensory receptors
  - Proprioceptors

Lesions

- Most CN lesions produce *ipsilateral* signs and symptoms.

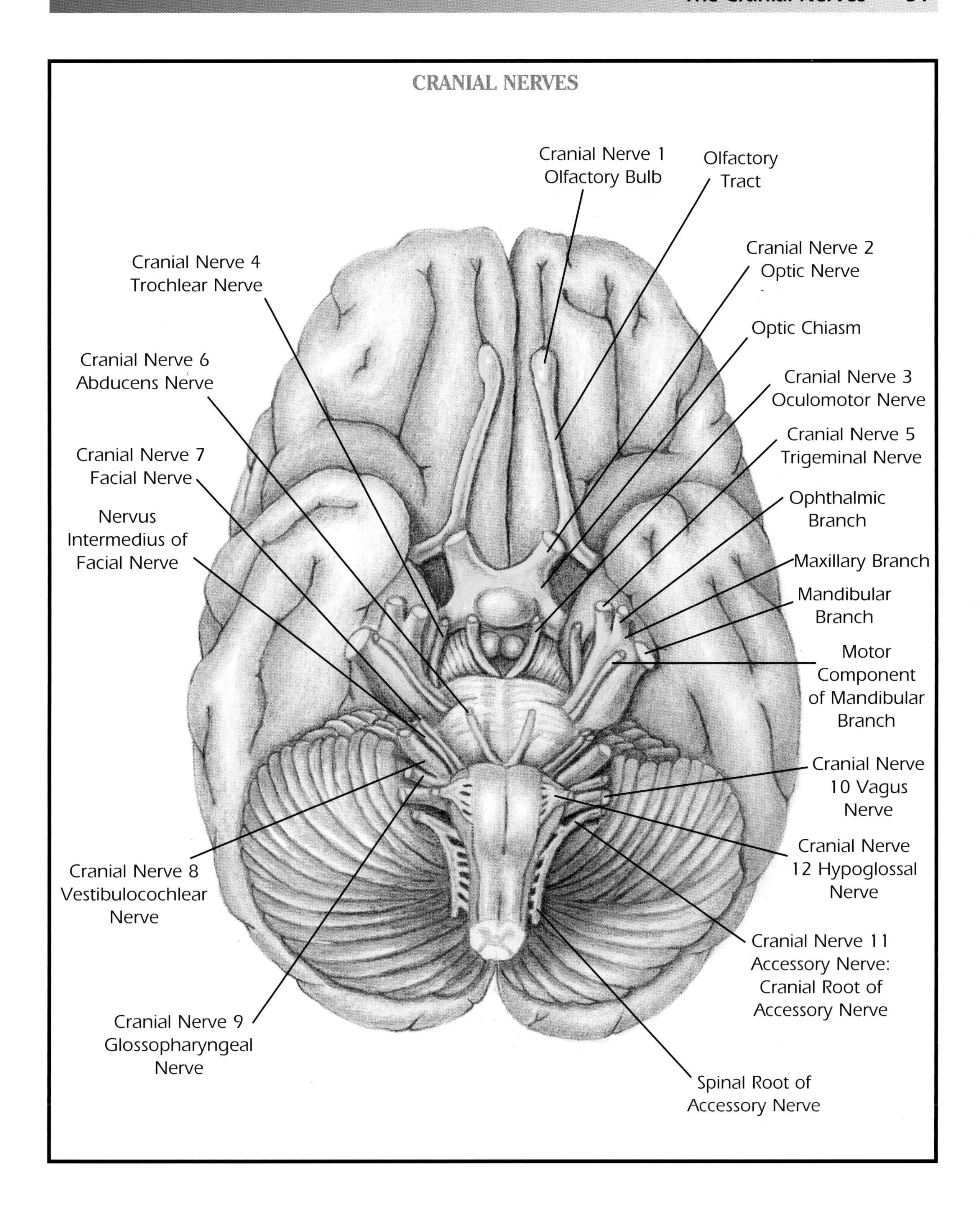
CRANIAL NERVES
Cranial Nerve 1
Olfactory Bulb
Olfactory
Tract
Cranial Nerve 4
Trochlear Nerve
Cranial Nerve 2
Optic Nerve
Optic Chiasm
Cranial Nerve 6
Abducens Nerve
Cranial Nerve 3
Oculomotor Nerve
Cranial Nerve 5
Trigeminal Nerve
Cranial Nerve 7
Facial Nerve
Ophthalmic
Branch
Nervus
Intermedius of
Facial Nerve
Maxillary Branch
Mandibular
Branch
Motor
Component
of Mandibular
Branch
Cranial Nerve
10 Vagus
Nerve
Cranial Nerve
12 Hypoglossal
Nerve
Cranial Nerve 8
Vestibulocochlear
Nerve
Cranial Nerve 11
Accessory Nerve:
Cranial Root of
Accessory Nerve
Cranial Nerve 9
Glossopharyngeal
Nerve
Spinal Root of
Accessory Nerve

Cranial Nerve 1: Olfactory Nerve

Carries

- Sensory information

Nuclei Location

- Chemoreceptors in the nose

Function

- Olfaction (smell)

Pathway

- Chemoreceptors in the nose send olfactory messages to the inferior frontal lobes—where the olfactory bulb is located.
- The olfactory information then travels from the olfactory bulb down the olfactory cranial nerve to the hippocampal formation in the temporal lobe.
- The hippocampus is responsible for long-term storage of odor memories. This is why odors can elicit old memories with an efficiency far greater than any other sense.
- Olfactory messages are then sent to the hypothalamus, thalamus, and finally to the orbitofrontal cortex where they are interpreted and eventually integrated with gustatory information.

Lesion Symptoms

- If the lesion is *unilateral* (only occurs in one olfactory nerve):
  - There are *no symptoms*—because the opposite olfactory nerve compensates for the lost sense of smell on one side.
- If the lesion is *bilateral*:
  - The individual *loses* the *sense of smell—anosmia.*
  - Often, anosmia occurs as a result of head injury.
  - There is also a lack of olfactory function.

Test

- Test one olfactory nerve at a time.
- Occlude vision.
- Block the patient's nostril on the opposite side being tested.
- Present one odor at a time.
- Provide the patient with a verbal choice of specific odors if he or she has difficulty identifying the odors but can easily smell them or if the patient has word-finding difficulties.

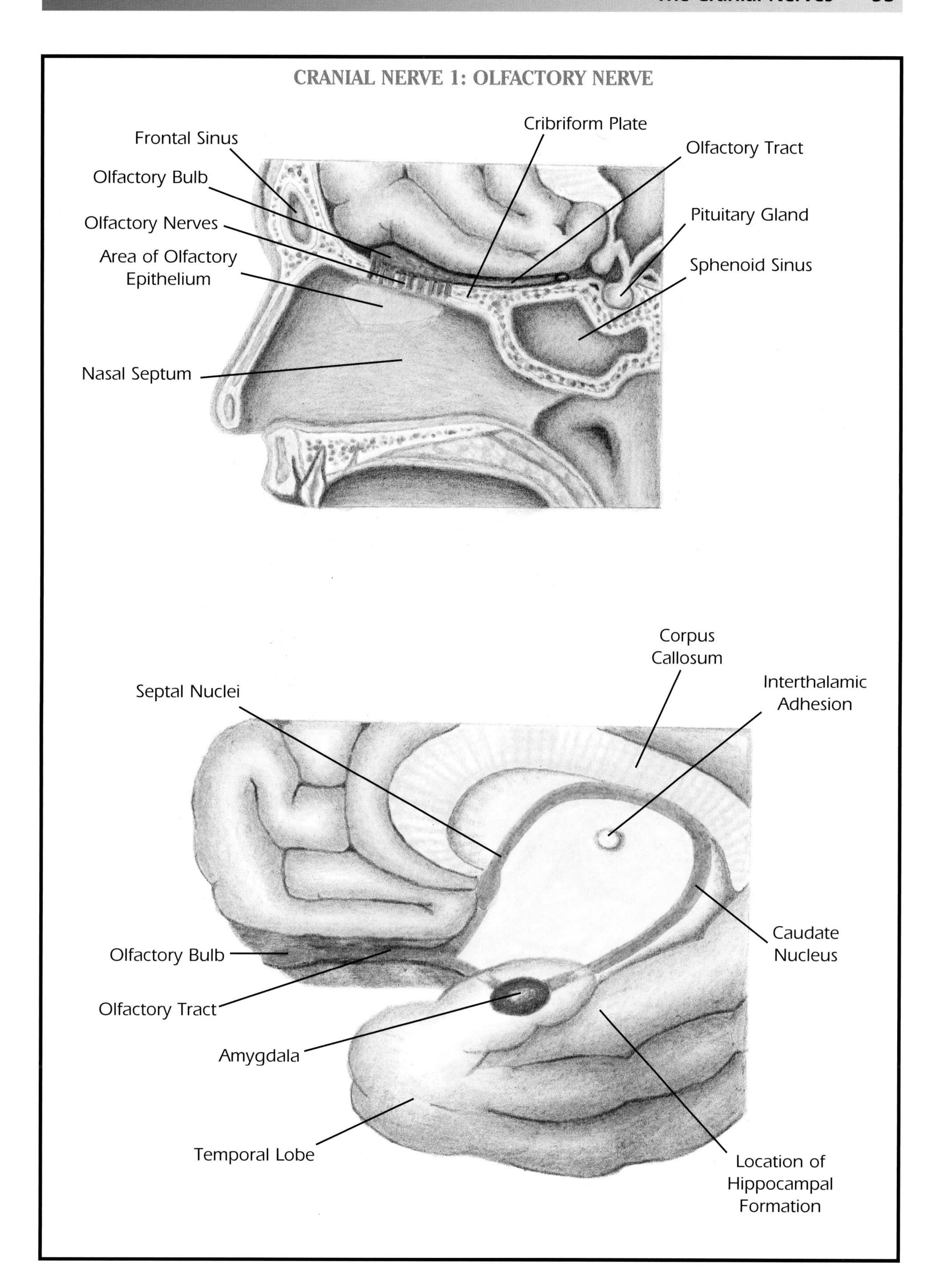
CRANIAL NERVE 1: OLFACTORY NERVE
Frontal Sinus
Olfactory Bulb
Olfactory Nerves
Area of Olfactory Epithelium
Nasal Septum
Cribriform Plate
Olfactory Tract
Pituitary Gland
Sphenoid Sinus
Corpus Callosum
Septal Nuclei
Interthalamic Adhesion
Olfactory Bulb
Olfactory Tract
Amygdala
Temporal Lobe
Caudate Nucleus
Location of Hippocampal Formation

Cranial Nerve 2: Optic Nerve

Carries

- Sensory information

Nuclei Location

- Photoreceptors of the retina (rods and cones)

Function

- Visual acuity (the *accuracy* of sight; *not* the *interpretation* of visual information).
- Visual messages that travel from the thalamus to the superior colliculi (and back) are involved in the following:
  - Pupillary reflexes
  - Awareness of light and dark
  - Orientation of head and eye movements

Pathway

- The *rods* and *cones* of the retina send visual information down the *optic nerves* to the *optic chiasm*.
- The visual information then travels from the optic chiasm through the *optic tracts*.
- Visual information travels from the optic tracts to the *lateral geniculate bodies* of the thalamus.
- A branch carries visual messages from the lateral geniculate bodies of the thalamus to the *superior colliculi* of the midbrain (the information is then processed back through the thalamus). This pathway allows certain visual information to be detected on an unconscious level—even in the case of cortical blindness, when the person's brain cannot interpret visual data but the visual anatomy remains intact.
- Visual messages then travel to the *occipital lobes* for visual detection and interpretation.

Lesion Symptoms

- A *unilateral lesion* (in only one optic nerve) produces *ipsilateral blindness*.
- A *bilateral lesion* (in both optic nerves) produces *bilateral blindness*.

Test

- When testing optic nerve function, three kinds of data should be collected (or the examination is not complete).
  - Results from a *Visual Acuity Test* (Snellen Eye Chart)
  - Results from a *Visual Field Test*
  - Results from a *Funduscopic Exam*
- Generally, therapists perform Visual Acuity Tests and Visual Field Tests. Ophthalmologists perform Funduscopic Exams.

Visual Acuity Test (Snellen Eye Chart)

- Test one eye at a time. Then test both eyes together.
- If the patient normally wears corrective lenses, test the patient with glasses on.

Visual Field Test

- Test one eye at a time. Occlude vision in the opposite eye.
- Test vertical and temporal peripheral vision.
- Normal *vertical vision* is *45 degrees*.
- Normal *temporal vision* is *85 degrees*.

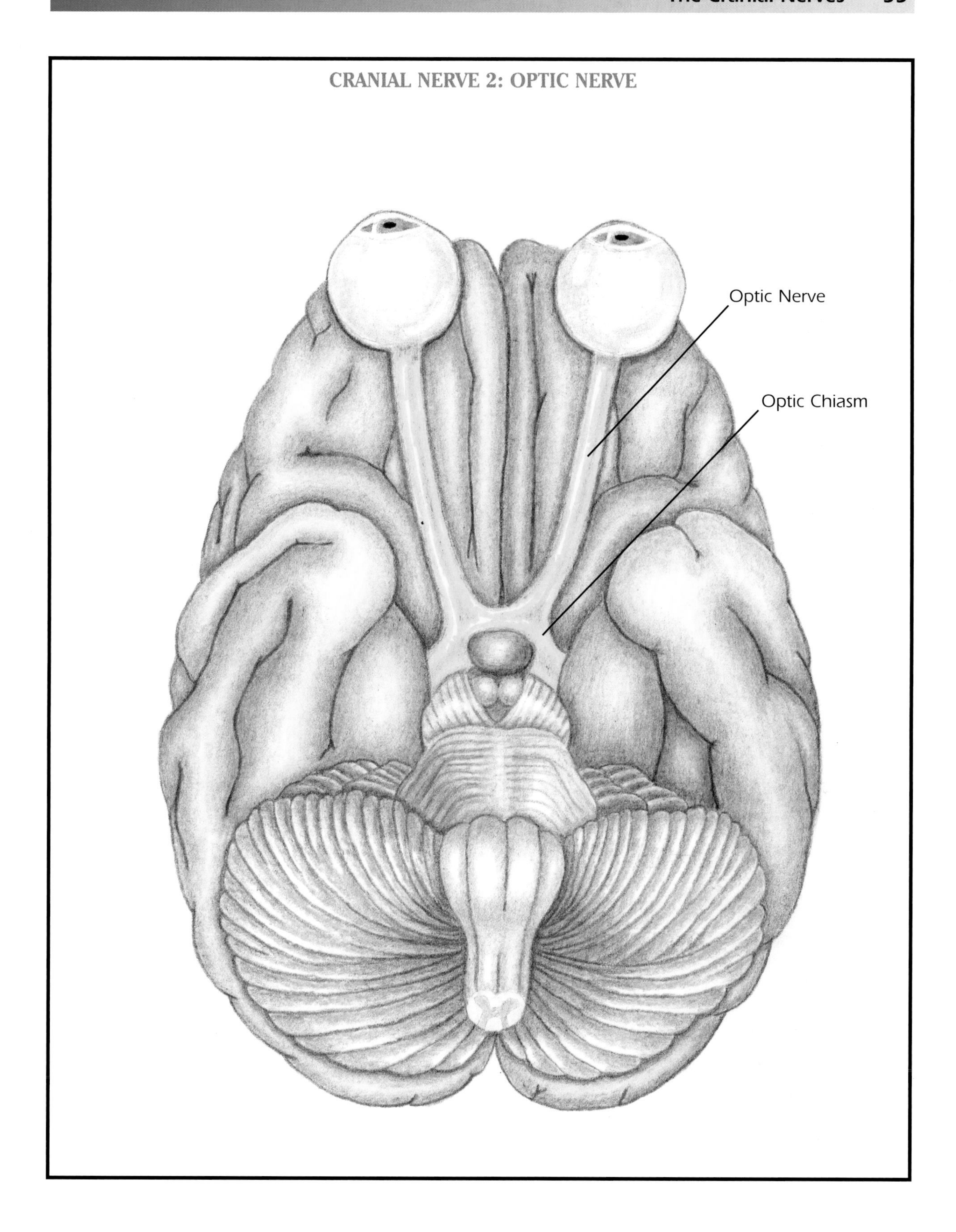
CRANIAL NERVE 2: OPTIC NERVE
Optic Nerve
Optic Chiasm

Cranial Nerve 3: Oculomotor Nerve

Carries

- Motor information

Nuclei Location

- Midbrain at the level of the superior colliculi

Function

- Extraocular eye movements.
- CN 3 is responsible for the *eyeball movements up, down, medially,* and *laterally.*
- The *oculomotor nerve innervates* the *eye muscles* that control these movements.
- The oculomotor nerve is considered to be one of the *extraocular motor nerves,* along with the *trochlear* and *abducens* nerves.
- The three extraocular motor nerves use the *medial longitudinal fasciculus* to communicate with each other and with the *vestibular system.*
- The medial longitudinal fasciculus is a brainstem tract that *coordinates head and eye movements* by providing bilateral connections among the *vestibular nerve nuclei, extraocular nerve nuclei,* and the *accessory nerve nuclei* in the brainstem.

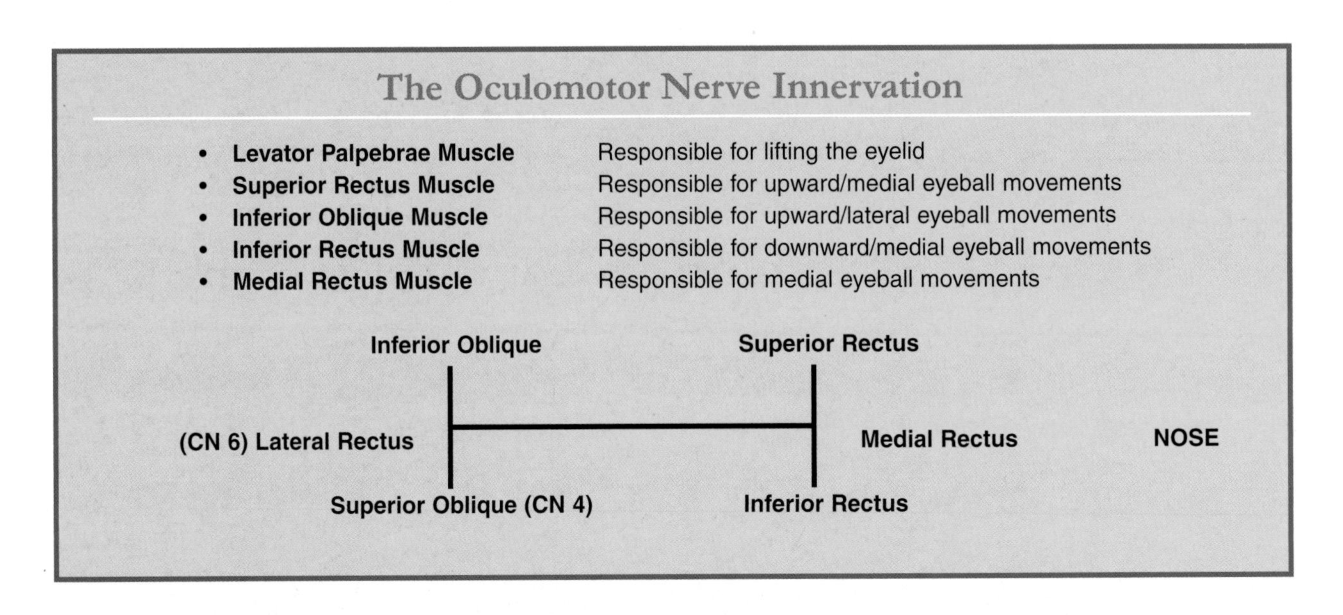

Oculomotor Nerve Reflexes

Pupillary Reflex	Pupil of the eye constricts when light is shined into it.
Accommodation	Lens of the eye adjusts to focus light on the retina.
Convergence	Pupils move medially when viewing an object at close range.

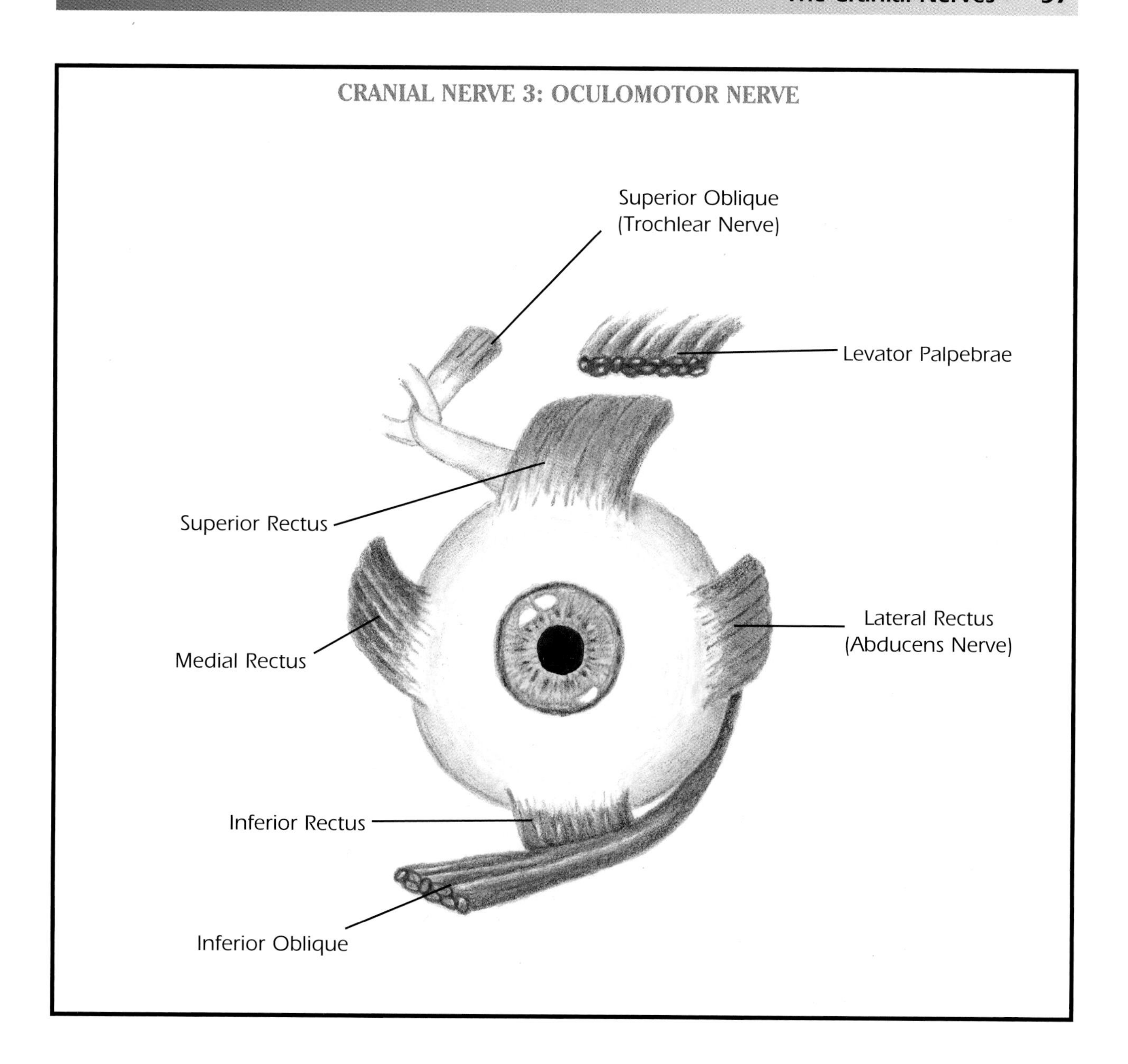
CRANIAL NERVE 3: OCULOMOTOR NERVE
Superior Oblique
(Trochlear Nerve)
Levator Palpebrae
Superior Rectus
Lateral Rectus
(Abducens Nerve)
Medial Rectus
Inferior Rectus
Inferior Oblique

Lesion Symptoms

- Lateral Strabismus (External Strabismus or Exotropia)
  - The *eyeball deviates outward* (or laterally) because the medial rectus is lost and the lateral rectus is working unopposed.
  - This can cause *diplopia*, or double vision.

Ptosis

- Ptosis is the drooping of a body region; in this case, the *ipsilateral eyelid droops*.

Nystagmus

- Nystagmus is involuntary back and forth movements of the eye in a quick, jerky, oscillating fashion when the eye moves laterally or medially to either the temporal or nasal extremes. Nystagmus can occur in the center of each visual field as well.
- Nystagmus can be normally elicited in an intact CNS using rotational or temperature stimulation of the semicircular canals.
- Pathological nystagmus is a sign of CNS abnormality and can occur with or without external stimulation.

Test

- Therapists *test* the *extraocular motor nerves simultaneously*—the oculomotor, trochlear, and the abducens nerves.
- Test one eye at a time. Occlude the eye not being tested.
- Instruct the patient to maintain head in a fixed position while visually scanning a moving stimulus.
- The moving stimulus can be a colored pen cap.
- The therapist moves the visual stimulus in the shape of an H (see "The Oculomotor Nerve Innervation" text box on p 56). This allows the therapist to determine if the eyeball muscles are functioning adequately (ie, if the extraocular muscles are adequately innervated by their extraocular motor CNs).
- Observe *symmetry* of *pupil size*.
- In a dimly lit area, shine pen light at the bridge of the nose and observe for symmetrical corneal reflection. If corneal reflection is asymmetrical, *strabismus* is indicated.
- Shine pen light into one eye at a time for a 2-second duration. Check for constriction of stimulated pupil.

SYMPTOMS OF OCULOMOTOR NERVE (CRANIAL NERVE 3) DAMAGE

Lateral Strabismus

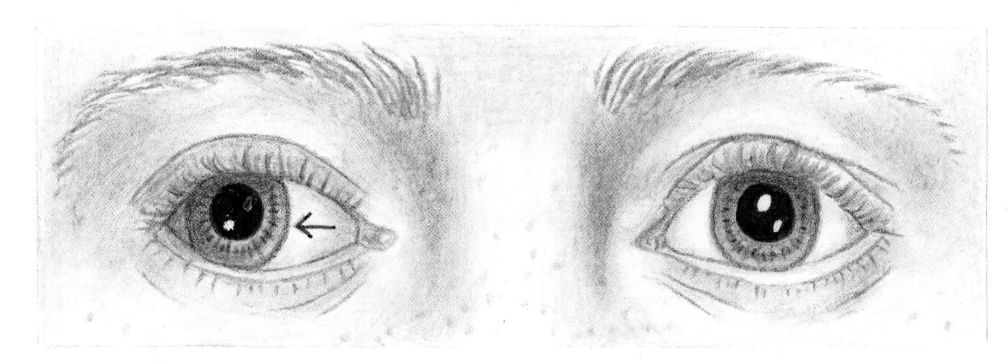

Ptosis

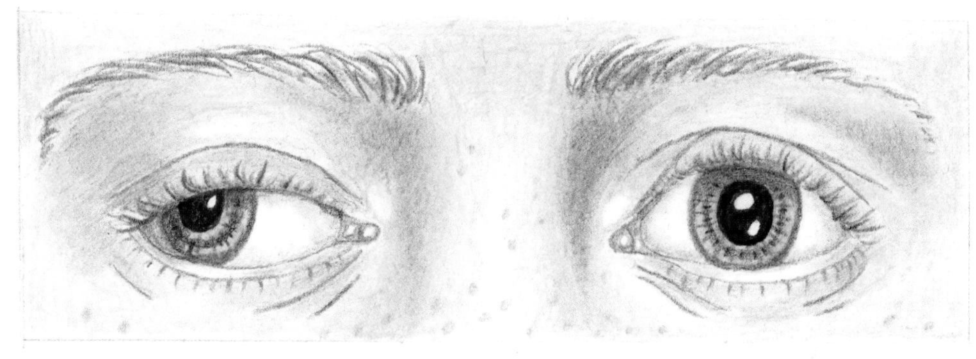

Nystagmus

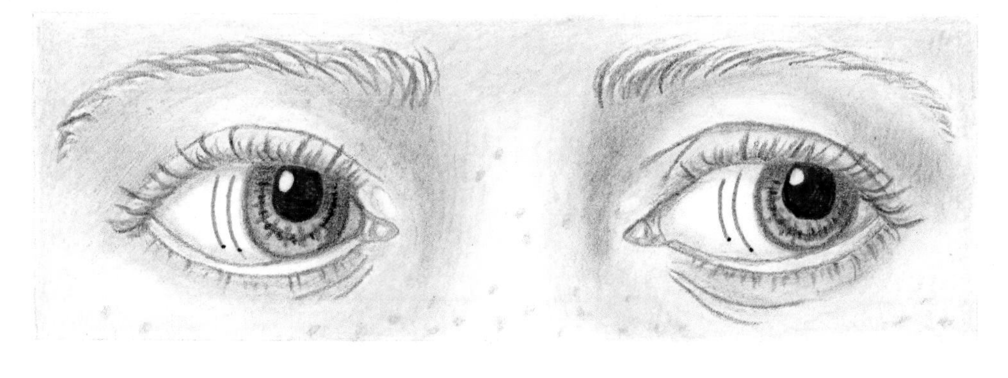

Cranial Nerve 4: Trochlear Nerve

Carries

- Motor information

Nuclei Location

- Midbrain at the level of the inferior colliculi

Function

- Extraocular eye movements
- Responsible for *downward* and *lateral eyeball movements*
- Considered to be one of the *extraocular motor nerves*

Innervates

- Superior oblique muscles

Lesion Symptoms

- The patient will experience difficulty moving the eyeball down and laterally.
- This occurs because the *superior oblique muscle* is *lost*. The medial rectus and the superior rectus muscles are working unopposed to pull the eyeball up and medially.
- This results in a subtle *vertical, medial strabismus*.
- The patient may display difficulty walking down steps.
- *Vertical diplopia* is often reported at both near and far distances.
- *Nystagmus*

Test

- Test the trochlear nerve function simultaneously with the oculomotor and abducens nerves (see Oculomotor Nerve Testing on p 58).

SYMPTOMS OF TROCHLEAR NERVE DAMAGE (CN 4)

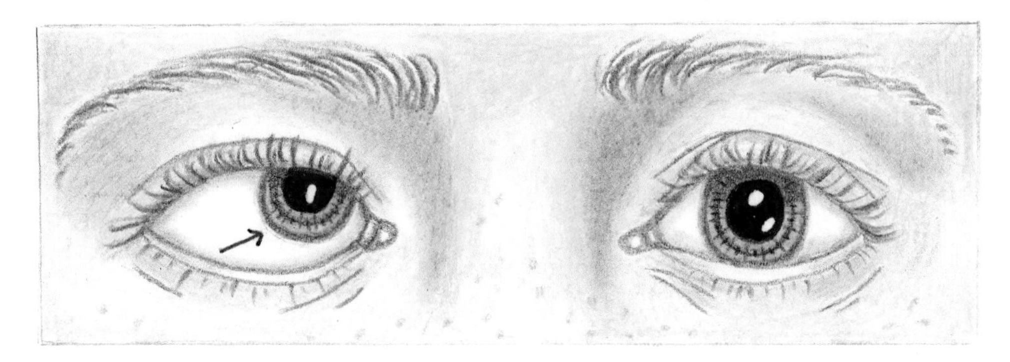

Vertical, Medial Strabismus

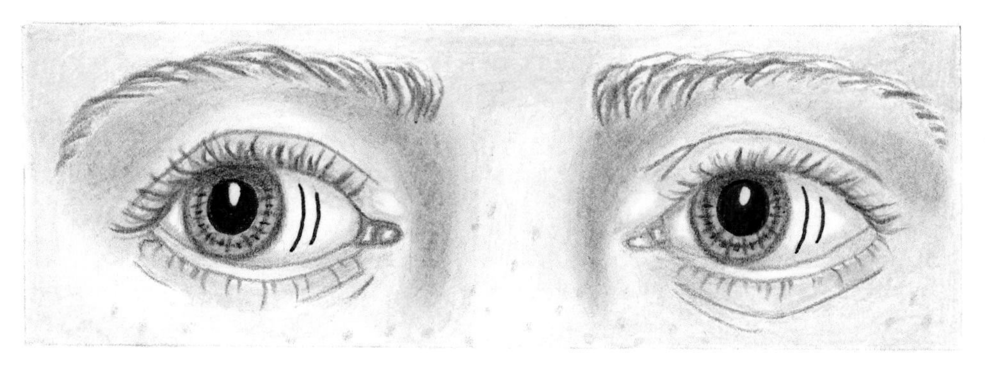

Nystagmus

Cranial Nerve 5: Trigeminal Nerve

Carries

- Sensory and motor information

Nuclei Location

- Mid pons

Function

Sensory

- The sensory half of the trigeminal nerve mediates sensation of the face, head, cornea of eye, and the *inner oral cavity* (the ophthalmic, maxillary, and mandibular regions).
- Sensation includes pain, temperature, and discriminative touch.

Motor

- The motor half of the trigeminal nerve *innervates the jaw muscles* that control *chewing* (or mastication).

Lesion Symptoms

Sensory

- Damage to the sensory half of the trigeminal nerve causes *ipsilateral loss of sensation* to the *head*, *face*, and *inner oral cavity*.
- *Trigeminal neuralgia* occurs when half of the face loses sensation.

Motor

- Damage to the motor half of the trigeminal nerve causes *weakness in chewing* (mastication).
- The *jaw* also *deviates* to the *affected side*.

Test

Sensory Half of the Trigeminal Nerve

- Evaluate the patient's *sensory* abilities on the *face*, *head*, and *inner oral cavity*.
- Occlude vision.
- Use a cotton swab to stroke the inner oral cavity. Assess the intact side first. Then evaluate the involved side.
- Use a cotton swab to stroke the patient's forehead, cheek, jaw, and chin. Apply stimulus to the unaffected side first. Then proceed to the involved side.
- Touch the patient's cornea lightly with a cotton swab to check for *corneal reflex*—eyelid should close.

Motor Half of the Trigeminal Nerve

- Occlude vision.
- Ask the patient to open his or her mouth. Check for *deviation of the jaw* to the affected side.
- Check for *asymmetry* of the size of the *mouth opening*—the patient will likely exhibit a decreased ability to open the mouth on the affected side.
- Ask the patient to move jaw from side to side. Check for asymmetry of *jaw movement*.
- Instruct the patient to bite down on a tongue depressor. Ask the patient to resist attempts, made by the therapist, to pull the tongue depressor out. Check for asymmetry between right and left jaw strength.

Trigeminal Neuralgia

- Trigeminal neuralgia involves sudden, excruciating pain of short duration along the second (maxillary region) and third (mandibular region) divisions of the trigeminal nerve.
- The etiology involves chronic compression of the trigeminal nerve as a result of a vessel position or tumor. This causes demyelination and impaired nerve signaling.
- Patients describe the pain of trigeminal neuralgia as sharp and shooting.
- Pain-free periods can be experienced between episodes, or a dull ache may continuously be felt in the affected region.
- Pharmacologic treatment includes antiseizure medications (eg, Tegretol, Dilantin, or Neurontin).
- Surgical intervention is attempted when drug therapy is unsuccessful. Such procedures involve surgical decompression of the trigeminal nerve.

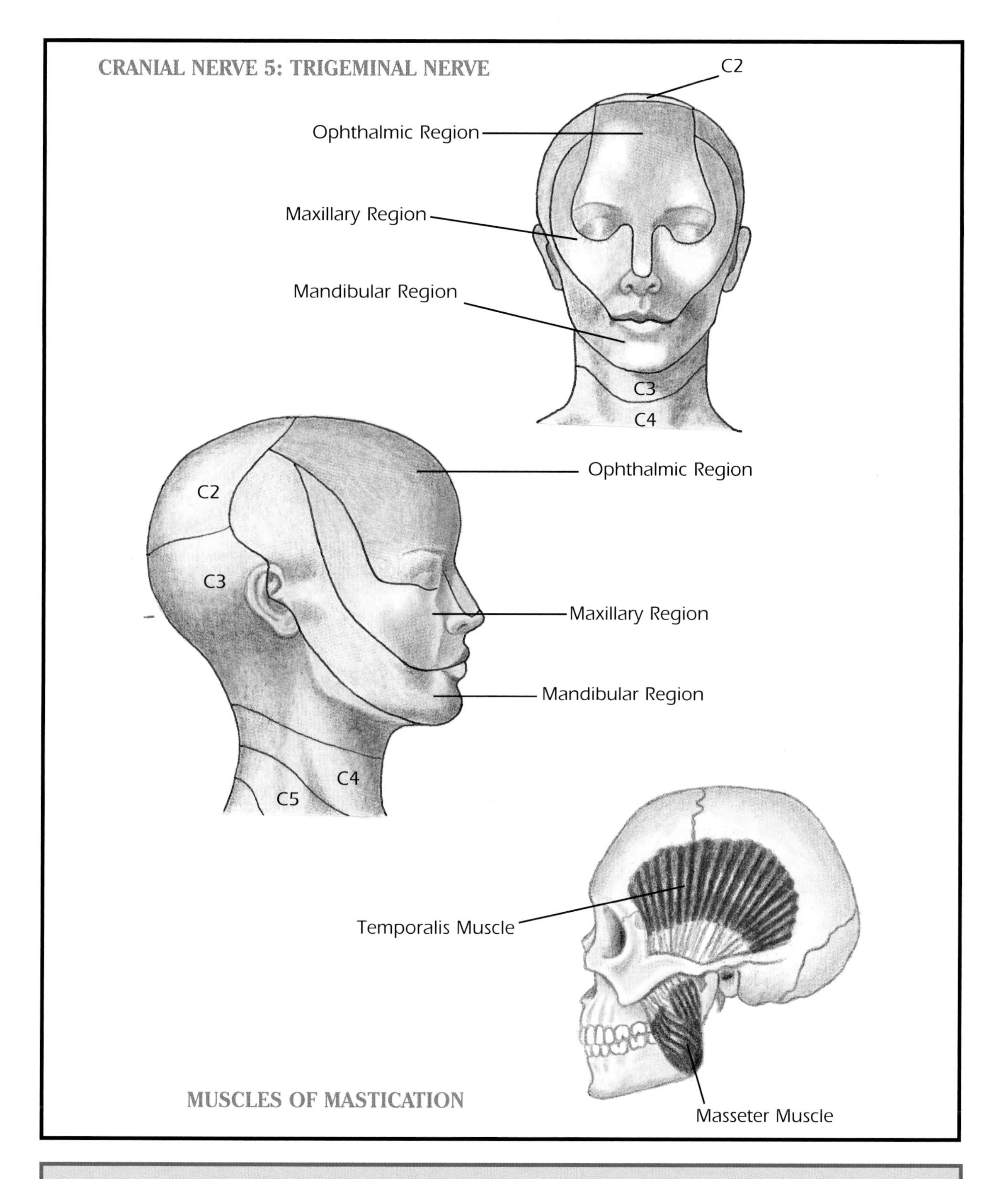

Trigeminal Nerve Reflexes

Masseter Reflex

- When the masseter muscle is lightly tapped with a reflex hammer, the masseter contracts.

Corneal Reflex

- When the cornea is touched, the eyelids close.

Cranial Nerve 6: Abducens Nerve

Carries

- Motor information

Nuclei Location

- Low pons

Function

- Extraocular eye movements
- Responsible for *lateral deviation* of the *eyeball—looking laterally*

Innervates

- Lateral rectus muscle

Lesion Symptoms

Medial Strabismus (Internal Strabismus or Esotropia)

- *Turning inward* of the *eyeball*
- This occurs because the *lateral rectus muscle is lost*. The medial rectus works unopposed to pull the eyeball medially.
- Can cause *double vision* or *diplopia*
- *Nystagmus*

Test

- The abducens nerve is simultaneously tested with the oculomotor and trochlear nerves (see Oculomotor Nerve Testing on p 58).

SYMPTOMS OF ABDUCENS NERVE DAMAGE (CRANIAL NERVE 6)

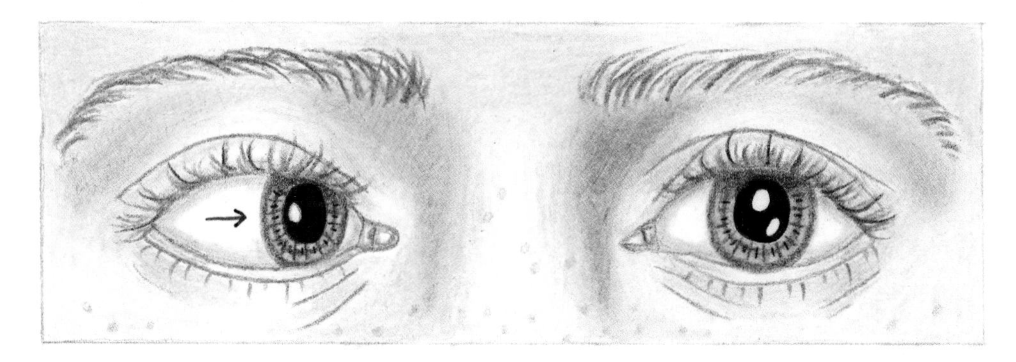

Medial Strabismus

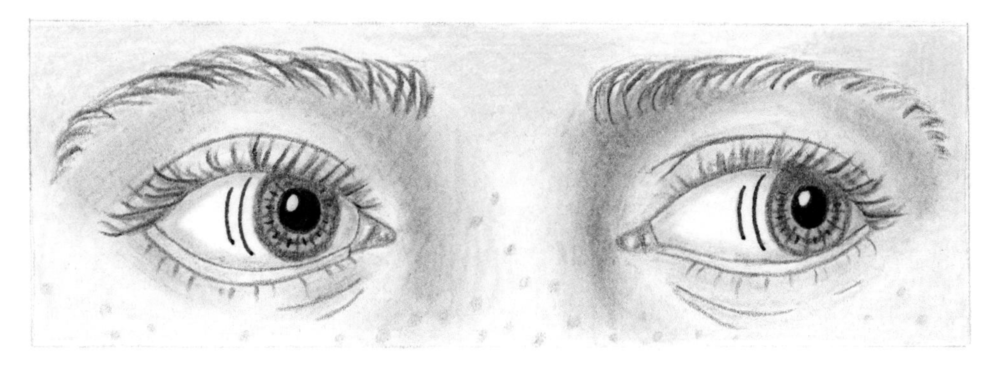

Nystagmus

Cranial Nerve 7: Facial Nerve

Carries

- Sensory and motor information

Nuclei Location

- Mid to low pons

Function

Sensory

- The sensory portion of the facial nerve comes off of a separate branch of the facial nerve called the *nervous intermedius*.
- It innervates the *taste receptors* on the *anterior tongue*.

Motor

- The motor portion of the facial nerve innervates the following:
  - Muscles of *facial expression*
  - Muscles for *eyelid closing*
  - *Stapedius muscle* (controls the stapes of the middle ear)

Lesion Symptoms

- *Decreased taste* on the anterior of the *tongue*
- *Decreased corneal reflex*

Test

Sensory Portion of the Facial Nerve

- Test the sense of taste on the anterior tongue.
- Occlude vision.
- Present *sweet*, *salty*, and *sour* solutions to the outer and lateral portions of the anterior tongue.
- Present each taste substance one at a time. Ask the patient to indicate if he or she can taste the substance and identify whether it is sweet, sour, or salty.

Motor Portion of the Facial Nerve

- Test the *strength* and *symmetry* of facial muscles.
- Ask the patient to *elevate eyebrows* and *forehead*.
- The *ability* to *wrinkle the forehead* is used to distinguish an upper motor neuron (UMN) lesion from a lower motor neuron (LMN) lesion:
  - In an UMN lesion, the *muscles* of the *forehead* will be spared (remain intact) even while the *lower facial regions are not*. This commonly occurs in *cerebrovascular accident*.
  - *Bell's palsy* is a LMN disorder (the facial nerve has been lesioned). *Both* the *forehead* and the *lower face* are *involved* in the paralysis of facial muscles.
  - Ask the patient to smile, frown, and pucker lips. Check for asymmetry on the right and left sides of the face.
  - Ask the patient to blow cheeks up with air. Gently push on cheeks while asking the patient to resist. Check for asymmetry of facial muscle strength.

Facial Nerve Reflex

- **Corneal (or Blink Reflex)** When the cornea is touched, the eyelids close.

Bell's Palsy

- Occurs when the facial nerve swells and becomes compressed as it passes through the petrous temporal bone.
- The herpes simplex 1 virus is implicated in 60% to 70% of cases.
- Other viruses that may cause Bell's palsy include cytomegalovirus, Epstein-Barr, rubella, and mumps.
- Trauma to the facial nerve may also play a role in the some cases.
- Bell's palsy is characterized by the following:
    - Drooping of the ipsilateral side of the face
    - Sagging eyebrow
    - Inability to close the affected eye completely
    - The mouth is drawn down toward the affected side.
    - The ear on the affected side becomes hypersensitive to noise.
- The acute phase is characterized by marked edema causing tension to the facial nerve.
- An inflammatory process begins; this induces edema with secondary vascular compromise.
- Inflammation and vascular compromise cause anoxia to the facial nerve.
- Anoxia then leads to vasodilation, transudation of fluid (oozing of fluid through the pores), and further pressure that confines the pathway of CN 7.
- Patients with diabetes are four times more likely to develop Bell's palsy. It is also more common in the third trimester of pregnancy and in people who are immunocompromised (such as in AIDS).

Hyperacusis in Bell's Palsy (Increased Sensitivity to Sound)

- Because the facial nerve travels through the internal auditory meatus and later gives off branches to the stapedius muscle, patients with Bell's palsy may complain of ipsilateral hyperacusis.

Cranial Nerve 8: Vestibulocochlear Nerve

Carries

- Two sensory branches

Nuclei Location

- Pons–medulla junction

Function

Auditory Branch

- The auditory branch of the vestibulocochlear nerve transmits sensory impulses that result from the vibrations of the fluid in the cochlea.
- The function of the auditory branch is *audition* (hearing).

Vestibular Branch

- The vestibular branch receives sensory stimulation from the semicircular canals of the inner ear.
- It is concerned with *balance* and the sensations of *vertigo* (dizziness).
- The functions of the vestibular branch are *balance*, *equilibrium*, and the *position of the head in space*.

Pathway

Auditory Branch

- The auditory branch of the vestibulocochlear nerve runs from the *hair cells* (or receptors) of the *organ of Corti* (in the inner ear) to the *vestibular nucleus* in the brainstem.

Vestibular Branch

- The vestibular branch runs from the *semicircular canals*, *utricles*, and *saccules* (of the inner ear) to the *vestibular nuclei* in the brainstem.
- The semicircular canals, utricles, and saccules detect changes in head position.

Lesion Symptoms

Auditory Branch Lesions

- *Deafness* or *tinnitus*

Vestibular Branch Lesions

- *Nystagmus* (due to connections to the extraoculomotor nerve nuclei)
- *Vertigo*
- *Decreased balance*
- *Decreased protective responses*
- *Changes in extensor tone* (because the vestibulospinal tract is responsible for mediating extensor tone)

Test

Auditory Branch of the Vestibulocochlear Nerve

- An audiologist must first distinguish between two possible types of hearing impairment:
  - *Sensorineural*—involves the inner ear, vestibulocochlear nerve, and the brain
  - *Conductive*—involves the outer ear and the middle ear structures

Vestibular Branch of the Vestibulocochlear Nerve:

- Test for *nystagmus*.
- Patient should assume a seated position.
- Have patient track a moving object (at a distance of 15 inches) in an H and X pattern.
- Check for nystagmus both within and at end ranges of visual fields.
- Test *balance* and the presence of *protective responses* (Romberg Test).
- Have patient stand with eyes open, then closed.
  - Check for increased sway and loss of balance.
  - Gently displace patient's balance. Check for protective responses.
  - Test for the presence of *extensor tone* in the *lower extremities*.

Vestibular Neuritis

- Vestibular neuritis is characterized by an acute onset of vertigo, nausea, vomiting, disequilibrium, and nystagmus.
- Improvement usually occurs within 1 to 2 weeks; however, some patients develop recurrent episodes.
- A large percentage of patients report having had an upper respiratory tract infection 1 to 2 weeks prior to the onset of symptoms. This suggests a viral origin.
- In some patients, vestibular neuritis can recur over months or years. There is no way to determine whether a first attack will be followed by repeated occurrences.
- Drug treatment may involve anticholinergic drugs (eg, scopolamine, atropine), monoaminergic drugs (eg, amphetamine, ephedrine), and antihistamines (eg, Antivert, Marezine, Dramamine, and Phenergan).
- Vestibular rehabilitation usually involves balance training and habituation exercises that can retrain the brain's response to motion-induced vertigo.

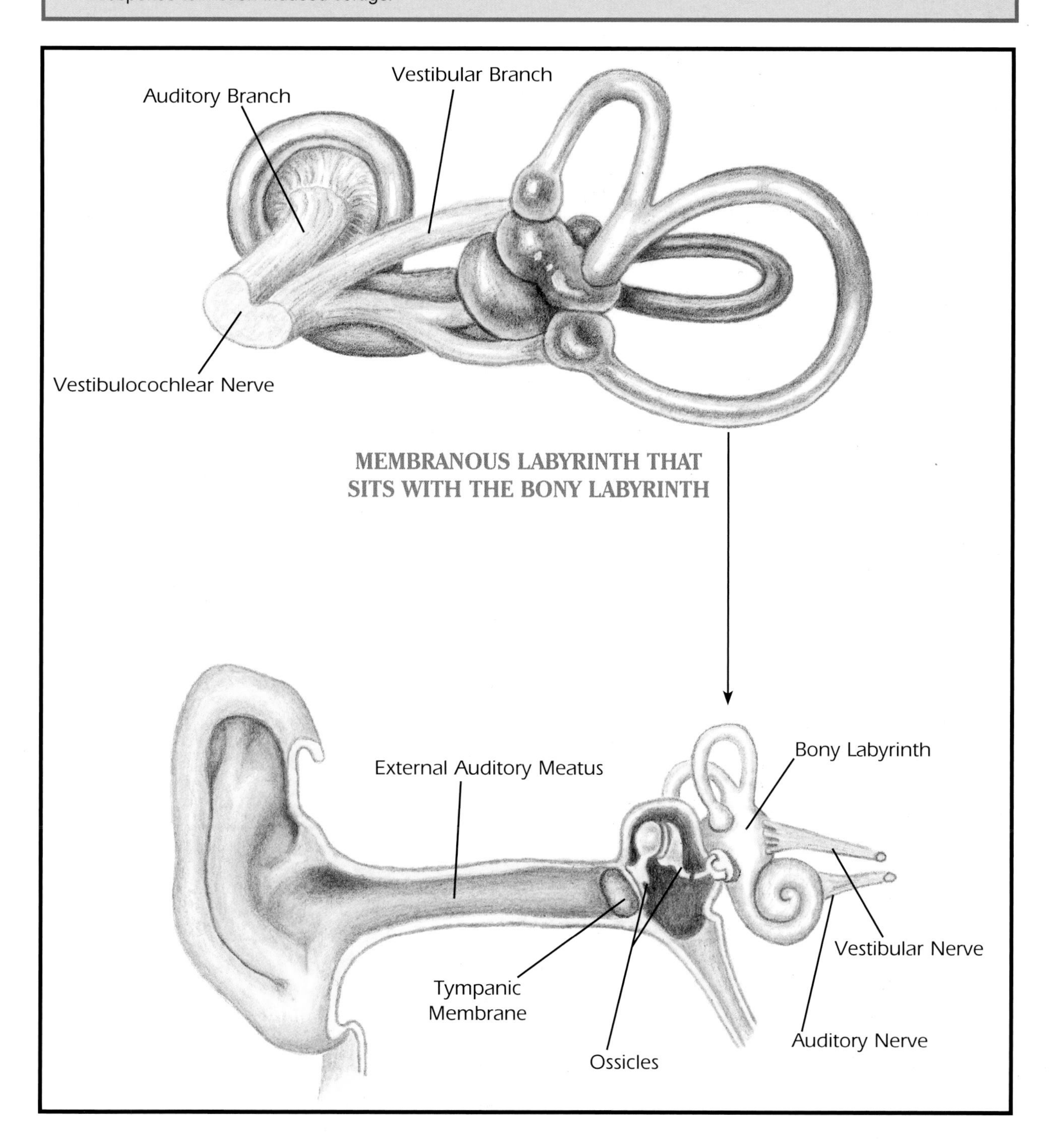

Cranial Nerve 9: Glossopharyngeal Nerve

Carries

- Sensory and motor information

Nuclei Location

- Nucleus ambiguous in the medulla

Function

Sensory

- Taste on the posterior aspect of the tongue

Motor

- Swallowing

Lesion Symptoms

Sensory

- *Loss of taste* sensation on *posterior* aspect of *tongue* (loss of *bitter* taste modality)

Motor

- Loss of the *gag* and *swallowing reflexes*
- *Dysphagia*—difficulty swallowing, leading to choking or food aspiration

Test

- Because they mediate similar functions, the glossopharyngeal and vagus nerves are tested simultaneously.

Sensory

- Use the same testing procedures described for the sensory half of the facial nerve, except apply the procedure to the posterior aspect of the tongue (where bitter tastes are detected).
- Ask the patient to chew on a lemon rind to determine if bitter tastes can be detected.

Motor

- Attempt to elicit the *gag reflex* by swiping a tongue depressor or cotton swab at the back of the throat.
- Observe the patient's ability to *swallow different consistencies* of food.
- Present different consistencies of food one at a time (eg, solid foods, pureed foods, thick liquids, thin liquids).
- Ask the patient to consume each presented food type. Check for food aspiration, coughing, throat clearing, or a wet vocal quality (indicating that the food is pocketing in the larynx).
- If aspiration precautions have been indicated previously, modify the above procedure accordingly.

Glossopharyngeal Nerve Reflexes

Gag Reflex

- Touching of the pharynx elicits contraction of the pharyngeal muscles.

Swallowing Reflex

- Food touching the pharynx elicits movement of the soft palate and contraction of the pharyngeal muscles.

Cranial Nerve 10: Vagus Nerve

Carries

- Sensory and motor information

Nuclei Location

- Dorsal vagal nuclei in the medulla
- Nucleus ambiguous in the medulla

Function

Visceral Branches (to Hollow Organs)

- Visceral branches carry sensory and motor information. Sensory carries *taste information* from the *palate* and *epiglottis*.
- Motor carries parasympathetic information to and from the *heart*, *pulmonary system*, *esophagus*, and the *gastrointestinal tract*.

Skeletal Muscle Branches

- Skeletal muscle branches carry motor information to the *muscles* of the *larynx*, *pharynx*, and *upper esophagus*.
- These muscles are responsible for *swallowing* and *speaking*.

Lesion Symptoms

Visceral Branch Lesions

- Transient tachycardia (irregular rapid heart beat)
- Dyspnea (difficulty breathing)

Bilateral Visceral Branch Lesions

- Asphyxia (suffocation)

Skeletal Muscle Branch Lesions

- Dysphonia (hoarse voice)
- Dysphagia (difficulty swallowing)
- Dysarthria (difficulty articulating words clearly—slurring words)

Test

- The *glossopharyngeal* and *vagus nerves* are *tested simultaneously*. See testing procedures under Glossopharyngeal Nerve section (p 70).
- Observe the patient's ability to speak clearly without slurring words.
- Check for decreased *phonal volume* and *hoarse voice*.

Vagus Nerve Reflexes

Gag Reflex

- Touching of the pharynx elicits contraction of the pharyngeal muscles.

Swallowing Reflex

- Food touching the pharynx elicits movement of the soft palate and contraction of the pharyngeal muscles.

Cranial Nerve 11: Accessory Nerve

Carries

- Motor information

Accessory Nerve Has Two Roots

Cranial Nerve Root

- Emerges from the *nucleus ambiguous* and *joins the vagus nerve*
- Innervates the intrinsic muscles of the *larynx*

Spinal Nerve Root

- Emerges from the *ventral horn* of the *upper cervical spinal cord*
- Innervates the *sternocleidomastoid* (SCM) and *upper trapezius* muscles

Nucleus Location

Cranial Nerve Root

- Medulla

Spinal Nerve Root

- C1 – C5 spinal cord levels in the ventral horn.

Function

Cranial Nerve Root

- Controls *elevation* of the *larynx* during *swallowing*

Spinal Nerve Root

- Innervation of the *sternocleidomastoid* muscle allows for the following:
  - *Head rotation* to contralateral side
  - *Head flexion/extension*
- Innervation of the *upper trapezius* muscle allows for the following:
  - *Shoulder elevation* and *shoulder flexion* above 90 degrees
- The accessory and vagus nerves are the only CNs that *innervate organs* and *glands below* the *neck*.

Lesion Symptoms

Cranial Nerve Root Lesions

- *Dysphagia* secondary to decreased laryngeal elevation

Spinal Nerve Root Lesions

- *Weakness rotating* the *head* to the contralateral side (because the ipsilateral SCM muscle that is affected functions to rotate the head to the opposite side)
- *Weakness flexing* the *head* laterally and forward (SCM)
- *Weakness extending* the *head* (SCM)
- *Weakness elevating* the *shoulder* (shrugging the shoulder) on the ipsilateral side (upper trapezius muscle)
- *Weakness flexing* the *arm* above 90 degrees (upper trapezius muscle) on the ipsilateral side

Test

Cranial Nerve Root

- Place index and middle fingers over patient's Adam's apple (laryngeal muscles).
- Ask patient to swallow.

- Check for normal *rise* and *fall* of *larynx*.
- Check for presence of *dysphagia* using procedures described in the Glossopharyngeal Nerve section (p 70).

Spinal Nerve Root

- Test SCM muscle as follows:
  - Ask patient to *flex head laterally* and *forward*, and *rotate head* to the opposite side (on both the involved and uninvolved sides). Instruct the patient to resist attempts to prevent the desired movement.
  - Compare both sides of the body to *observe symmetry of movement* and *strength*.
  - Check for *atrophy* in involved SCM muscles.
- Test upper trapezius muscles as follows:
  - Ask patient to *shrug both shoulders* toward ears. Instruct the patient to resist attempts to depress the elevated shoulders.
  - Check for *symmetry* in *shoulder movement* and *strength*.
  - Check for *atrophy* of involved *upper trapezius muscles*.

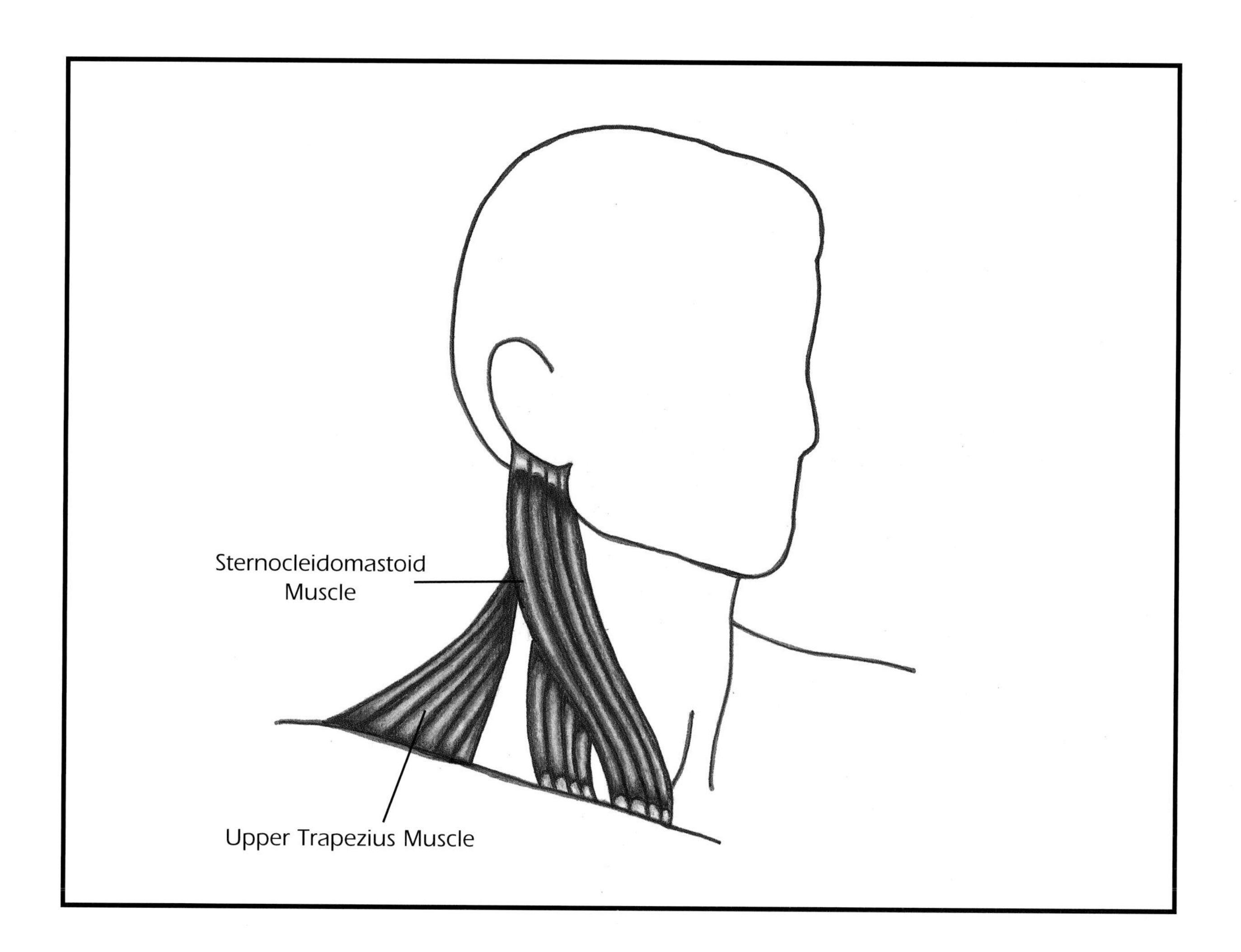

Cranial Nerve 12: Hypoglossal Nerve

Carries

- Motor information

Nuclei Location

- Medulla

Function

- Innervates the muscles of the tongue. The hypoglossal cranial nerve is responsible for *tongue movement.*

Lesion Symptoms

- *Dysarthria* secondary to impaired tongue musculature—inability to produce required movements for sound and word formation.
- *Ipsilateral deviation* of the *tongue*
- *Dysphagia* (because the tongue muscles are needed to manipulate food into a bolus in the mouth and propel bolus to the pharynx)
- *Ipsilateral atrophy* and *paralysis* of tongue

Test

- Ask patient to protrude tongue.
- Note whether tongue *deviates* to the lesion side.
- Note whether there is unilateral or bilateral *atrophy* of the tongue muscles.
- Check for *tongue tremors* or *involuntary tongue movements.*
- Ask patient to move tongue from side to side.
- Check for *asymmetry* in *movement* and *weakness.*
- Ask patient to push tongue against both cheeks.
- Ask patient to resist attempts to depress the cheek as patient pushes the cheek outward.
- Note whether there is *asymmetry* in *tongue strength.*

The Stages of Swallowing: Cranial Nerves that Mediate the Function of Swallowing

Three Stages of Swallowing

- Oral
- Pharyngeal/laryngeal
- Esophageal

Cranial Nerves Involved in Swallowing

- CN 5 Trigeminal Nerve
- CN 7 Facial Nerve
- CN 9 Glossopharyngeal Nerve
- CN 10 Vagus Nerve
- CN 11 Accessory Nerve
- CN 12 Hypoglossal Nerve

Stages of Swallowing

Stage	*Function*	*CN*
Oral	1. Food is brought into the mouth; the lips close.	CN 7
	2. Jaw, cheek, and tongue movements manipulate food into a bolus.	CN 5, 7, 12
	3. Tongue moves bolus to pharynx.	CN 12
	4. Larynx closes.	CN 10
	5. Swallow reflex is triggered.	CN 9
Pharyngeal/ Laryngeal	1. Food moves into pharynx.	CN 9
	2. Soft palate rises to block food from entering the nasal cavity.	CN 10
	3. Epiglottis covers trachea to prevent food from entering the lungs.	CN 10
	4. Pharynx rises and falls during swallowing.	CN 11
	5. Peristalsis moves food to the esophageal entrance, the sphincter opens, and food moves into the esophagus.	CN 10
Esophageal	Peristalsis moves food into the stomach.	CN 10

SECTION 9

Sensory Receptors

- A *sensory receptor* is a specialized nerve cell that is designed to respond to a specific sensory stimulus (eg, touch, pressure, pain, temperature, light, sound, position in space).
- Normally, sensory receptors only accept molecules that have a *complimentary receptor site organization*. Exceptions occur in states of pathology.

Three Types of Sensory Receptors

Exteroceptors

- Exteroceptors are sensory receptors that are adapted for the reception of stimuli from the *external world* (outside of the body).
- Example: visual, auditory, tactile, olfactory, and gustatory receptors.

Interoceptors

- Interoceptors receive sensory information from *inside* the body (eg, from the viscera—hollow organs and glands).
- Interoceptors detect *internal body sensations*—such as stomach pain, pinched spinal nerves, or inflammatory processes in the deep layers of the skin.

Proprioceptors

- Proprioceptors are sensory receptors located in the *muscles*, *tendons*, and *joints* of the body and in the *utricles*, *saccules*, and *semicircular canals* of the *inner ear* (ie, the labyrinths of the inner ear).
- These detect *body position and movement.*

Developmental Classification of Sensory Receptors

Protopathic

- The protopathic sensory system is considered to be old phylogenetically.
- Protopathic receptors are adapted to *identify gross bodily sensation* rather than specific regions of sensation. Thus, protopathic receptors cannot precisely locate the origin of pain.
- Protopathic receptors detect *crude touch* and *dull pain* rather than discriminative touch and sharp pain.
- Evolutionary function: enables the organism to detect possible (but not imminent) danger in the environment.

Epicritic

- With regard to phylogeny of the species, the epicritic sensory system is considered to have developed more recently than the protopathic system.
- Epicritic receptors can *detect sensation* with *precision*, *accuracy*, and *acuteness*. *Discriminative touch*, *sharp pain*, *exact joint position*, and the *exact localization of a stimulus* are within the functions of the epicritic system.

Children with Sensory Processing Problems

- The epicritic and protopathic sensory systems may not function optimally in children with sensory processing disorders.

A Child with a Dominant Protopathic Sensory System

- This child may not be able to receive adequate sensory stimulation from the environment, or the sensory data that enter the central nervous system (CNS) are experienced by the child as if they had been dulled. It would be as if the child received all sensory data filtered through gloves, earmuffs, and a heavy winter coat that decreased the child's direct sensory stimulation from the environment.
- Some children who display self-stimulation behaviors (eg, rocking, biting oneself, humming continuously) may have a dominant protopathic system. They may seek a great deal of sensory stimulation in an attempt to compensate for their inability to receive adequate sensory stimulation from the environment.
- Such sensory stimulation is necessary for the development of the CNS.

A Child with a Dominant Epicritic Sensory System

- These children may be hypersensitive to sensation from the environment. Some sensations—that may be innocuous to others—may be experienced as painful or intolerable to children with heightened epicritic systems.
- Tactile defensiveness is a condition in which certain normal sensations are experienced as highly noxious—such as the feel of water hitting the skin during bathing and showering or the feel of clothing labels rubbing against one's neck. Children with hypersensitive epicritic sensors may refuse to take showers and baths or may insist on wearing shorts in the winter because they cannot tolerate the feel of clothing against their skin.
- Children with hypersensitive proprioceptors may be unable to tolerate movement caused by gravitational displacement. Gravitational insecurity is a condition in which children experience gravitationally induced movement as frightening. For example, such children may be abnormally fearful of using playground equipment such as a swing, sliding board, seesaw, or jungle gym. While such playground games commonly provide exhilarating sensory stimulation to many children, those with gravitational insecurity may experience these sensations as frightening.

- Evolutionary functionallows the organism to explore the environment with precise detail, thus allowing the ability to detect imminent danger.

Sensory Receptors Classified by Anatomical Location

Cutaneous Receptors

- Cutaneous receptors respond to *pain*, *temperature*, *pressure*, *vibration*, and *discriminative* touch.
- They are found in the superficial and deep layers of skin.
- Cutaneous receptors are also classified as *exteroceptors* (coming from an external source) or *interoceptors* (eg, inflammatory process in the deep layers of the skin).

Muscle, Tendon, and Joint Receptors

- These sensory receptors are located in the *muscles*, *tendons*, and *joints* of the body.
- They detect *muscle length*, *muscle tension*, *joint position*, *deep muscular and joint pain*, and *tendonitis*.
- These sensory receptors are also classified as *proprioceptors*.

Visceral Receptors

- Visceral receptors respond to *pressure* and *pain* from the internal organs.
- They are also considered to be *interoceptors*.

Special Sense Receptors

Visual Receptors

- Rods and cones of the retina. Considered an exteroceptor

Olfactory Receptors

- Hair cells located in the mucous lining of the nasal canal. Considered an exteroceptor

Auditory Receptors

- Hair cells of the cochlea. Considered an exteroceptor

Gustatory Receptors

- Taste buds on the tongue. Considered an exteroceptor

Equilibrium

- Semicircular canals, utricles, and saccules of the inner ear. Considered proprioceptors

Sensory Receptors Classified by Structural Design

Mechanoreceptors

- Stimulated by *mechanical deformity*
- Example: Hair cells of the labyrinth system, some receptors of the skin, and stretch receptors of the skeletal muscles
- Detect *touch*, *pressure*, *vibration*, *proprioception*, *equilibrium*, and *audition*

Thermal Receptors

- Detect changes in *temperature*
- Located all over an organism's external and internal body

Chemoreceptors

- Respond to the presence of a particular *chemical*
- Example: Chemoreceptors are the sensory receptors involved in *olfaction* and *gustation*.

Photoreceptors

- Detect light on the retina of the eye

Receptor Fields

- A receptor field is a body area that contains specific types of sensory receptor cells.
- When stimulated, the receptors on a specific receptor field become activated.

Location of Small and Large Receptor Fields

Small Receptor Fields

- Located on body areas with the *greatest sensitivity—lips, hands, face, soles of feet*

Large Receptor Fields

- Located on *legs, abdomen, arms, back.*

Function of Small vs Large Receptor Fields

Small Receptor Fields

- *Fine discrimination*
- Necessary for the exploration of the environment

Large Receptor Fields

- *Gross discrimination*

Cortical Representation of Receptor Fields (The Homunculus)

- Small receptor fields (eg, those of the mouth, tongue, hands, feet) are given a *large amount of cortical representation* on both the sensory and motor homunculi.
- Large receptor fields (eg, legs, abdomen, arms, back) are given *smaller amounts of cortical representation on the homunculi.*

SECTION 10

Neurons and Action Potentials

Neuron

- A neuron is the electrically excitable nerve cell and fiber of the nervous system.
- It is composed of the following:
  - Cell body (soma), nucleus, dendrites, main axon branch (with possible axon collaterals), and terminal boutons.

Cell Body

- The cell body contains the *nucleus* of the neuron, which stores the genetic codes of the organism.

Dendrites

- Dendrites are the *tree-like processes* that attach to the cell body.
- They receive messages from the terminal boutons of a presynaptic neuron.
- Dendrites can *bifurcate*, or produce additional dendritic branches.
- Bifurcation increases the neuron's receptor sites.

Axon Hillock

- The axon hillock is the region where the cell body and axon attach.

Axon

- An axon is the fiber emerging from the axon hillock and extending to the terminal boutons.
- Axons transmit *action potentials*, or nerve signals, to the terminal boutons.

Myelin

- Axons are covered by a cellular sheath called myelin.
- The myelin is composed of lipids and proteins.
- It acts as an *insulating substance* of the axon and serves to conduct *nerve signals*.
- The more myelin the axon has, the faster its conduction rate.
- There are two primary types of myelin:
  - *Schwann cells*
    - In peripheral nervous system (PNS)
    - *Able to regenerate* if damaged
  - *Oligodendrites*
    - In central nervous system (CNS)
    - *Cannot regenerate* if damaged

Nodes of Ranvier

- Nodes of ranvier are spaces between the myelin where nerve signals jump from one node to the next in the process of conduction.

Multiple Sclerosis: Disease of the Myelin

- Characterized by random demyelination of the CNS.
- The disease is characterized by periods of exacerbation and remission over many years.
- In the early stages of multiple sclerosis (MS), there is normal or near-normal neurologic function between exacerbations. As the disease progresses, remissions grow shorter and are marked by less improvement.
- Signs and symptoms can be variable because demyelination can occur in a wide variety of locations in the CNS.
- Sensory symptoms:
  - Numbness, paresthesias, and Lhermitte's sign (causalgia radiating down the back and lower extremities—elicited by neck flexion)
- Motor symptoms:
  - Abnormal gait, bladder and sexual dysfunction, vertigo, nystagmus, fatigue, and speech disturbance
  - MS of the spinal cord produces asymmetrical weakness secondary to plaques that interfere with the descending motor tracts.
  - Ataxia of the limbs occurs secondary to an interruption of nerve signal conduction in the dorsal columns.
  - MS is considered to be a disease process that affects the upper motor neurons (UMNs).
- Commonly affected areas:
  - Optic nerve ⟶ visual field acuity
  - Corticospinal tracts ⟶ muscle strength
  - Corticobulbar tracts ⟶ speech and swallowing functions
  - Cerebellar tracts ⟶ gait and coordination
  - Spinocerebellar tracts ⟶ balance
  - Medial longitudinal fasciculus ⟶ conjugate gaze of the extraocular eye muscles
  - Dorsal columns ⟶ discriminative touch, pressure, vibration, proprioception, kinesthesia

Amyotrophic Lateral Sclerosis: Death of Upper and Lower Motor Neurons

- Amyotrophic lateral sclerosis (ALS) is a severe degenerative neurologic disorder affecting motor function.
- ALS is a disease of both the CNS and the PNS. The UMNs of the cerebral cortex and the lower motor neurons (LMNs) of the ventral horn of the spinal cord are both affected.
- Sensory and cognitive functions remain intact.
- The death of UMNs and LMNs leads to denervation, causing muscle atrophy and spasticity.
- UMN lesions result in the following:
  - Muscle weakness
  - Spasticity
  - Loss of fine motor control
- Both UMN or LMN lesions can account for the following:
  - Dysphagia (difficulty swallowing)
  - Dysarthria (difficulty articulating words clearly)
  - Dysphonia (difficulty projecting one's voice audibly)
- LMN lesions result in the following:
  - Fasciculations
  - Muscle weakness
  - Muscle atrophy
  - Hyporeflexia
- Common early symptoms:
  - Muscle cramps involving the distal legs
  - A slow, progressive weakness and atrophy of the distal muscle groups of one upper extremity
- The above early symptoms are then followed by a wider spread of muscle weakness in surrounding body areas.
- Eventually UMNs and LMNs involving multiple limbs and the head are affected.
- In advanced stages of the disease, muscles of the palate, pharynx, tongue, neck, and shoulders are affected.
- Death commonly occurs from denervation of the respiratory musculature.

Myasthenia Gravis: Disorder of the Neuromuscular Junction

- Myasthenia gravis is a chronic autoimmune disorder that affects the neuromuscular junction of voluntary muscles.
- The disease process is characterized by production of acetylcholine receptor antibodies that destroy acetylcholine receptors at the neuromuscular junction. Essentially, acetylcholine receptor antibodies block the transmission of acetylcholine across the neuromuscular junction. The production of acetylcholine receptor antibodies is thought to be an autoimmune response.
- This results in severe muscular weakness and fatigue.
- Typically, the disease first affects eye and head musculature. Progression continues to the limbs and sometimes to respiratory muscles.
- Anticholinesterase agents are commonly used to reduce the breakdown of acetylcholine in the neuromuscular synaptic space.
- Increased availability of acetylcholine in the synaptic space enhances muscular contraction.
- Corticosteroid drugs are used to suppress the immune response if anticholinesterase agents are ineffective.

Axon Collaterals

- Axon collaterals project from the main axon structure.
- They serve to transmit nerve signals to several parts of the nervous system simultaneously.

Terminal Boutons (Synaptic Boutons)

- Terminal boutons emerge from the end branches of the axon and contain the *neurotransmitter* substances.

Synaptic Cleft

- The synaptic cleft is the space between a presynaptic neuron's terminal boutons and a postsynaptic neuron's dendrites.
- The *terminal boutons* release their *neurotransmitter* substances into the *synaptic cleft*.

Neurotransmitter

- The neurotransmitter is a chemical stored in the *terminal boutons*.
- It is released into the *synaptic cleft* to transmit messages to another neuron.

Presynaptic Neuron

- The presynaptic neuron is the *first order neuron* that releases its neurotransmitter into the synaptic cleft.

Postsynaptic Neuron

- The postsynaptic neuron is the *second order neuron*.
- It receives the presynaptic neuron's neurotransmitter substance from the synaptic cleft but only if the neurotransmitter possesses the specific molecules that can bind to the postsynaptic neuron's receptor sites.

Re-Uptake Process

- This is the process by which a neurotransmitter is reabsorbed into the presynaptic neuron's terminal boutons.

Down Regulation

- Down regulation is a decrease in the number of receptors on the postsynaptic neuron (for a specific neurotransmitter), often due to long-term exposure to the neurotransmitter.
- In response to too much neurotransmitter, the receptor sites on the second order neuron will decrease.

Synaptic Delay

- Synaptic delay is the time required for the neurotransmitter to diffuse across a postsynaptic neuron's membrane.
- Average time is 1 to 2 milliseconds.

Antitransmitter

- An antitransmitter is a chemical substance that *breaks down a neurotransmitter* so that the postsynaptic neuron can *repolarize* in order to fire again.
- Antitransmitters terminate the postsynaptic neuron's response.
- They are located in the *synaptic cleft.*
- Example: Acetylcholinesterase (AChE) is the antitransmitter for acetylcholine (ACh). ACh is a neurotransmitter important for activating neuromuscular–joint movements.

Post-Tetanic Potentiation

- This occurs in synapses that are frequently used.
- When the presynaptic bouton becomes excited, it releases greater amounts of neurotransmitter substance.
- The *postsynaptic neuron then has prolonged and repetitive discharge after firing* due to too much neurotransmitter release or too slow antitransmitter work.

Decreases in Synaptic Transmission

Anoxia

- Anoxia is lack of oxygen.
- Synaptic transmission begins to fail within 45 seconds of anoxia.

Paralysis Due to Poison

- Paralysis due to poison—such as snake venom or curare—occurs because the *poison blocks ACh at the neuromuscular junction*.
- This causes a decreased postsynaptic potential at the cell membrane of the muscle.

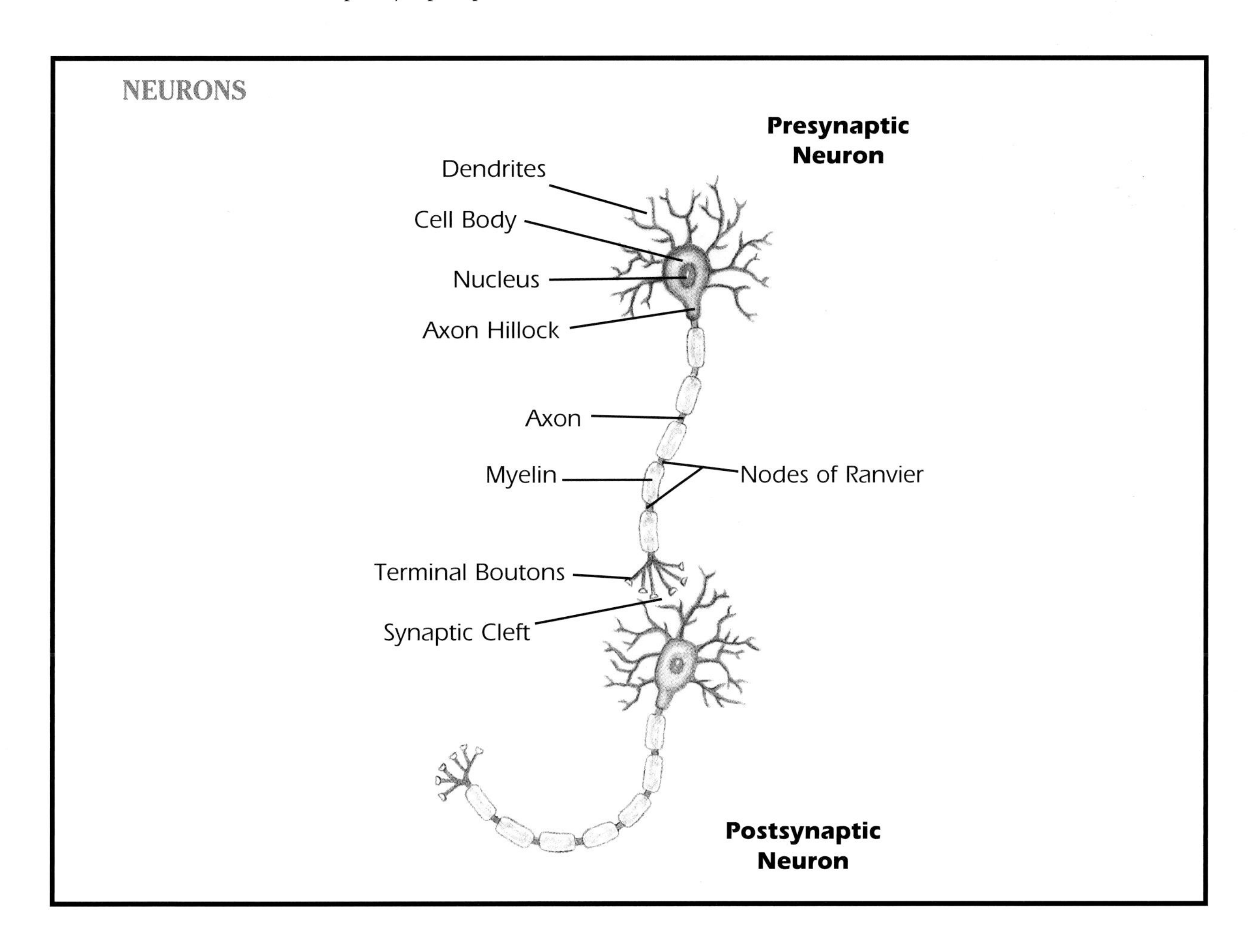

- Example: *Botulism* (food poisoning) is a condition in which the release of ACh at the presynaptic bouton is blocked. This causes fatigue of the neurotransmitters at the neuromuscular junction, resulting in skeletal paralysis.

Spasms Due to Cholinergic Drugs

- Cholinergic drugs are chemicals that bind with and *activate the same receptors as ACh.*
- These drugs increase the effects of ACh at the neuromuscular junction, thus producing spasms in the muscle.
- *Tardive dyskinesia* is a condition caused by cholinergic drugs. Patients may experience full body spasms, lip smacking, and tongue protrusion.

Synaptic Fatigue

- Synaptic fatigue occurs as a result of a *neurotransmitter depletion* due to the repetitive stimulation of a presynaptic neuron.
- Synaptic fatigue underlies the addiction process. Every time an individual uses an addictive substance, the CNS is abnormally flooded with the neurotransmitter dopamine. This causes the brain to decrease its own natural production of dopamine, leading to cravings for the addictive substance.

Electrical Potentials at Synapses

- The release of neurotransmitters into the synaptic cleft results in the stimulation of receptors on the postsynaptic neuron's cell membrane.
- The chemical stimulation of these receptors can cause the *membrane ion channel to open.*
- The flux of ions across the postsynaptic membrane *generates a postsynaptic potential.*

Postsynaptic Potentials

- Postsynaptic potentials are changes in ion concentration on the *postsynaptic membrane.*
- When a neurotransmitter binds to a receptor site on a postsynaptic membrane, the result may be either *depolarization* or *hyperpolarization* of the postsynaptic membrane.
- Depolarization of the cell membrane causes the neuron to become excited. An *excitatory postsynaptic potential* (EPSP) is generated.
- Hyperpolarization of the cell membrane causes inhibition of the neuron. An *inhibitory postsynaptic potential* (IPSP) is generated.

Membrane Potentials

- The membrane potential is the electrical charge that travels across the cell membrane.
- It is the difference between the chemical composition inside and outside the cell—in other words, the *sodium potassium balance inside and outside the cell.*
- If the membrane potential is strong enough, it causes an action potential.

Action Potentials

- An action potential is the brief electrical impulse that provides the basis for conduction of nerve signals along the axon.
- It results from the brief changes in the cell's membrane permeability to sodium and potassium ions.
- A strong enough action potential will cause the neuron to become excited and start the conduction process.

Sequence of Events of Synaptic Transmission

- An action potential arrives at the presynaptic terminal boutons.
- The membrane of the presynaptic terminal boutons depolarizes, thus causing the opening of the voltage-gated CA++ channels.
- The influx of CA++ (in the terminal boutons) triggers the release of neurotransmitters into the synaptic cleft.
- If the neurotransmitter has compatible molecules, it will bind to the receptor sites of the postsynaptic neuron.

Events of Excitatory Postsynaptic Potentials

- Na+ channels remain closed during the state of a resting membrane.
- The Na+ channels open when a neurotransmitter, released into the synaptic cleft, binds to the membrane receptors of the postsynaptic cell membrane.
- The resulting influx of Na+ depolarizes the membrane and causes excitation of the postsynaptic neuron.

Events of Inhibitory Postsynaptic Potentials

- Cl– channels remain closed during the state of a resting membrane.
- Cl– channels open when a neurotransmitter is released into the synaptic cleft and binds to receptor sites of the postsynaptic cell membrane.
- The resulting influx of Cl– hyperpolarizes the cell membrane, thus causing inhibition of the postsynaptic neuron.

SECTION 11

Special Sense Receptors

- The *special sense receptors* are designed to transmit sensory information for olfaction, gustation, vision, audition, and equilibrium.

OLFACTION

Olfactory Receptors

- *Cilia*, or hair cells, located in the nostrils and nasal membranes

Physiology

- Some chemical stimulus (an odor) is received by the *hair cell receptors*.
- The odor molecules dissolve in the mucous that bathes the *receptors* on the cilia.
- The dissolution of the molecules causes a *membrane potential* in the receptor endings.
- If the membrane potential is strong enough, an *action potential* will be generated.
- The action potential is then propagated down cranial nerve (CN) 1, the *olfactory nerve*.

Olfactory Pathway

- The olfactory pathway leads from the *nasal membrane* to the *olfactory bulb* and *tract*, and travels to the *olfactory cortex* (*pyriform cortex* in the temporal lobe).
- The olfactory cortex is part of the *limbic system*, along with the amygdala and hippocampus.
- The amygdala and hippocampus play roles in the storage of long-term memories.
- The olfactory cortex takes information from the *hippocampus* and projects it to the hypothalamus.
- From the *hypothalamus*, the olfactory information is then projected to the *dorsomedial thalamus* and finally to the *orbitofrontal cortex* for conscious association of the odor with previously stored memories.
- The connection between the olfactory pathway and the hippocampus is phylogenetically old. This is the connection that triggers the *remembrance of old memories in response to odors*. For example: the memories of one's grandmother may be elicited in response to the smell of the kind of cookies she baked, or the memories of playing softball as a child may be elicited in response to the smell of freshly cut grass on a summer day.
- The memories that are elicited as a result of the connection between the olfactory tract and hippocampus have a *strong emotional component*. This results because the limbic system regulates emotion.

Therapeutic Significance

- Olfactory stimulation is used with comatose patients to facilitate CNS arousal and activity.
- Aroma therapy is based on the principle that odors can elicit enhanced moods as a result of the limbic–olfactory connection.

Olfactory Lesions

A Bilateral Lesion of the Olfactory Nerve

- This will result in the loss of smell (anosmia). This type of lesion occurs frequently in patients with traumatic brain injury (TBI) secondary to brainstem involvement.

A Lesion of the Olfactory Cortex (Pyriform Cortex)

- This results in the loss of smell.

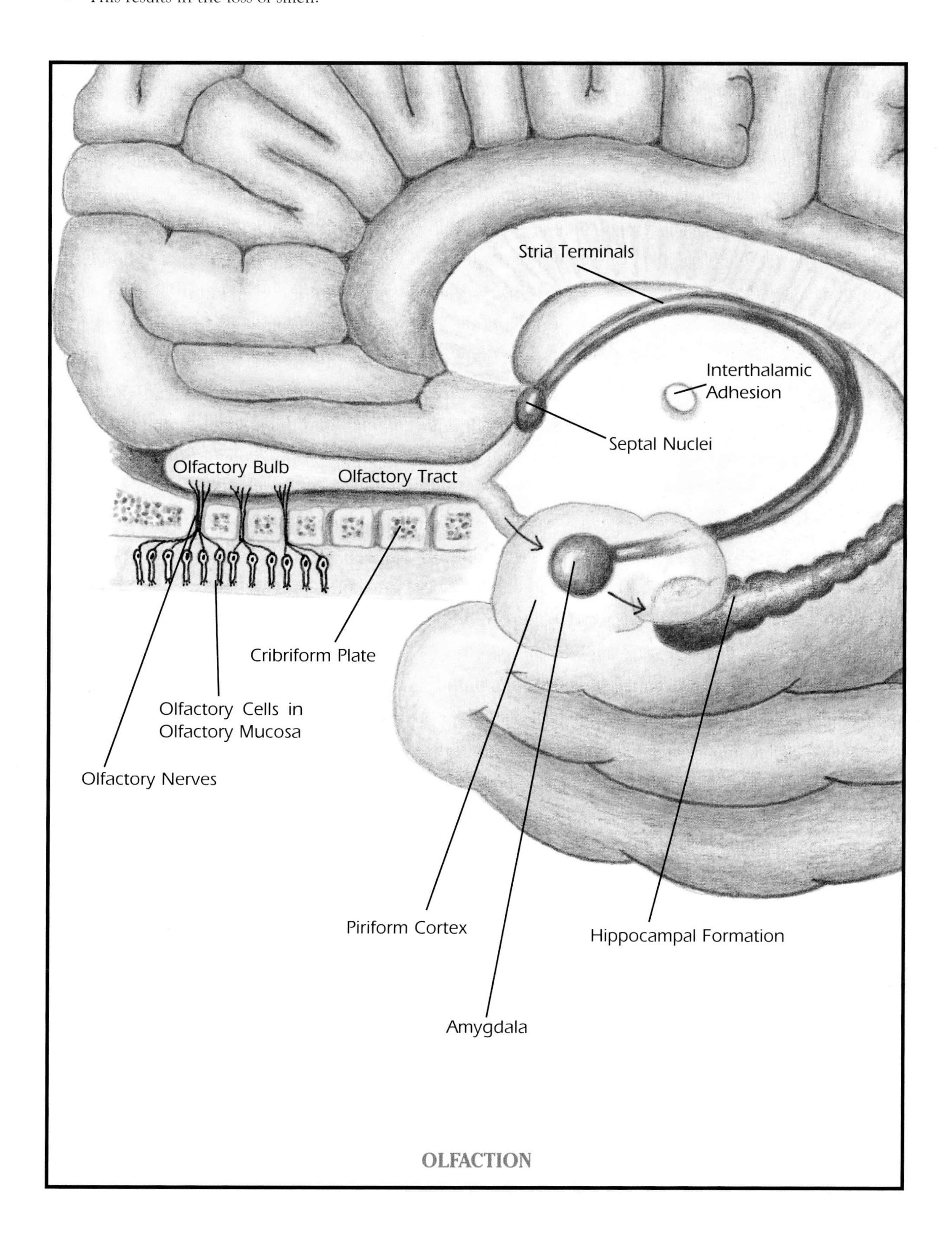

OLFACTION

GUSTATION

Gustatory Receptors

- The gustatory receptors are the *taste buds* found on the folds of *papillae* located on the surface of the tongue.
- Papillae are the small protuberances on which the taste buds lie.
- The taste buds form synapses with the sensory neurons that convey taste information to the brain.

Physiology

- Saliva dissolves food in the mouth.
- Dissolved ions enter the pores of the papillae and taste buds.
- This causes a *membrane potential.*
- If the membrane potential is strong enough, it will cause an action potential.

Gustatory Pathway

- The gustatory pathway leads from the *taste bud receptors* to the solitary nucleus in the *medulla.*
- Gustatory information is then propagated to the *amygdala, hypothalamus,* and *thalamus.*
- From the thalamus, gustatory information then travels to the *primary gustatory cortex* (located in the *frontal insula*) for conscious interpretation.

The Connection Between Taste and Smell

- The sense of smell is critical for the function of taste.
- If the sense of smell is lost, the sense of taste will also be lost.
- The *orbitofrontal cortex* seems to be a brain region that integrates gustatory and olfactory information.
- When the orbitofrontal cortex cannot integrate olfactory and gustatory information, *gustation is lost.*
- Example: Gustation is lost during *common colds* when olfaction is severely diminished due to respiratory congestion.
- Damage to the orbitofrontal region is common in *TBI* and may account for the lost sense of smell and taste that many individuals with TBI experience.

Therapeutic Significance

- Gustatory stimulation is used in the sensory stimulation treatment of comatose patients.
- It is also used to facilitate oral motor function in children and adults with oral musculature dysfunction.

Gustatory Information is Received from Cranial Nerves 7, 9, and 10

- The taste receptors on the anterior of the tongue are mediated by CN 7.
- The taste receptors on the posterior of the tongue are mediated by CN 9.
- The taste receptors on the palate and epiglottis are mediated by CN 10.

Super Tasters vs Nontasters

- Humans fall into at least two primary categories of tasters—super tasters and nontasters.
- Super tasters can detect phenylthiocarbamide—a pungent substance that can be synthesized in a lab—while nontasters cannot detect the substance.
- Super tasters are more sensitive to the four basic tastes—sweet, sour, salty, and bitter. They are also able to detect a proposed fifth taste called umami. Umami is a Japanese word meaning meaty or hearty.
- While the tongue normally contains 100 to 200 papillae per square centimeter, super tasters may have twice as many. Nontasters have half the normal number of papillae but each papillae seems to be much larger.
- Other studies have found that super tasters tend to be thinner than nontasters. Because they can detect taste more acutely than nontasters, they may be able to feel satiated more quickly and eat less than nontasters.

Specific Taste Regions of the Tongue

- The tip of the tongue can detect all tastes but predominantly detects sweet tastes.
- Salty tastes are detected on the sides of the tongue.
- Bitter and sour tastes are detected on the posterior of the tongue.

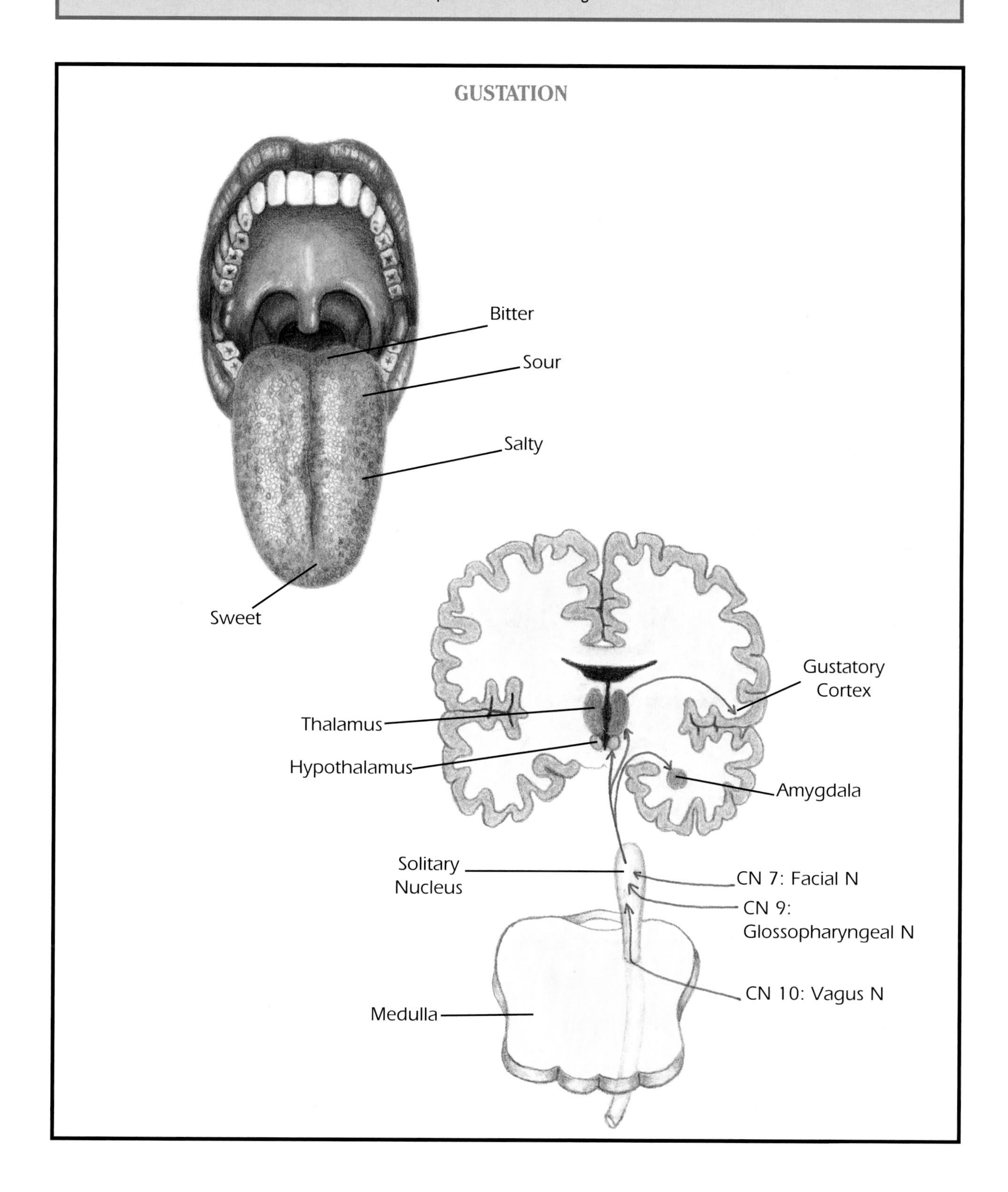

Vision (Acuity)

Anatomy of the Eyeball

Sclera

- The sclera is the outer layer of the eyeball.

Cornea

- The transparent anterior portion of the sclera is called the cornea.
- The cornea lies in front of the iris and pupil.

Choroid

- The choroid is the middle layer of the eyeball.
- It contains the *iris* and *lens*.
- It extends behind the retina.
  - *Iris*
    - The iris is the circular structure that forms the colored portion of the eye.
    - It controls the *size* of the *pupil* opening.
  - *Lens*
    - The lens is the structure that focuses *light rays* on the *retina*.

Anterior Chamber

- The anterior chamber contains the gelatinous fluid between the cornea and lens called *aqueous humor*.

Posterior Chamber

- The posterior chamber is located directly behind the lens and extends to the retina.
- It contains both aqueous *humor* and *vitreous humor*.

Pupil

- The pupil is the *circular black opening* in the center of the iris.
- Light enters through the pupil.
- The amount of light permitted is controlled by the constriction of the iris.

Fovea Centralis

- The fovea is the *central region* of the *retina*.
- It primarily contains *cone receptors*.
- The fovea is the location upon which incoming visual images are focused.

Retina

- The retina is the photosensitive layer at the posterior of the eyeball.
- It contains the rods and cones—the *photoreceptors*.
  - *Rods*
    - Specialized receptors for *peripheral vision*
    - Function optimally in *dim light*
  - *Cones*
    - Specialized receptors for *color vision* and *acuity*
    - Function optimally in *bright light*
    - In dim light, humans are color blind and lack acuity.

Visual Receptor Pathway

- Light waves enter the eye and are refracted by the *cornea* and *lens*.
- The light waves are then focused upon either the *fovea* (where the cones are located) or the *peripheral retina* (where the rods are located).
- A chemical reaction occurs in the rods and cones, and a *membrane potential* is generated.
- If the membrane potential is strong enough, an *action potential* is then propagated along CN 2—*the optic nerve*.
- Visual signals are then propagated down the *optic nerve* to the *optic chiasm*.

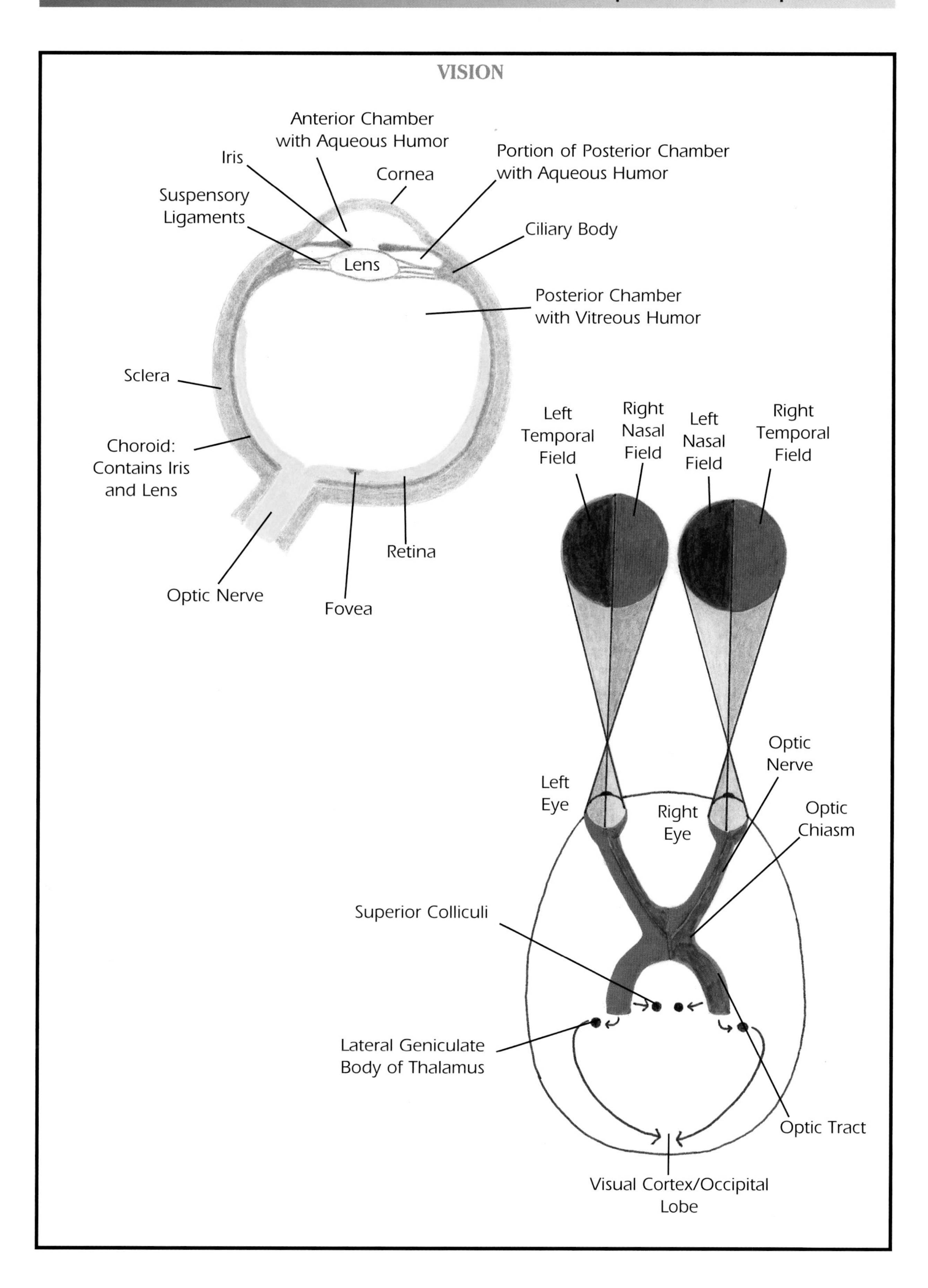
VISION
Anterior Chamber with Aqueous Humor
Iris
Cornea
Portion of Posterior Chamber with Aqueous Humor
Suspensory Ligaments
Ciliary Body
Lens
Posterior Chamber with Vitreous Humor
Sclera
Choroid: Contains Iris and Lens
Retina
Optic Nerve
Fovea
Left Temporal Field
Right Nasal Field
Left Nasal Field
Right Temporal Field
Optic Nerve
Left Eye
Right Eye
Optic Chiasm
Superior Colliculi
Lateral Geniculate Body of Thalamus
Optic Tract
Visual Cortex/Occipital Lobe

Light-Sensitive Receptors in the Eye: Role in Setting the Internal Clock

- A new type of receptor cell has been found in the retina and appears to be comprised of photoreceptive ganglion cells.
- These cells contain a light-sensitive protein called melanopsin—a protein that helps convert light into an electrochemical signal that is sent to the pineal gland and hypothalamus.
- The hypothalamus and pineal gland regulate the body's circadian rhythms, or internal clock.
- Melatonin is produced by the pineal gland and plays a role in the ability to fall asleep and maintain sleep. Melatonin production ceases when the photoreceptive ganglion cells in the eye detect light.
- Researchers suggest that the photoreceptors may help regulate the body's circadian rhythms. Without the ability to detect light, the internal clock quickly begins to dysfunction. This may underlie such conditions as seasonal affective disorder and sleep disturbances.
- Researchers have begun to experiment with functional applications that use certain types of light to enhance health and cognitive stimulation. For example, some hospitals are now using warm, reddish light in patient rooms for its calming effect. Similarly, bluer, alertness-enhancing light is being used in work settings where employees need to maintain a high level of vigilance for lengthy periods of time.

- The optic chiasm is where the optic nerves join together at the base of the brain. The optic chiasm is a midline structure. The chiasm joins the optic tracts.
- Visual signals then project from the optic tracts to two locations: *superior colliculi* of the midbrain and the *lateral geniculate nucleus* of the thalamus.
- The fibers that project to the superior colliculi then travel along the *medial longitudinal fasciculus* back to the *thalamus*.
- All visual fibers that travel through the thalamus then continue on to the *occipital lobe*.
- In the occipital lobe, visual signals first travel to the *primary visual cortex* (V1). V1 is responsible for the *detection* of a visual stimulus.
- Visual signals then travel from V1 to the *visual association cortices* for *interpretation* of the visual stimulus.

Visual Field Pathways

- Visual information from the *nasal fields* is refracted by the lens. As a result, visual information from the nasal fields travel along the *lateral aspect* of the optic nerves and tracts.
- Visual information from the *temporal fields* is also refracted by the lens. As a result, visual information from the temporal fields travel along the *medial aspect* of the optic nerves and tracts.
- Visual information from the *temporal fields cross* at the *optic chiasm*.
- The lens also inverses the visual image on the retina. As a result, visual information from the right visual field is focused on the left region of the retina.
- Visual information from the left visual field is focused on the right region of the retina.
- Each cerebral hemisphere receives visual information from the contralateral visual field.
- The right cerebral hemisphere receives visual information from the left visual field.

Primitive Visual System

The visual pathways that travel from the optic tracts to the superior colliculi (of the midbrain) and lateral geniculate nucleus (of the thalamus) allow the brain to process and interpret visual data on an unconscious (noncortical) level. These pathways form a primitive visual system that enables organisms to respond quickly to visual stimuli without conscious interpretation. For example, it has been documented that patients with cortical blindness—in which the occipital lobe cannot interpret visual data but the external visual anatomy remains intact and can process visual stimuli—are able to identify images with greater accuracy than chance. The evolutionary significance of these pathways may have been to allow organisms the ability to respond to danger more quickly than the time required for the occipital lobes to consciously detect and interpret visual data and send that information to the frontal lobe for conscious decision-making.

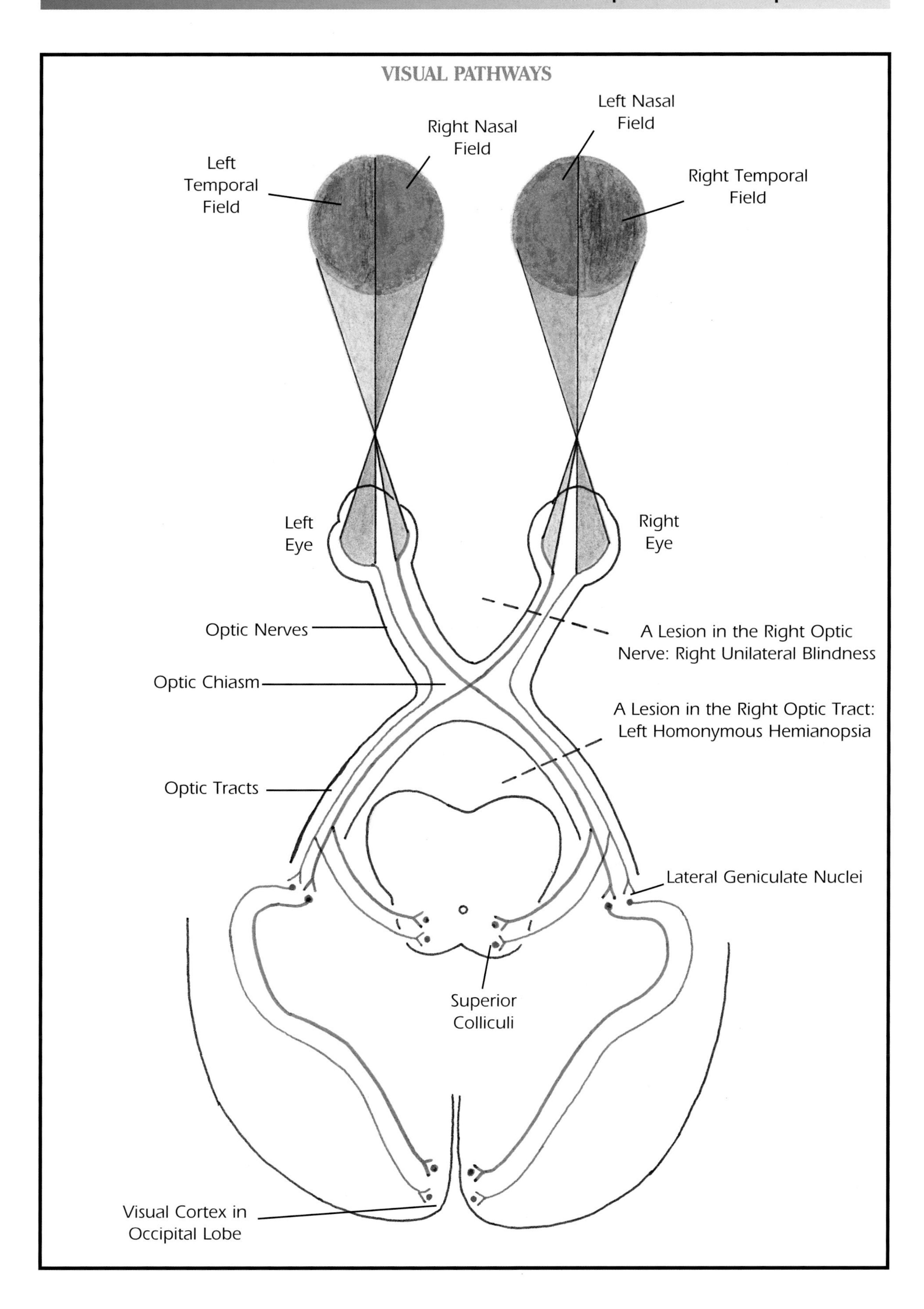
VISUAL PATHWAYS
Left Nasal Field
Right Nasal Field
Left Temporal Field
Right Temporal Field
Left Eye
Right Eye
Optic Nerves
A Lesion in the Right Optic Nerve: Right Unilateral Blindness
Optic Chiasm
A Lesion in the Right Optic Tract: Left Homonymous Hemianopsia
Optic Tracts
Lateral Geniculate Nuclei
Superior Colliculi
Visual Cortex in Occipital Lobe

- The left cerebral hemisphere receives visual information from the right visual field.
- Thus, if the left visual field is lost, it is due to a lesion in the right optic tract or right cerebral hemisphere. If the right visual field is lost, it is due to a lesion in the left optic tract or left cerebral hemisphere.

Pathology of the Visual Pathways

Contralateral Homonymous Hemianopsia

- A contralateral homonymous hemianopsia is a loss of the visual field on the *opposite side of the lesion.*
- A left visual field cut, or *left contralateral homonymous hemianopsia,* results from a lesion in the *right optic tract.*
- A right visual field cut, or *right contralateral homonymous hemianopsia,* results from a lesion in the *left optic tract.*

Bitemporal Hemianopsia

- A bitemporal hemianopsia occurs when the *temporal fields* in *both eyes* have been *lost.*
- It results from a *lesion* to the *central optic chiasm.*
- Bitemporal hemianopsia results in *tunnel vision.*

Blindness as a Result of Optic Pathway Damage

- *Unilateral* blindness can result from a *lesion* to the *ipsilateral optic nerve.*
- *Bilateral* blindness can result from a large lesion that knocks out the entire *optic chiasm.*

A Blind Spot in the Visual Field

- A blind spot inside a visual field suggests retinal damage (not optic nerve or tract damage).

Audition

Auditory Receptor Anatomy

Cochlea

- The cochlea is a *fluid filled structure* located in the *inner ear.*
- It contains a structure called the *organ of Corti.*

Organ of Corti

- The organ of Corti contains the *auditory receptors,* or hair cells that are attached to the basilar membrane at the base of the organ of Corti.

Auditory Pathway

- Sound, or vibration, is the stimulus for the organ of Corti.
- Sound waves travel through the *external auditory meatus* (the ear canal) to the *tympanic membrane* (the eardrum).
- When sound waves reach the tympanic membrane, the membrane vibrates and causes the *ossicles* to vibrate.
- The ossicles are the bones of the middle ear—the *malleus* (hammer), *incus* (anvil), and *stapes* (stirrup).
- The stapes presses against the *oval window* causing it to vibrate.
- Cochlear vibration then causes the *basilar membrane* to flex back and forth.
- Pressure changes of the cochlear fluid are then transmitted to the *round window.*
- In response, the round window flexes back and forth in a manner opposite to the movement of the oval window.
- The hair cells of the basilar membrane begin to bend causing a *membrane potential* to be generated.
- If the membrane potential is strong enough, an *action potential* will be propagated down *CN 8, the vestibulocochlear nerve.*
- The auditory message is then sent to the *cochlear nucleus* in the medulla. From here, fibers project to the *superior olivary nucleus* (medulla level), the *inferior colliculus* (midbrain level), and to the *medial geniculate nucleus* of the thalamus.
- From the thalamus, auditory fibers project to the *primary auditory cortex* (A1) in the temporal lobe. A1 is responsible for the *detection* of auditory stimuli.
- The auditory messages are then sent to the *auditory association areas* (A2+) for *interpretation.*

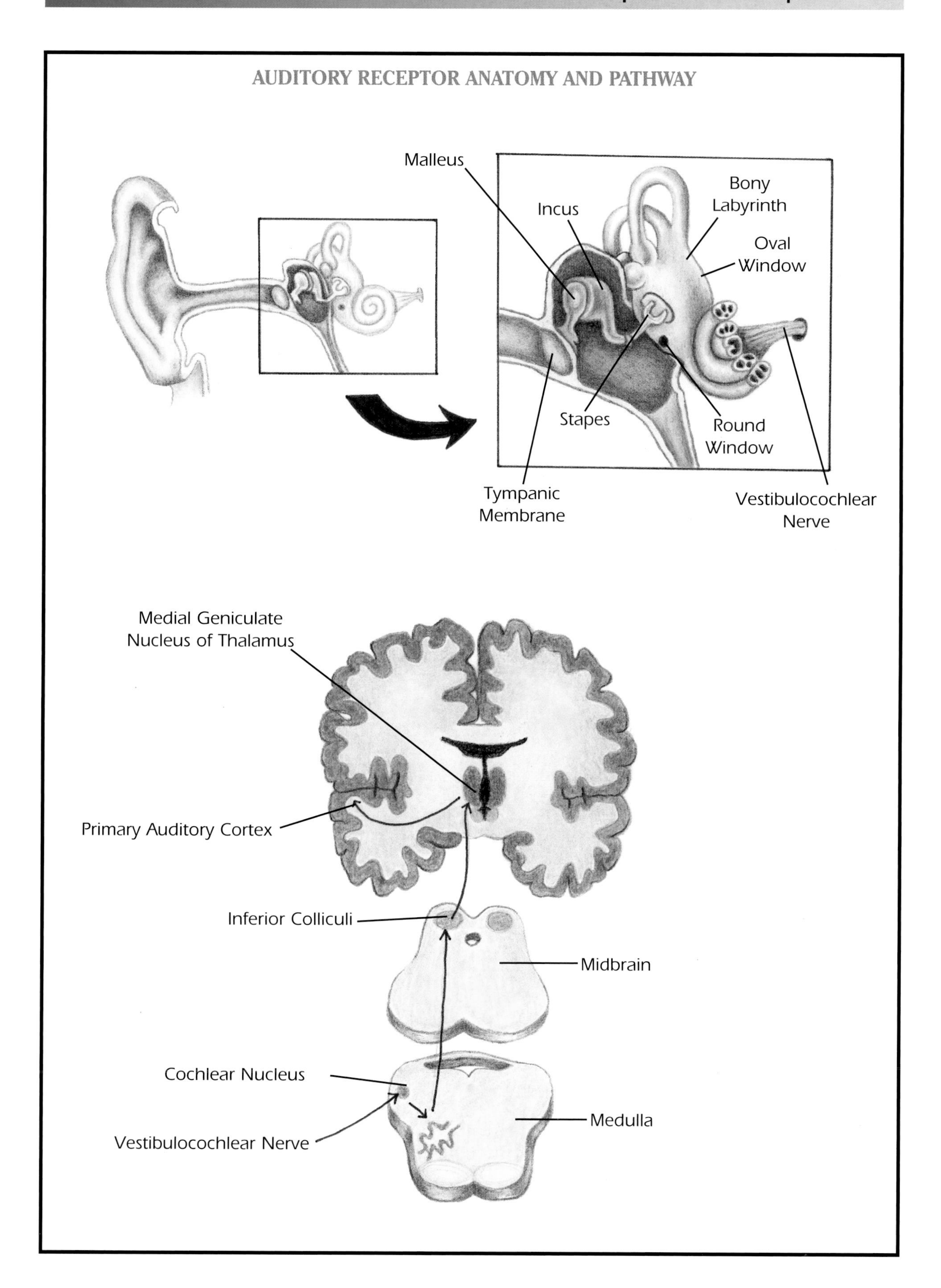
AUDITORY RECEPTOR ANATOMY AND PATHWAY
Malleus
Incus
Bony Labyrinth
Oval Window
Stapes
Round Window
Tympanic Membrane
Vestibulocochlear Nerve
Medial Geniculate Nucleus of Thalamus
Primary Auditory Cortex
Inferior Colliculi
Midbrain
Cochlear Nucleus
Vestibulocochlear Nerve
Medulla

Hearing Impairment

Sensorineural

- Sensorineural hearing impairment involves damage to the inner ear, vestibulocochlear nerve, and/or the brain.

Conductive

- Conductive hearing impairment involves damage to the outer and/or middle ear structures.

Lesion to the Auditory Portion of the Vestibulocochlear Nerve

- This causes deafness or tinnitus.

Lesion to the Primary Auditory Cortex

- This causes cortical deafness.

Lesion to the Auditory Association Cortices

- This may cause auditory agnosia—inability to interpret sounds.

EQUILIBRIUM

Equilibrium Receptor Anatomy

- The receptors of equilibrium are the *hair cells* of the *semicircular canals*, *utricle*, and *saccule* located in the inner ear.

Three Semicircular Canals

- The three semicircular canals follow:
  - *Superior Semicircular Canal*
  - *Posterior Semicircular Canal*
  - *Horizontal Semicircular Canal*
- These semicircular canals respond to *movement of the head*.
- The semicircular canals are a system of canals called the *bony labyrinth*.
- Within the bony labyrinth is a *membranous labyrinth*.
- Both labyrinths contain *fluid*.
- *Perilymph* is the fluid located within the bony labyrinth.
- *Endolymph* is the fluid located within the membranous labyrinth.

Ampulla

- Located at the *end* of each semicircular canal is a small enlargement called the ampulla.
- Each ampulla contains hair cells.

Utricle

- The utricle is located just medially to the semicircular canals.

Saccule

- The saccule is located just medial and inferior to the utricle.

Together, the Utricle and Saccule Respond to:

- Gravity
- Changes in head position

The Semicircular Canals Respond to:

- Angular acceleration and deceleration (rotation)
- Linear acceleration and deceleration (moving forward on a bike or on skis)

Postrotary Nystagmus

- Nystagmus is an involuntary, rapid, repetitive, jerky oscillating movement of the eyeballs in either a horizontal, vertical, or rotary direction.
- Nystagmus can occur in the absence of pathology in response to the following:
  - Rotation
  - Optokinetic nystagmus (eg, being on a train and watching sequential telephone poles pass by)
  - Caloric testing
- Nystagmus can also occur as a result of pathology:
  - Damage to the labyrinths of the inner ear
  - Vestibular nuclei damage at the pons–medulla junction
  - Vestibulocochlear nerve damage
  - Extraocular nuclei damage in the midbrain and pons
  - Extraocular cranial nerve damage
  - Cerebellar damage
- *Postrotary nystagmus* occurs normally after rotation. Patients with neurologic deficits will experience postrotary nystagmus for longer periods than normal after the rotation has stopped.
- The individual is rotated to the right (on a rotation board/swing).
- Fluid inside the labyrinths displaces towards the left—due to the sudden rotation of the head to the right.
- This causes the *hair cells* to *bend* and become excited.
- An *action potential* is generated along the vestibular portion of the *vestibulocochlear nerve*.
- While the individual is rotating to the right, the fluid in the labyrinths eventually catches up to the labyrinths and begins to move at the same rate as the labyrinths.
- This causes the hair cells to *stop firing*—they are no longer being bent.
- The rotation board/swing is stopped.
- The fluid in the labyrinths now displaces to the right—because the head has stopped but the fluid has not yet stopped.
- This again *bends* the *hair cells* and causes them to become excited.
- An *action potential* is generated and travels along the vestibular portion of the *vestibulocochlear nerve*.
- The individual experiences nystagmus to the right and has a sensation of the room rotating to the right.
- If asked to point at a fixed object, the individual will overshoot it by pointing further right (past pointing to the right).
- The individual also experiences a tendency to fall to the right.

Nystagmus Testing

- There are several standardized tests to assess whether nystagmus is pathological or normal.
- All tests must be done with caution as they can induce nausea due to the connection between the vestibular system and the vagus nerve.

Postrotary Nystagmus Test (Section of the Sensory Integration and Praxis Test)

- This is a standardized test for children and adults.
- The individual sits on a rotating swing. The rotation is stopped and the therapist examines the length of the nystagmus.
- A normal range for nystagmus to continue after rotation has stopped is 8 to 14 seconds.
- One disadvantage of this method is that the vestibular apparatus on both sides must be tested simultaneously.

Caloric Testing

- Caloric testing is usually performed by physicians.
- The patient's head is elevated to 30 degrees.
- To induce nystagmus, 30 to 50 ml of warm or cold water is squirted into the external auditory canal.
- This procedure produces convection currents in the optic fluid that mimic the effects of angular acceleration.
- The advantage of caloric testing is that the vestibular apparatus on each side can be tested separately.

Optokinetic Testing

- Optokinetic testing can be induced from prolonged, recurrent stimulation of the extraocular cranial nerves.
- Examples: The person observes a black and white rotating drum or watches telephone poles pass by while seated on a train.

Electronystagmograph

- Electronystagmograph (ENG) is a precise and objective method of assessing nystagmus.
- The test is usually performed by physicians.
- Electrodes are placed on the canthus (angle) of each eye and above and below each eye. A ground electrode is placed on the forehead. This electrode placement accesses the extraocular eye muscles.
- The individual is rotated; the rotation is stopped.
- The ENG records the nystagmus on paper (like an electromyograph).
- The velocity, frequency, and amplitude of nystagmus can be evaluated.
- Advantages of ENG are that it is easily administered, is noninvasive, and does not interfere with vision.

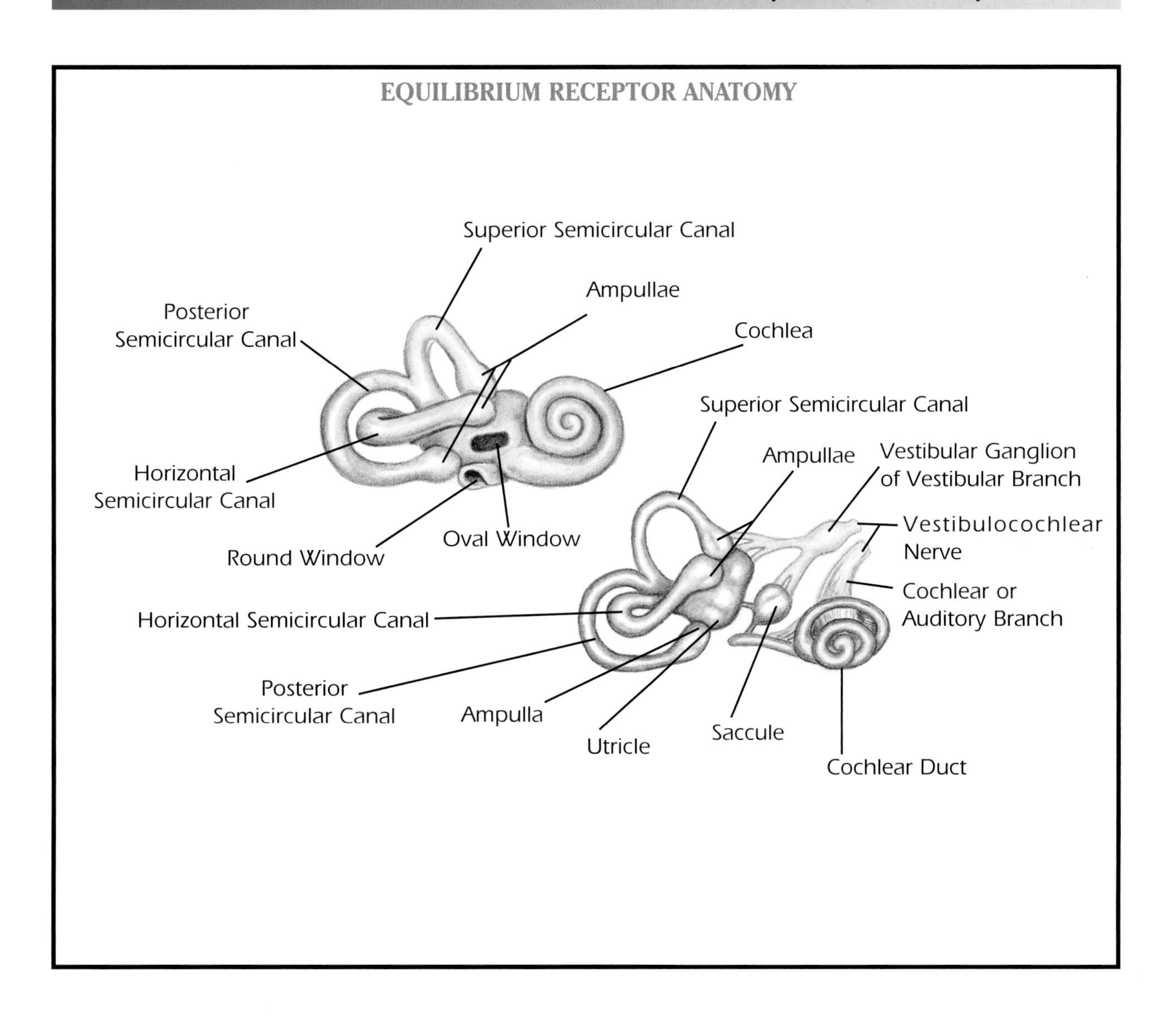
EQUILIBRIUM RECEPTOR ANATOMY
Superior Semicircular Canal
Ampullae
Posterior Semicircular Canal
Cochlea
Horizontal Semicircular Canal
Round Window
Oval Window
Superior Semicircular Canal
Ampullae
Vestibular Ganglion of Vestibular Branch
Vestibulocochlear Nerve
Cochlear or Auditory Branch
Horizontal Semicircular Canal
Posterior Semicircular Canal
Ampulla
Utricle
Saccule
Cochlear Duct

SECTION 12

Vestibular System

Function of the Vestibular System

- Maintains *equilibrium* and *balance*
- Maintains the *head* in an *upright vertical position*
- Has a role in the *coordination* of *head* and *eye movements*
- Influences *tone* through the alpha and gamma motor neurons and the medial and lateral vestibulospinal tracts
- The *vestibular*, *visual*, and *proprioceptive systems* work collaboratively to maintain an understanding of the *body's position in space*. When one of these systems becomes compromised, the others are relied on to compensate.
- When the vestibular system is impaired, patients rely more heavily on their proprioceptive and visual systems to maintain an awareness of their body's position in space.
- Disorders causing a progressive loss of vestibular function (from a degenerative disorder) may go unnoticed by the patient as the person grows increasingly reliant upon the visual and proprioceptive systems.
- They will experience difficulty, however, on uneven ground surfaces (such as sand) or in dim light—when visual and proprioceptive cues cannot be accurately used.

The Vestibular System Receives Input From the Following

- Vestibulocochlear nerve
- Vestibular nuclei in the pons–medulla junction
- Vestibular apparatus in the inner ear (semicircular canals, utricle, and saccule)
- Cerebellum
- Extraocular cranial nerves and nuclei

Vestibular Pathway

- The *vestibular nuclei* in the *pons–medulla junction* receive vestibular information about balance and equilibrium from the following:
  - Vestibular apparatus in the inner ear
  - Extraocular nuclei in the midbrain and pons
  - Cerebellum

- The vestibular nuclei send the information they receive to the *alpha* and *gamma motor neurons* in the ventral horn of the spinal cord.
- This information is sent via the *vestibulospinal tracts.*
- The motor neurons in the ventral horn synapse with *spinal nerves* in the *periphery.*
- The spinal nerves in the periphery project to *antigravity muscles*—the muscles that maintain the body's upright position against gravity. These are the extensors in the legs, trunk, and back.
- The *muscle spindles, Golgi tendon organs,* and *joint receptors* then send information about the body's position in space back to the cerebellum in a feedback loop.
- The *cerebellum* uses this information to make ongoing decisions about the modification of muscular activity in order to enhance balance.
- There are also *motor neurons* in the *ventral horn* that project to the *head* and *neck musculature* via the *medial longitudinal fasciculus.*
- These structures mediate the *head righting* and *tonic neck reflexes*—reflexes that remain present throughout the lifespan to maintain the head in an upright position.
- Eye and head movements are also integrated by the *medial longitudinal fasciculus.*
- This tract allows feedback about the head's position to be continuously sent back to the extraocular nuclei in the midbrain and pons.

Relationship Between the Reticular Formation and the Vestibular System

- The *reticular formation* is diffusely located in the brainstem.
- It is composed of the *reticular activating* and *inhibiting systems.*
- The *reticular activating system screens sensory information* to alert the brain to attend to important incoming sensory data. It also responds to excitatory vestibular stimulation by *arousing* the *brain* and *body.*
- The *reticular inhibiting system* acts as a mechanism to calm the brain/body in response to inhibitory sensory information.
- The reticular formation in the brainstem integrates information from the vestibular system via the *medial longitudinal fasciculus tract* and the *vestibulospinal tracts.*
- Vestibular sensory data—such as *slow rocking*—are integrated by the reticular formation and calm the individual.
- Excitatory vestibular sensory data—such as *fast dancing* (rotary acceleration) and *roller coasters* (linear acceleration)—are integrated by the reticular formation and arouse the brain/body.
- Children seek this kind of inhibitory and excitatory vestibular sensation to facilitate the development and organization of their nervous systems.
- The *need for intense vestibular stimulation decreases as humans age.* Often, adults cannot tolerate the same kind of intense vestibular stimulation they craved as children and adolescents.

Relationship Between the Autonomic Nervous System and the Vestibular System

- The autonomic nervous system (ANS) mediates *cranial nerve (CN) 10, the vagus nerve.*
- The *visceral branches* of the vagus nerve conduct signals to and from the *gastrointestinal (GI) tract.*
- Because the vagus nerve has connections to the vestibular pathways, these connections may explain why *overexcitation* of the vestibular system can induce *nausea* and *vomiting.*

Two Categories of Vestibular Dysfunction

Peripheral (PNS)

- Involves the vestibular apparatus in the inner ear:
  - Semicircular canals
  - Utricle
  - Saccule

Central (CNS)

- Involves the vestibular pathways and structures of the CNS
  - Vestibulocochlear nerve
  - Vestibular nuclei in the pons–medulla junction
  - Cerebellum
  - Extraocular cranial nerve nuclei

Signs and Symptoms of Vestibular Impairment

- Nystagmus
- Tinnitus
- Vertigo
- Hearing loss
- Loss of balance and possible falls
- Broad-based stance (to accommodate for imbalance)
- Sweating, nausea, and vomiting (due to ANS involvement

Motion Sickness: Disorder of the Vestibular System

- Motion sickness is a disorder of the vestibular system that is caused by repeated rhythmic stimulation—for example, car, plane, or boat travel.
- Symptoms include vertigo, nausea, vomiting, lowered blood pressure, tachycardia, and sweating.
- Pooling of blood in the lower extremities commonly leads to postural hypotension and fainting.
- Motion sickness can sometimes be alleviated by obtaining a match between visual and motion signals reaching the vestibular system.
- For example, observing the upcoming traffic conditions in a car—rather than reading a book—can suppress feelings of motion sickness.
- Motion sickness usually decreases in severity with repeated exposure.
- Anti-motion sickness drugs suppress the activity of the vestibular system.

Romberg Test

- The Romberg test is used to assess disequilibrium.
- The patient is asked to stand with feet together and shoulders flexed to 90 degrees (and positioned in front of the patient).
- The patient is then asked to close his or her eyes. When visual cues are removed, the patient's postural stability is based on vestibular and proprioceptive information.
- The therapist observes the patient's degree of postural sway, balance, and arm stability.
- Impaired vestibular function is indicated by postural sway and a tendency for the arms to drift toward the affected side.
- If the vestibular system is severely impaired, the patient will fall toward the affected side.

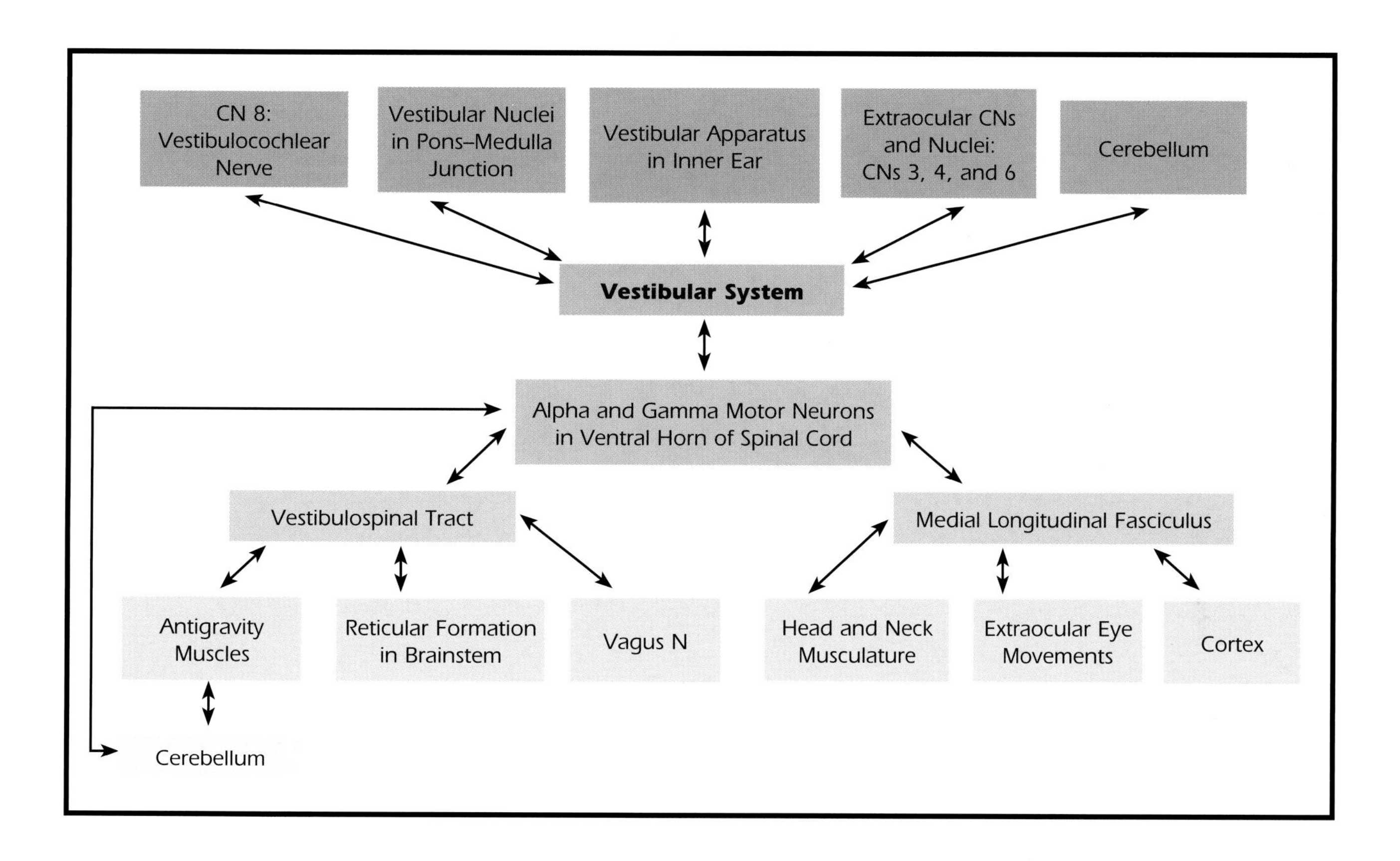

SECTION 13

Autonomic Nervous System

Function of the Autonomic Nervous System

- Innervates the *internal organs*, *blood vessels*, and *glands*
- Regulates *cardiac* and *smooth muscle* (muscle of glands and organs)
- Regulates *secretion* from *glands*
- Controls *vegetative functions*:
  - Temperature
  - Digestion
  - Heart rate
  - Respiration
  - Metabolism
  - Maintains homeostasis of internal organs
  - Blood pressure
- Influences *muscle tone* through the vestibulospinal, rubrospinal, and reticulospinal tracts

Central Components of Autonomic Nervous System

- The central portion of the ANS consists of parts of the *cerebral cortex*, *hypothalamus*, *thalamus*, *limbic system*, *cerebellum*, and *spinal cord*.
- Efferent fibers that originate in the cortex project descending fibers through the *thalamus* and *hypothalamus*.
- These fibers end on a *cranial nerve nuclei* to influence involuntary muscles, vessels, and glands.

Anterior and Posterior Hypothalamus

- The hypothalamus is a major control center of the autonomic nervous system (ANS).
- It regulates *temperature*, *thirst*, *feeding behaviors*, and *endocrine functions* (eg, the secretion of glands).
- The *anterior hypothalamus* projects pathways to the *parasympathetic nervous system*.
- The *posterior hypothalamus* projects pathways to the *sympathetic nervous system*.
- The hypothalamus exerts autonomic regulation over various brainstem centers that control *vegetative functions* (ie, life sustaining functions such as cardiovascular and respiratory functions).

Brainstem Centers: The Reticular Formation

- The reticular formation consists of interconnected neurons throughout the brainstem.
- It is diffusely located throughout the *midbrain*, *pons*, and *medulla*.
- It sends and receives projections to and from the *diencephalon*, *cortex*, and *spinal cord*.
- The reticular formation is involved in the *control of posture*, *visceral motor function*, *sleep*, and *arousal/wakefulness*.
- This brainstem center is an ancient part of the nervous system. The brains of primitive vertebrates are made up primarily of a reticular-type formation designed for survival functions.
- There are *respiratory* and *cardiovascular* centers in the reticular formation. These control vital functions and reflexes such as the *gag*, *cough*, *sneeze*, *swallow*, and *vomit reflexes*.
- Severe injury to the brainstem often results in death or poor prognosis.
- If the brainstem remains intact but the cortex is no longer active, one can still survive—the patient is considered to be in a *persistent vegetative state* (because the brainstem controls vegetative functions).

- The reticular formation also receives sensory fibers from the *somatic* and *visceral* systems (including vision and olfaction). This sensory information is then relayed to the *thalamus* and *cortex*.
- The reticular formation's motor fibers synapse with (a) motor neurons of the pyramidal and extrapyramidal systems and (b) motor neurons that synapse with preganglionic autonomic motor neurons.
- The reticular formation is the origin of the *reticulospinal tracts*—descending extrapyramidal tracts of the spinal cord that inhibit and facilitate the *antigravity muscles* and thereby *influence muscle tone*.

Reticular Activating System

- The reticular activating system (RAS) is the portion of the reticular formation that is responsible for *arousal*, *alertness*, and *wakefulness*.
- It *filters* all *incoming sensory information* and alerts the cortex to attend to important sensory input.
- This results in a sharpening of attention to important sensory information from the environment.
- The RAS is diffusely located throughout the brainstem but is believed to be primarily located in the *midbrain*.

Lesions to the Reticular Activating System

- These result in a *stuporous state*. When the RAS is lost, the reticular inhibitory system becomes dominant and causes heightened somnolence.

Reticular Inhibitory System

- The reticular inhibitory system (RIS) is responsible for *calming* the organism. Certain types of sensory input—slow rocking, deep pressure—will activate the RIS to calm the body.
- It is believed that the RIS diffusely extends from the *caudal midbrain* to the *caudal medulla*.

Lesions to the Reticular Inhibitory System

- These result in *constant wakefulness*. When the RIS is lost, the RAS becomes dominant and causes heightened arousal.

Limbic Lobe

- The limbic lobe has a role in the relationship between our emotions and the ANS.
- *Heightened emotional states* will cause the sympathetic nervous system to become activated and dampen parasympathetic nervous system activity.
- When humans are upset or excited, they may experience *loss of appetite*—because the sympathetic nervous system is geared up and has shut down the digestive system (regulated by the parasympathetic nervous system).
- When humans or animals are frightened, *loss of bladder control* can occur—fear can interfere with ANS regulation.
- When individuals are in great pain, *nausea* may occur as a result of the connections between the ANS, the vagus nerve, and the limbic lobe.
- Similarly, *blushing*, *heart palpitations*, *clammy hands*, and *dry mouth* are emotional responses that are mediated by the limbic system and the ANS.

Spinal Cord

- The spinal cord contains important ANS reflexes that are modulated by higher CNS centers.
- When there is lost communication between *spinal cord reflexes* and *higher CNS centers*—as in spinal cord injury—such reflexes function in an unmodified manner.
- For example, *uncontrolled sweating*, *vasomotor instability*, and *reflex bowel* and *bladder functions* occur when regulation of ANS spinal cord level reflexes are lost.

PERIPHERAL COMPONENTS OF AUTONOMIC NERVOUS SYSTEM

- The peripheral components of the ANS consist of *pre-* and *postganglionic fibers* that innervate the viscera.

Preganglionic Fibers

- Preganglionic fibers are a collection of neurons in the peripheral nervous system (PNS).
- They are also called *presynaptic neurons* (or first order neurons).
- Preganglionic fibers have their *cell bodies* in the *brainstem* (CNs 3, 7, 9, 10, 11) and *spinal cord* (in the *intermediolateral horn* of the thoracic sections and first two lumbar sections).

Postganglionic Fibers

- Postganglionic fibers are also called postsynaptic or second order neurons.
- These ganglionic fibers are located in the periphery.
- Their *cell bodies* are located in the *autonomic ganglia* (groups of ANS organs and tissues).

Sympathetic Nervous System

- The sympathetic nervous system's *preganglionic neurons* have their *cell bodies* located in the *intermediolateral horn* of the *thoracic* and *first two lumbar sections* of the spinal cord.

Parasympathetic Nervous System

- The parasympathetic nervous system's *preganglionic fibers* emerge from the *brainstem* and *sacral spinal cord*.

SYMPATHETIC NERVOUS SYSTEM

Function

- The sympathetic nervous system activates the *fight/flight response*. This response occurs during situations of stress and involves the following:
  - Accelerated heart rate
  - Increased blood pressure
  - Shift of blood flow from the skin and gastrointestinal (GI) tract to the skeletal muscles and brain
  - Increased blood sugar level
  - Dilation of the bronchioles and pupils
  - Constriction of the stomach, intestine, and internal sphincter of the urethra

Location of Cell Bodies

- The sympathetic nervous system is also called the *thoracolumbar division* because its *preganglionic neurons* have their *cell bodies* located in the *intermediolateral horn* of the *thoracic* and *first two lumbar sections* of the spinal cord.

Sympathetic Chain Ganglia

- The sympathetic chain ganglia are a *series of interconnected sympathetic ganglia* that lie adjacent to the vertebral column (on both sides).
- The chain ganglia receive input from the *preganglionic sympathetic fibers*.
- The chain ganglia *project postganglionic sympathetic fibers* to *specific target organs and tissues*.

Preganglionic Sympathetic Fibers

- Preganglionic sympathetic fibers extend from their cell bodies (in the intermediolateral horn of T1 – L2) to the chain ganglia.

Postganglionic Sympathetic Fibers

- Postganglionic sympathetic fibers extend from the chain ganglia to (one of three) *collateral ganglia* (located outside of the chain).
- A collateral ganglion is a *collection of cell bodies located outside* of the *sympathetic chain ganglia*.
- There are three main collateral ganglia in the body (three on each side of the vertebral column):
  - Celiac Ganglion
  - Superior Mesenteric Ganglion
  - Inferior Mesenteric Ganglion
- The postganglionic sympathetic fibers ascend or descend through the chain ganglia before synapsing on one of the collateral ganglion.
- After synapsing on one of the collateral ganglion, the postganglionic sympathetic fibers project to a target or end organ.

Neurotransmitters of the Sympathetic Division

- The preganglionic fibers use *acetylcholine (ACh)*.
- The postganglionic fibers use *norepinephrine (noradrenalin)*.

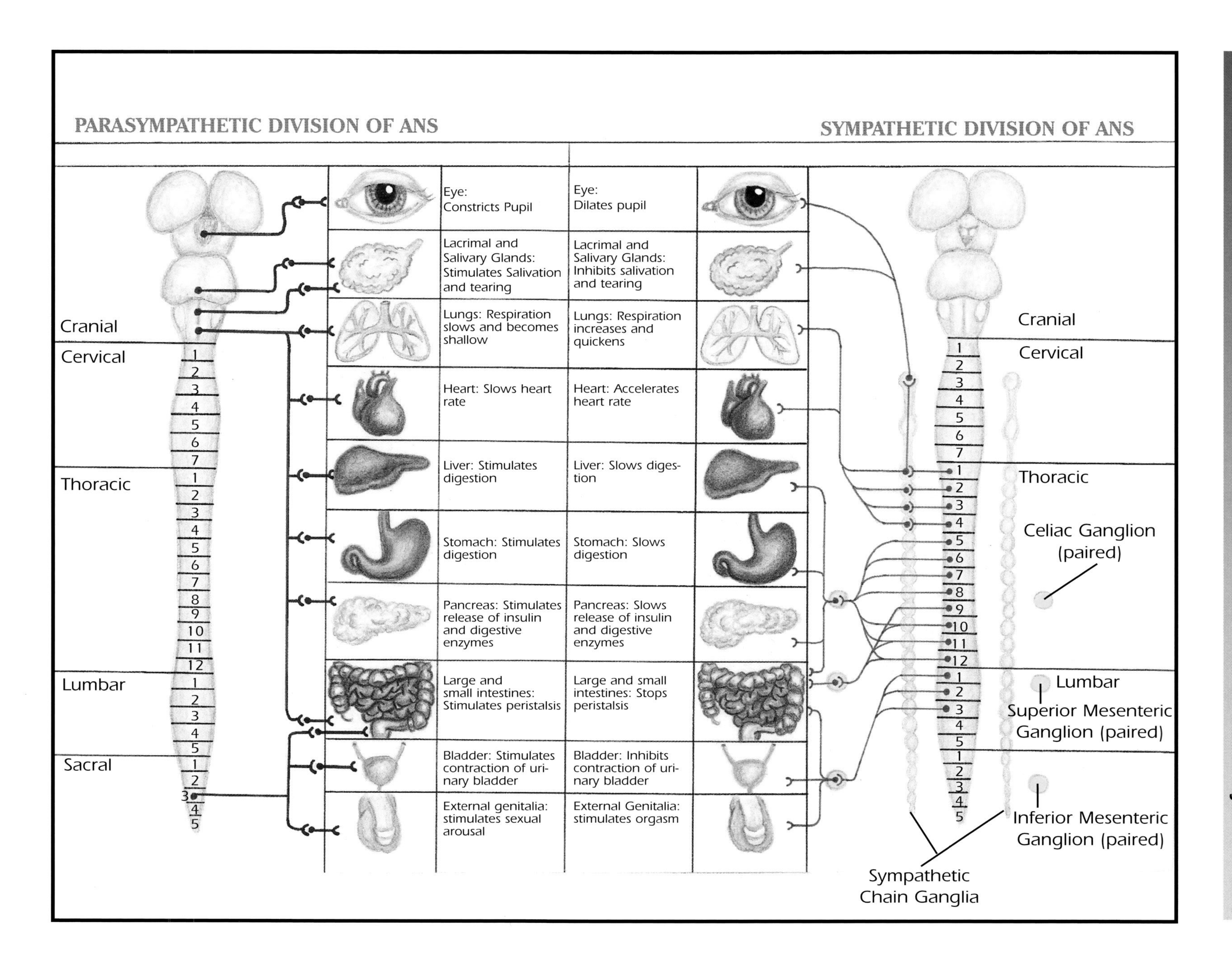

PARASYMPATHETIC DIVISION OF ANS
SYMPATHETIC DIVISION OF ANS
Cranial
Cervical
Thoracic
Lumbar
Sacral
Eye: Constricts Pupil
Eye: Dilates pupil
Lacrimal and Salivary Glands: Stimulates Salivation and tearing
Lacrimal and Salivary Glands: Inhibits salivation and tearing
Lungs: Respiration slows and becomes shallow
Lungs: Respiration increases and quickens
Heart: Slows heart rate
Heart: Accelerates heart rate
Liver: Stimulates digestion
Liver: Slows digestion
Stomach: Stimulates digestion
Stomach: Slows digestion
Pancreas: Stimulates release of insulin and digestive enzymes
Pancreas: Slows release of insulin and digestive enzymes
Large and small intestines: Stimulates peristalsis
Large and small intestines: Stops peristalsis
Bladder: Stimulates contraction of urinary bladder
Bladder: Inhibits contraction of urinary bladder
External genitalia: stimulates sexual arousal
External Genitalia: stimulates orgasm
Cranial
Cervical
Thoracic
Celiac Ganglion (paired)
Lumbar
Superior Mesenteric Ganglion (paired)
Inferior Mesenteric Ganglion (paired)
Sympathetic Chain Ganglia

Parasympathetic Nervous System

Function

- Conservation and energy storage and replenishment
- Maintains heart rate, respiration, metabolism, and digestion in a state of *homeostasis*

Location of Cell Bodies

- Also called the *craniosacral division* because the parasympathetic nervous system's cell bodies are located in the *brainstem* and *sacral spinal cord*

Preganglionic Parasympathetic Fibers

- The preganglionic parasympathetic fibers extend directly from their cell bodies to the terminal ganglia located within or very near the organs that they supply.

Postganglionic Parasympathetic Fibers

- The postganglionic parasympathetic fibers are short in length. They extend from the *terminal ganglia* to the *target end organ*.

Neurotransmitters of the Parasympathetic Division

- Both the pre- and postganglionic fibers use ACh.

Comparison of the Parasympathetic and Sympathetic Nervous Systems

Function

- The sympathetic nervous system activates the *fight/flight response* and prepares the *body for action*.
- The parasympathetic nervous system regulates *homeostasis* and *slows the body down*.

Cell Bodies

- Sympathetic nervous system cell bodies are located in the *intermediolateral horn* of the *thoracic* and *first two lumbar segments* of the spinal cord.
- Parasympathetic nervous system cell bodies are located in the *brainstem* and *sacral* sections of the spinal cord.

Pathways

- Preganglionic sympathetic fibers extend from their *cell bodies* to the *sympathetic chain ganglia*. Postganglionic sympathetic fibers then travel from the *chain ganglia* to a *collateral ganglion* and finally synapse on a *target organ*.
- Preganglionic parasympathetic fibers extend from their *cell bodies* to a *terminal ganglion*. Postganglionic parasympathetic fibers then travel from the *terminal ganglia* to the *target organ*.

Neurotransmitters

- Preganglionic sympathetic fibers use *ACh*. Pre- and postganglionic parasympathetic fibers use *ACh*.
- Postganglionic sympathetic fibers use *norepinephrine (noradrenalin)*.

Autonomic Nervous System and Disease/Illness

Stress and Heart Disease

- Individuals with chronic stress are more prone to *heart disease*. Stress activates the *sympathetic nervous system*.
- This causes constriction of the *blood vessels*, *increases heart rate*, and *heightens cholesterol production*.

Stress and Healing

- Stress promotes sympathetic nervous system activity.
- This causes decreased repair of cell structures.
- Parasympathetic nervous system activity promotes homeostasis and facilitates cellular repair and tissue restoration.
- Parasympathetic nervous system activity is heightened during restorative sleep—stage 4 sleep.

Symptoms of Autonomic Dysregulation

- Cold, dry, flaky skin with loss of hair over the denervated skin region.
- Alterations in sweating.
- The nails may become thickened.
- The denervated skin region may become smooth and glossy.

Horner's Syndrome

- Horner's Syndrome is an ANS disorder.
- Results from a transection of the Oculomotor Sympathetic Pathway (CN 3).
- Causes:
  - Ipsilateral Mioses (constriction of the pupil)
  - Partial Ptosis (drooping of the ipsilateral eyelid)
  - Flushed Dry Skin on the Ipsilateral Face
  - Ipsilateral Sunken Eyeball

- Some research has shown that *fibromyalgia*—an illness characterized by muscular and joint pain—may result in part from a lack of stage 4 restorative sleep.

Denervation

- Denervation is an interruption of neuronal innervation.
- It causes hypersensitivity—the injured neuron experiences hypersensitivity to its own neurotransmitter.
- This results in dysregulation of the ANS.

Autonomic Nervous System Characteristics

Sympathetic

- Activated in short bursts when individual is stressed, angry, frightened, or excited. May be activated continuously if individual has chronic stress.
- Fight/flight response
- Pupils dilate
- Increased focus on event/environment
- Heart rate increases
- Blood pressure increases
- Respiration increases and quickens
- Vasoconstriction of arteries to bring blood to heart faster
- Saliva thickens
- Blood is diverted away from the GI tract and shifts to skeletal muscles and brain
- Digestive juices stop production
- Peristalsis stops
- Activation of all muscle groups
- Cell destruction
- Body temperature changes

Parasympathetic

- Activated continuously to maintain homeostasis of body systems. Will shut down when sympathetic nervous system takes over.
- Homeostasis
- Pupils contract
- Decreased awareness of environment
- Heart rate slows
- Blood pressure decreases
- Respiration slows and becomes shallow
- Vasodilation of arteries
- Saliva thins
- Blood returns to viscera in GI tract
- Digestive juices begin to be secreted again
- Peristalsis is continuous unless sympathetic nervous system shuts parasympathetic nervous system down
- Relaxation of most muscle groups
- Cell repair
- Body temperature is maintained at a constant level (98.6 degrees Fahrenheit)

SECTION 14

Enteric Nervous System

- The enteric nervous system (ENT) is an independent circuit of ganglionic cells that is loosely connected to the central nervous system (CNS) but can function alone without instruction from the CNS.

Location

- The ENS is located in sheaths of tissue that line the esophagus, stomach, small intestine, and colon.

Composition

- The ENS is composed of a network of *neurons*, *neurotransmitters*, and *proteins*.
- The ENS contains *100 billion neurons*—one-hundredth of the number of neurons in the brain, and significantly more than the number in the spinal cord (SC).

Chemical Substances in the Brain and Enteric Nervous System

- Every chemical substance that helps to control the brain has been found in the intestines:
  - Major neurotransmitters such as *serotonin*, *dopamine*, *glutamate*, *norepinephrine*, and *nitric oxide*
  - Two dozen small brain proteins called *neuropeptides*
  - *Mast cells*—cells of the immune system
  - *Enkephalins*—one class of the body's natural opiates
  - *Benzodiazepines*—the family of psychoactive chemicals that include Valium and Xanax

Development of the Enteric Nervous System

- A formation of tissue called the *neural crest* develops early in embryogenesis.
- One section of the neural crest turns into the *CNS* (brain and SC).
- Another piece migrates to become the *ENS*.
- Only later in fetal development are the two systems connected via the *vagus nerve (CN 10)*.

Communication Between the Central and Enteric Nervous Systems

- The CNS sends signals to the ENS through a small number of *command neurons*.
- These command neurons then send signals to the ENS's interneurons (small neurons that connect two major neurons).
- Both command neurons and interneurons are located in two layers of intestinal tissue called the:
  - Myenteric Plexus
  - Submucosal Plexus
- *Command neurons* appear to control the pattern of activity in the gastrointestinal (GI) tract. These neurons form an *independent system* of the ENS.
- The *CNS* appears to primarily control the firing rate of the command neurons.
- The *vagus nerve* only alters the firing rate of the command neurons.

The Myenteric and Submucosal Plexus

- The plexus have sensors for sugar, protein, and acidity levels.
- These sensors monitor the progress of digestion and determine how the intestines should mix and propel its contents.

Drug Interactions and the Connection Between the Enteric and Central Nervous Systems

- When pharmacologists design a drug to have effects on the brain, it very likely has concomitant effects on the GI tract that are undesired (ie, negative side effects).
- The intestine contains a substantial amount of *serotonin*. When pressure receptors in the GI tract's lining are stimulated, serotonin is released and begins the reflexive motion of *peristalsis*.
- Twenty-five percent or more of individuals taking a selective serotonin re-uptake inhibitor (SSRI—a class of antidepressants including Prozac, Paxil, Zoloft, and Celexa) experience *accompanying GI problems*—nausea, diarrhea, and constipation.
- These drugs act on serotonin, preventing its re-uptake by receptor cells. It is beneficial if higher levels of serotonin remain in the CNS.
- But when high levels of serotonin are released and remain in the GI tract, it causes bowel problems.
- The SSRIs also double the speed at which food is passed through the colon, explaining why some people taking SSRIs experience diarrhea.
- Sometimes, too much antidepressant drugs can have the effect of producing constipation.
- Some antibiotics—like erythromycin—act on the GI receptors to produce oscillations, causing cramping and nausea.
- Drugs like morphine and heroin attach to the intestine's opiate receptors and can produce constipation.
- The ENS can become addicted to drugs just like the CNS does.
- People with Alzheimer's and Parkinson's disease often experience constipation because the pathology of their CNS disorder also causes dysregulation of the intestine's functioning.

Similarities Between the Central and Enteric Nervous Systems

- Both the CNS and ENS act similarly when deprived of input from the external world.
- During sleep, the CNS produces 90-minute cycles of *slow wave sleep* punctuated by periods of *rapid eye movement* (REM) sleep in which dreams occur.
- During the night when the ENS has no food, it produces 90-minute cycles of *slow wave muscle contractions* punctuated by short bursts of rapid muscle movements.
- The two systems influence each other while in this state. Patients with bowel problems have been shown to have abnormal REM sleep.
- This finding seems consistent with the folk wisdom that indigestion can cause nightmares.

Enteric and Central Nervous Systems' Response to Fight/Flight Situations

- When individuals encounter frightening situations, the CNS releases stress hormones that prepare the body to fight or flee (via the sympathetic nervous system).
- The GI tract also contains many sensory nerves that are stimulated by this chemical surge. These produce the common sensation of butterflies in the stomach.
- The CNS instructs the GI tract to shut down during fight/flight situations.
- Fear and chronic stress also cause the *vagus nerve* to increase the firing rate of serotonin circuits in the GI tract.
- When the GI tract becomes overstimulated with high levels of serotonin, bowel problems can occur (eg, colitis, irritable bowel syndrome).
- Similarly, when nerves in the *esophagus* are highly stimulated by an increase in the release or production of serotonin or norepinephrine, the esophagus reacts by constricting—thus making swallowing difficult. This is probably where the phrase "choked up with emotion" is derived from.

SECTION 15

Pain

- Pain is the sensory experience that is unpleasant and is associated with possible tissue damage.
- The detection of pain indicates that a pathological condition may be occurring in the organism.
- The detection of pain is called *nociception.*
- Nociceptors are receptors that detect harmful stimuli.

THE PROCESS OF PAIN: FOUR STAGES

Nociception can be divided into four stages:

Transduction

- Transduction occurs when *free nerve endings in the periphery* (ie, nociceptors) become *stimulated.*
- Nociceptors are located in the skin, muscles, connective tissue, circulatory system, and the viscera.
- Stimulation results from damage to the nerve endings or from the release of chemicals at the injury site.

Transmission

- Transmission involves the *conduction of pain signals* along afferent pathways in the periphery to the *spinal cord (SC)* and *brain.*
- Two primary fibers are involved in the transmission process: A delta and C fibers.
  - *A delta fibers* are large, unmyelinated fibers that transmit signals quickly in response to tissue damage. Pain signals propagated along A delta fibers are sharp, stinging, highly localized, and short-lasting.
  - *C fibers* are small, unmyelinated, and conduct pain signals more slowly. Pain signals carried along C fibers are poorly localized, dull, aching, and longer-lasting.

Perception

- Perception is the process whereby the *cortex attaches meaning to,* or interprets, pain signals.
- The perception of pain involves (a) pain threshold and (b) pain tolerance.
  - *Pain threshold* refers to the amount of pain stimulation required before pain is perceived. Pain thresholds are generally similar among all people.
  - *Pain tolerance* refers to the amount of pain one is able to tolerate before seeking medical intervention. Pain tolerance varies widely from person to person.
- The primary somatosensory area (SS1), secondary somatosensory area (SS2), posterior multimodal association area, and limbic system structures all have a role in the perception of pain.

Modulation

- Modulation involves the *modification of pain signals* by different CNS and PNS centers along the pain pathway.
- Generally, pain can be modified at the level of the peripheral nociceptor, the SC, the brainstem, and the cortex.

Types of Pain

Somatic Pain

- Somatic pain occurs from the body (eg, skin, skeletal muscles, bones).
- *Superficial somatic pain* is usually well-localized (eg, pin prick).
- *Deep somatic pain* is commonly poorly localized (eg, muscular ache).

Visceral Pain

- Visceral pain occurs from the *viscera* (internal organs, glands, smooth muscle).
- Visceral pain is dull or diffuse and not well-localized.
- This type of pain is usually accompanied by an *autonomic nervous system response* (eg, changes in heart rate, respiration, and blood pressure; nausea; dilated pupils; perspiration; pallor).

Qualities of Pain

- Dull ache—tends to be diffuse.
- Dull aching pain tends to last a long time because it is carried by slow conducting, small, unmyelinated C fibers.
- Sharp pain—tends to be well-localized.
- Sharp pain tends to last a short time because sharp pain is carried by fast-conducting, large A delta fibers.

Pain Receptors

- Pain receptors are believed to be *free nerve endings*.
- If stimulated intensely enough, other types of receptors may act as pain receptors as well.

Major Pain Pathways

Spinothalamic Spinal Cord Tracts

- The spinothalamic tracts are ascending somatic sensory pathways that receive pain information from the skin and skeletal muscles.
- Sensory nerves carry pain information from the skin and skeletal muscles in the periphery to the dorsal horn of the SC.
- When these spinal nerves synapse in the dorsal horn, they release a neurotransmitter called *Substance P*.
- The spinothalamic tracts travel from the SC to the thalamus and send projections to the cortex for conscious pain detection and interpretation.
- Substance P is transmitted via the spinothalamic tracts to the thalamus and cortex.

Reticulospinal Tracts

- The reticulospinal tracts are descending sensory tracts that receive pain information from the periphery through afferent spinal nerves that synapse in the reticular formation of the brainstem.
- The reticulospinal tracts have their origin in the *medullary reticular formation*.
- There, they travel to the *raphe nuclei of the brainstem*. The raphe nuclei are a group of nuclei located in the medulla. They are situated along the midline.
- When the raphe nuclei become excited, they *release endorphins* through a descending pathway to the place of pain origin to decrease the pain sensation.

Trigeminothalamic Tracts

- The trigeminothalamic tracts have their origin in the *trigeminal lemniscus* in the brainstem.
- The trigeminal lemniscus projects afferent fibers from the *trigeminal nerve* (CN 5) to the thalamus and then to the cortex.
- This tract carries *pain sensation from the face*.

Pathways to the Cortex for the Conscious Detection of Pain

- Pain messages from the SC tracts are projected to the thalamus and then to the cortex for conscious detection and interpretation.
- The *primary somatosensory area (SS1)*, located in the postcentral gyrus, detects incoming somatosensory data from the periphery.
- Pain messages are then projected to the secondary somatosensory area (SS2) for interpretation. SS2 is located just posterior to SS1.
- Pain messages are then projected to the *posterior multimodal association area* for the integration of pain information with other sensory data.
- Example: The multimodal association area integrates pain with smell. The individual will learn to associate the smell of spoiled food with abdominal pain—in order to refrain from eating spoiled food in the future.
- The multimodal association area sends projections to the limbic system for the integration of sensation, emotion, and memory.
- The individual is able to remember the smell of spoiled food at later dates and will remember the pain associated with eating spoiled food.

How the Body Controls Pain

Gate Control Theory

- One of the first theories of pain control was the Gate Theory, proposed by Melzack and Wall in 1965.
- Simplistically, the Gate Theory suggested that the transmission of pain information is blocked in the dorsal horn, closing the gate to pain.
- Lamina II, or the substantia gelatinosa in the dorsal horn, was suggested as the site of interference with pain message transmission.
- The Gate Theory suggested that afferent sensory fibers carry pain sensation from the periphery into the dorsal horn of the SC.
- This information synapses in the substantia gelatinosa.
- T-cells are specialized cells in the SC. They begin to fire and cause the release of Substance P—a neurotransmitter involved in the sensation of pain.
- If the substantia gelatinosa can be facilitated by another pathway (a collateral pathway), the T-cell firing will diminish causing a decrease in pain transmission.
- While portions of the Gate Theory have been disproved, the theory continues to provide the foundation for many subsequent theories describing the mechanisms of pain.

The Counterirritant Theory of Pain Cessation

- The Counterirritant Theory of Pain Cessation has incorporated research-based findings first suggested by the Gate Theory.
- The Counterirritant Theory suggests that non-nociceptors in the dorsal horn inhibit the excited nociceptors (also in the dorsal horn).
- For example, pressure (such as rubbing the painful area) stimulates mechanoreceptive afferent fibers.
- The proximal branches of the mechanoreceptors in the dorsal horn activate interneurons that synapse on the excited nociceptors (in the dorsal horn).
- These interneurons release the neurotransmitter *enkephalin*—a chemical in the family of endorphins.
- Enkephalin binds with the excited nociceptor and diminishes the release of Substance P.
- Enkephalin binding on the nociceptor inhibits the transmission of nociceptive signals, thus decreasing the sensation of pain.

Analgesic Inhibition of Pain

- Analgesia is an absence of pain in response to stimulation that would otherwise cause pain.
- Analgesic mechanisms can be activated by the following:
  - Endorphins—naturally occurring substances that diminish the sensation of pain
  - Pharmaceuticals that diminish the sensation of pain

- Endorphins include *enkephalin*, *dynorphin*, and *beta-endorphin*.
- *Opiates* are the family of analgesic drugs that block nociceptor signals without affecting other sensations.
- Both endorphins and analgesic drugs bind to the same receptor site.
- The inhibition of nociceptive information can also be inhibited by the supraspinal levels of the nervous system. These are *brainstem centers* that provide natural analgesia. They are referred to as *pain inhibiting centers*:
  - The raphe nuclei in the medulla
  - The periaqueductal gray in the midbrain
  - The locus ceruleus in the pons
- When the raphe nuclei are stimulated, axons projecting to the SC release the neurotransmitter *serotonin* in the dorsal horn. This release of serotonin inhibits the transmission of nociceptive signals.
- When the periaqueductal gray is stimulated, it also produces an analgesic effect by activating the raphe nuclei.
- The ceruleospinal tract originates at the locus ceruleus in the pons. When stimulated, it causes *norepinephrine* to bind to the spinothalamic tract in the dorsal horn. Binding of norepinephrine to the spinothalamic tract suppresses the release of Substance P, thus diminishing pain messages to the cortex.
- Narcotic drugs (derived from opium or opiumlike compounds) bind to receptor sites in the periaqueductal gray, the raphe nuclei, and the dorsal horn. By binding to these receptor sites, narcotic drugs induce analgesia and stupor (a state of reduced consciousness).

Stress-Induced Analgesia

- The pain inhibiting centers can be activated naturally by injury and athletic over-exertion.
- Often, people injured during accidents, disasters, or athletic competition may not feel pain until the event has passed.
- Stress occurring during the event may trigger the pain inhibition centers.
- Stress-induced analgesia involves activation of the following:
  - The raphe nuclei descending tracts
  - The release of the hormonal endorphins from the pituitary gland (particularly beta-endorphins)
  - The release of hormonal endorphins from the adrenal medulla (particularly enkephalins)
- Hormonal endorphins bind to the opiate receptors in the brain and SC.
- Beta-endorphins are the most potent endorphins and can trigger analgesic affects that last for hours.

Pain Transmission Can Be Diminished at Several Nervous System Levels

The Periphery

- *Non-narcotic analgesics* (eg, aspirin) decrease the synthesis of prostaglandins, thus preventing prostaglandins from sensitizing peripheral pain receptors. Prostaglandins are a large group of biologically activated, carbon-20, unsaturated fatty acids.
- *NSAIDs* (nonsteroidal anti-inflammatory drugs—such as ibuprofen and naproxen) and local anesthetic agents also produce analgesic effects by interrupting peripheral transmission at an early stage of the pain process. NSAIDs inhibit prostaglandin production, thus reducing the number of pain chemicals available to stimulate peripheral nociceptors.
- *Local anesthetics* can be administered to nerve endings at the site of injury or to the nerve plexus supplying the area. Peripheral transmission of pain signals is interrupted by localized or regional blockade.
- The application of *heat and cold* to a painful area similarly reduces peripheral pain signals by altering blood flow to the area and reducing swelling.

Dorsal Horn

- Inhibitory neurons in the dorsal horn release enkephalin or dynorphin. These can diminish pain sensation through interneurons that bind to the excited nociceptor. This is the principle of the Counterirritant Theory.

Supraspinal Descending Systems

- The raphe nuclei, periaqueductal gray, and the locus ceruleus can inhibit nociceptive information.

Hormonal System

- The hormonal system involves the release of hormonal endorphins from the *pituitary gland* (particularly beta-endorphins).
- It also involves the release of hormonal endorphins from the *adrenal medulla* (particularly enkephalins).

Cortical Level

- The cortical detection and interpretation of pain can be altered by an individual's expectations, distraction level, anxiety, and belief (particularly regarding placebo effects).

Pain Transmission Can Also Be Intensified at Several Nervous System Levels

- Edema and endogenous chemicals can sensitize free nerve endings in the periphery.
- For example, following a minor burn injury, sensory stimuli that would normally be innocuous can cause heightened pain.
- Fear and anxiety can also heighten the experience of pain.

Chronic Pain and Pain Tolerance

- Prostaglandins form as a result of damaged cells. An enzyme called phospholipase A breaks down phospholipids in the cell membrane and converts them to arachidonic acid.
- Another enzyme then breaks down arachidonic acid. As a result, prostaglandins are formed.
- Sensitization by prostaglandins lowers the threshold of pain fibers. This results in *allodynia*—a condition in which nonpainful stimuli now produce pain (ie, people who once had a higher pain tolerance now experience pain more easily).
- Some theorists suggest that a *synaptic memory of pain* in the nociceptive pathway can be formed when glutamate binds to certain pain receptors on the postsynaptic neuron. Then, when excessive or repeated stimulation of the small, unmyelinated C fibers occurs, SC neurons become sensitized, further creating the synaptic memory of pain.
- The synaptic memory becomes easier to trigger each time it is stimulated—so that a once non-noxious stimulus can now trigger the synaptic memory of pain.
- This phenomenon is referred to as *wind-up* and may underlie the development of *chronic pain*.

Referred Pain

- Referred pain is pain that is perceived to originate from one body region when it actually originates from a different body region.
- Usually referred pain occurs when visceral pain (from an internal organ or gland) is perceived as originating from a somatic area (such as the skin or skeletal muscles).
- For example, during a heart attack, the brain may misinterpret the nociceptive information as arising from the skin on the medial left arm.
- Similarly, gallbladder pain is often referred to the right subscapular region.
- The phenomenon of referred pain can be explained by the dermatomal distribution:
  - In a heart attack, nociceptive information from the heart projects to and from the SC segment T1. The dermatomal sensation of the medial left arm also projects to and from T1.
  - Because the cortex is unfamiliar with pain messages received from the heart, it initially interprets the pain as originating from the left arm—until the pain becomes excruciating.
  - Some dorsal root neurons have two peripheral axons—one that innervates the skin and skeletal muscle, and one that innervates the viscera. Stimulation of the visceral branch of a dual receptive neuron may be the source of cortical misinterpretation.

Noninvasive Pain Management Procedures

Stimulation of the Mechanoreceptors (Massage)

- Stimulation of the mechanoreceptors is based on the Counterirritant Theory and may involve rubbing or massaging the painful area.
- The proximal branches of the mechanoreceptors in the dorsal horn activate interneurons that synapse on the excited nociceptors (also in the dorsal horn).
- These interneurons release the neurotransmitter enkephalin—which binds with the excited nociceptor and diminishes the release of Substance P (released from the excited nociceptors).
- Enkephalin binding on the nociceptor inhibits the transmission of nociceptive signals, thus decreasing the sensation of pain.

Electrical Stimulation

- Electrical stimulation is a form of pain management that may work by (a) blocking the transmission of pain signals along the nerve, (b) promoting the release of endorphins, (c) causing vasodilation (widening of the blood vessels) allowing increased oxygenated blood to the painful area, or (d) some combination of the above.
- Electrical stimulation can be applied in various forms including (a) transcutaneous electrical nerve stimulations (TENS), (b) microcurrent electrical therapy (MET), and (c) interferential electrical stimulation (IFC).
- The administration of electrical stimulation commonly involves the placement of electrodes over the painful body part.
- Electrical stimulation is used to treat both acute and chronic pain conditions including back pain, headaches, and arthritis.

Transcutaneous Electrical Nerve Stimulation

- The therapist finds the dermatomal region of the pain area and applies TENS to activate the mechanoreceptors.
- This technique has marginal effectiveness.

Thermotherapy

- Thermotherapy involves the use of heat to treat chronic and acute pain syndromes, including back pain, arthritis, tendinitis, and muscular aches.
- Thermotherapy works through vasodilation—allowing the blood vessels in the painful area to widen, thus increasing blood flow to the area. It can also relax superficial muscles making them more elastic and decreasing joint stiffness.
- The most common forms of thermotherapy include hot packs, paraffin, ultrasound, and diathermy.

Hot Packs

- Hot packs contain a silicone gel that, when immersed in hot water, has the capacity to absorb and hold a great amount of heat.
- Hot packs are not able to provide deeply penetrating heat, as is ultrasound.
- They are recommended for large body areas, such as the back.
- They provide temporary relief of pain.

Ultrasound

- Involves conversion heating, or high frequency sound waves.
- As ultrasound is propagated through tissue, it is absorbed and converted into heat.
- Ultrasound may improve circulation, soften scar tissue and adhesions, reduce chronic inflammation, and reduce irritation of nerve roots.
- Ultrasound is one of the most effective heat modalities for deep structures; reported depths of penetration travel 5 to 6 cm below the superficial layers of the skin.

Paraffin

- Heated wax; a method of superficial heat conduction
- Recommended for specific joint tightness and pain—particularly in the hands.

Diathermy

- Also known as shortwave or microwave diathermy
- Microwaves or shortwaves are selectively absorbed by tissues with high water content.

Cryotherapy (Cold Packs)

- Cold packs are useful for anesthetizing sensory receptors to relieve pain. The analgesic effects of ice result from decreased nerve conduction along pain fibers.
- Cold therapy also produces vasoconstriction, which slows circulation, thus reducing inflammation and relieving pain.

Hydrotherapy

- Hydrotherapy is a method of convection heat using water and can be considered a type of thermotherapy.
- It can be performed in swimming pools, whirlpools, showers, or immersion tanks that allow patients to (a) exercise with reduced stress to joints and (b) relax tight or spastic muscle groups.
- This is particularly useful for large body areas such as the back.

Fluidotherapy

- Fluidotherapy is a dry superficial thermal modality that transfers heat to soft tissue through heated air and particles (corn husks ground to the size of sand grains).
- Fluidotherapy is much like a dry whirlpool with particles instead of water as the heating medium.
- The particles are circulated by hot air blown within the fluidotherapy machine. The particles agitate around the body part at a temperature of approximately 108 to 124 degrees.
- Fluidotherapy may work via vasodilation and mechanical stimulation. Vasodilation causes increased oxygenated blood flow to the painful area. Mechanical stimulation may help to desensitize skin and scar pain after injury.
- This technique is useful for pain relief in the extremities—particularly in the hand, wrist, forearm, and ankle.

Kinesiotape

- Kinesiotaping involves using a special elastic tape over the muscles to (a) assist function and provide support, (b) to prevent over-use, and (c) to reduce pain and inflammation.
- Kinesiotape is used for muscular disorders and lymphedema reduction.
- This technique involves a nonrestrictive type of taping that allows for full range of motion in functional activities—in contrast to traditional sports taping in which tape is wrapped fully around a joint for stabilization and limits both range of motion and vascular circulation.

Acupuncture

- Acupuncture is a part of traditional Chinese medicine (TCM) and is one of the oldest documented medical treatments in the world.
- In TCM, the body is believed to be in a state of health when its two forces—yin and yang—are balanced. When one force dominates the other, an imbalance occurs causing disease states. An imbalance between yin and yang is said to cause a blockage in the flow of qi—the vital life force. Such blockage is believed to result in pain and disease.
- In order to restore health and relieve pain, specific acupuncture points (called meridians) need to be stimulated. The most common form of acupuncture involves inserting hair-thin needles into the skin along the meridian points.
- A large body of research has been generated supporting the effectiveness of acupuncture.
- While the mechanism through which acupuncture works is unknown, it is suggested that it stimulates the release of naturally occurring opioids in the body.
- Acupuncture has been shown to provide lasting pain relief if the procedure is applied over time.

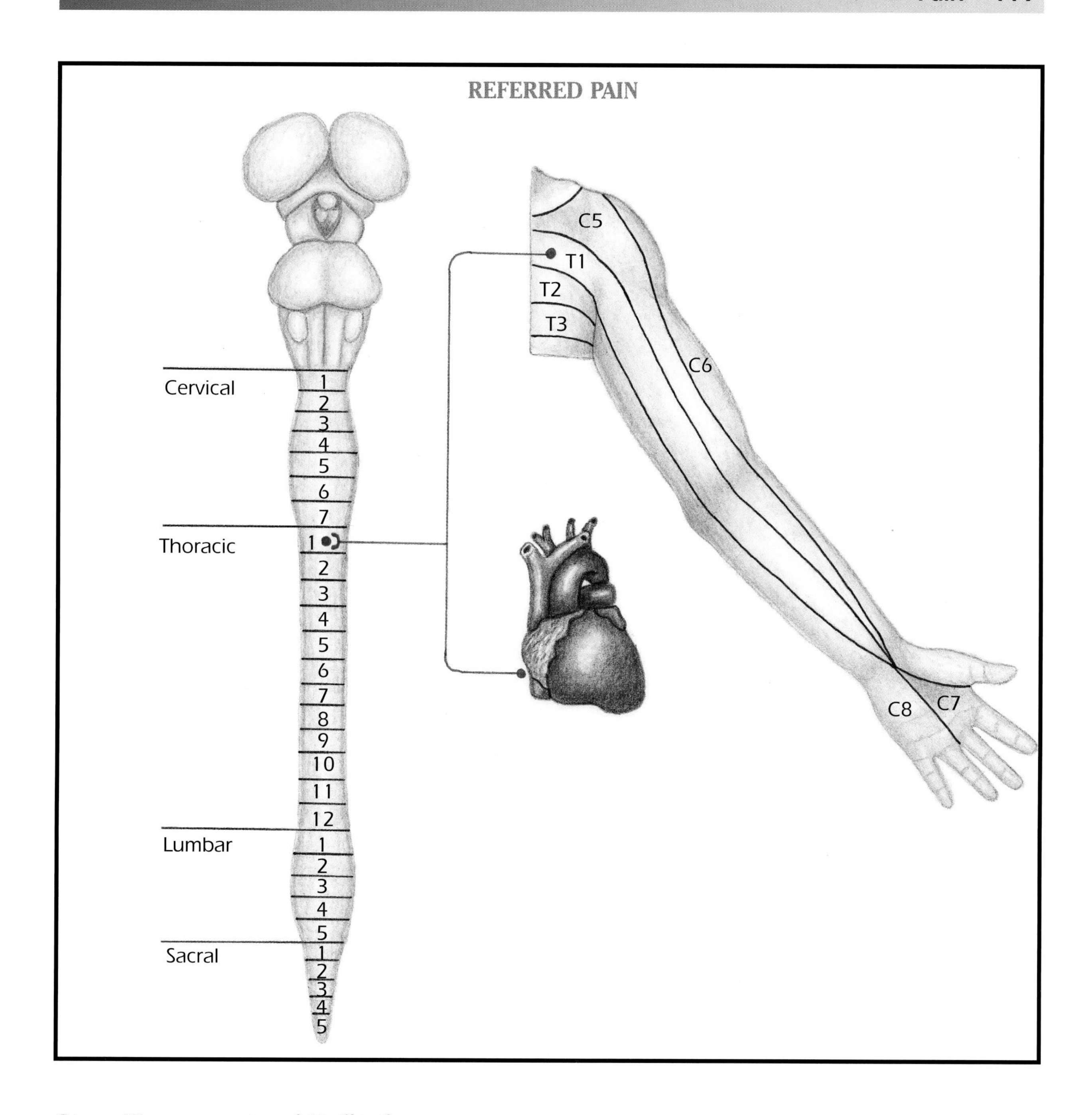

Stress Management and Meditation

- These techniques involve learning how to release the body's naturally occurring endorphins.
- They also involve learning how to raise pain thresholds by using visual imagery.

Biofeedback

- Using biofeedback, patients can learn to consciously control body functions that are typically involuntary—such as muscle tension and heart rate.
- Biofeedback involves learning to release tension in muscles that may be contracted because the individual is using those muscles to guard or protect the painful area.
- Biofeedback can offer patients an effective pain management strategy.

Invasive Pain Management Procedures

Nerve Blocks

- Nerve blocks involve injections of local anesthetics and steroids in the area of a spinal nerve in an attempt to decrease pain.
- There are several different types of nerve blocks:

Facet and Medial Branch Blocks

- A facet block involves an injection of local anesthetic and/or steroid—using x-ray guidance—into a joint space in the spine to relieve pain.
- A medical branch block is similar, but the anesthetic is injected outside of the joint space near the nerve that supplies the joint (called the medical branch).
- A series of injections over time is commonly required to provide effective and lasting relief of pain.

Root Blocks

- A root block involves the injection of anesthetic and/or steroid—using x-ray guidance—into the area where the nerve exits the SC.
- A root block is performed to alleviate pain in the extremities that follows the distribution of a single nerve.
- Root blocks will be repeated if good results are initially obtained.

Epidural Steroid Injections

- Epidural steroid injections involve the administration of a local anesthetic and steroid into the subdural space to provide pain relief by reducing the inflammation of the nerve roots as they exit the SC.
- The injections are commonly administered in a series of three injections, each approximately 1 month apart. If pain is substantially reduced after the first injection, further injections are not needed unless the pain returns at a later date.

Spinal Surgery

- Surgery of the spine may be considered as an option when back pain cannot be relieved through noninvasive methods.

Discogram

- A discogram is a diagnostic procedure used to determine the anatomical source of low back pain. The procedure is most frequently used to determine if degenerative disc disease is the cause of a patient's pain.
- A needle is inserted into the center of the suspected discs—one at a time. Radiographic dye is then injected into each disc. The patient is awakened at this time and asked to indicate present pain level. If the injection of dye recreates the patient's pain in a specific disc(s), that disc is determined to be the source of pain.
- Frequently, after the discogram is completed, a CT scan is performed to further examine the identified disc(s).

Discectomy

- Back pain is most commonly caused by herniation of the disc—in other words, the disc ruptures, moves out of place, and may impinge upon nearby structures (such as nerve roots).
- A discectomy involves the removal of a herniated disc and the replacement of the disc with synthetic material.

Laminectomy

- A laminectomy involves the partial or total removal of the lamina—the small bony plate that sits at the back of each vertebrae.
- This procedure is commonly performed for conditions of spinal stenosis—narrowing of the spinal canal causing impingement of the cord and nerves.

Foraminotomy

- A foraminotomy is a surgical procedure used to enlarge foraminal openings—the openings through which the nerves exit the SC.
- It is commonly performed to alleviate a compressed or impinged nerve.

Spinal Fusion

- Spinal fusion involves the use of synthetic and natural bone material when metal devices (such as cages, plates, screws, and rods) are used to stabilize the vertebrae.
- The bone grafts will grow around the metal instrumentation and strengthen the vertebral architecture.

Intrathecal Pumps (Indwelling Pain Pump)

- Intrathecal pumps are surgically implanted mechanical devices used to deliver medications to the SC region. The instrumentation consists of a computerized pump, a reservoir, and a catheter.
- Intrathecal pumps must be implanted surgically—usually under the skin of the lower abdomen. The catheter is placed into the spinal fluid space and connected to the reservoir.
- The pump is programmed to deliver pain medication in a controlled fashion.
- Intrathecal pumps are considered for patients who have not responded to oral opioid treatment.

Pharmaceutical Management of Pain

Nonsteroidal Anti-Inflammatory Drugs

- Nonsteroidal anti-inflammatory drugs (NSAIDs) can reduce swelling and the secondary damage that can occur as a result of swelling.
- These drugs do not alter cognitive functions, cause respiratory depression, or cause nausea. Long-term side effects include gastrointestinal problems.
- They include aspirin and ibuprofen.

Acetaminophen

- Acetaminophen is a group of analgesic medications that are most commonly used as alternatives to NSAIDs. They can be used alone or in combination with NSAIDs.
- Potential long-term side effects include liver and kidney damage.
- They include Tylenol, Anacin, and Valadol.

Opioids

- Opioids are indicated only in the case of severe pain due to their addictive potential.
- They include morphine, codeine, Demerol, and Oxycontin.

Muscle Relaxants

- Muscle relaxants are indicated in cases of severe pain.
- They include Flexeril, Valium, and Robaxin.

Fibromyalgia

- Fibromyalgia is a chronic pain syndrome that does not have a clear etiology.
- Some studies suggest that changes occur in the CNS leading to heightened sensitivity of pain fibers.
- Other studies suggest that deficient restorative sleep periods may contribute to the disorder. Restorative sleep occurs in deep stage 4 sleep—the period of rapid eye movement (REM). Humans generally experience two 90 minute cycles of stage 4 sleep each night. Cellular repair is believed to occur most efficiently in this stage. People who are chronically deficient of stage 4 sleep appear more likely to develop the symptoms of fibromyalgia.
- Primary symptoms include tenderness of muscles and adjacent soft tissues, stiffness of muscles, and aching pain. Other symptoms can include fatigue, sleep dysfunction, numbness and tingling, headache, and short term memory and concentration difficulties.
- Patients may have no accompanying disease; it is also likely for patients to concomitantly have rheumatoid arthritis, osteoarthritis, irritable bowel syndrome, Lyme disease, or sleep apnea.
- Multiple tender points can be found on palpation. Such painful areas follow a regional rather than a dermatomal or peripheral nerve distribution.
- Increased restorative sleep and appropriate exercise have been found to benefit patients with fibromyalgia.

SECTION 16

Peripheral Nerve Injury and Regeneration

Neuropathy

- Neuropathy is a general term for pathology involving one or more peripheral nerves.

Dermatomal Distribution

- The dermatomes are *skin segments* that are *innervated* by *specific peripheral nerves*.
- The dermatomal skin distribution corresponds closely to the skeletal muscles that are innervated by a specific peripheral nerve.
- When a therapist evaluates sensation, the therapist will test each dermatomal skin segment to identify possible loss of sensation.
- Identification of a specific dermatomal skin segment that has lost sensation will indicate the lesion level.
- For example, a loss of sensation on the lateral forearm and the lateral hand may result from C6 peripheral nerve loss.

Complete Severance of a Peripheral Nerve

- A completely severed peripheral nerve results in a *loss of sensation*, *loss of motor control*, and a *loss of reflexes* in the structures innervated by a specific peripheral nerve.

Nerve Compression

- Nerve compression results in a *loss of proprioception* and *discriminative touch*.
- Pain and temperature initially remain intact.
- This occurs because compression of a nerve affects the large myelinated fibers first—these are fibers that carry proprioceptive and discriminative touch information.
- There is initial relative sparing of the smaller fibers that carry pain and temperature.
- When compression of a nerve occurs, sensory loss proceeds in the following order:
  1. Conscious proprioception and discriminative touch
  2. Cold
  3. Fast pain or sharp pain
  4. Heat
  5. Slow pain or dull, diffuse, aching pain

When Compression Resolves

- When the compression is relieved, abnormal sensations called *paresthesias* occur as the blood supply increases.
- Paresthesias include *burning*, *pricking*, and *tingling sensations*.
- After compression is resolved, sensation returns in the reverse order in which it was lost:
  1. Slow pain or dull, diffuse, aching pain
  2. Heat
  3. Fast pain or sharp stinging sensations
  4. Cold

5. Conscious proprioception and discriminative touch

- This process can be observed when one of the body's limbs "falls asleep" as a result of nerve compression.

Flaccidity

- When a peripheral nerve is injured, the muscles that are innervated by that nerve become flaccid and gradually atrophy.
- Paralysis and loss of sensation occur just distal to the lesion.

Schwann Cells

- Schwann cells form the myelin in the peripheral nervous system (PNS).
- Schwann cells produce *nerve growth factor* (NGF).
- This allows peripheral nerve damage to resolve (unlike damage in the CNS).
- Nerve regrowth occurs approximately 1 mm per day or 1 inch per month.
- A C6 peripheral nerve injury in the upper arm could take a year or more to fully regenerate.
- Compression injuries recover more quickly than complete severance of a nerve.

Terminology of Sensory Pathology

Hypoesthesia

- A decrease in sensory perception. Also referred to as hypesthesia

Hyperesthesia

- An increase in sensory perception (eg, heightened perception of pain and temperature)

Paresthesia

- The occurrence of unusual feelings such as pins and needles

Dysesthesia

- Unpleasant sensation such as burning. Causalgia is an intense burning pain accompanied by trophic skin changes. Causalgia usually runs along the distribution of a nerve.

Thermesthesia

- The ability to perceive temperature (hot and cold)

Thermohyperesthesia

- An increase in temperature perception (ie, hot and cold sensations become heightened)

Thermohypoesthesia

- A decrease in temperature perception

Analgesia

- Loss of pain sensation

Hypalgesia

- A decrease in the ability to perceive pain

Hyperalgesia

- An increase in the ability to perceive pain (ie, pain sensation becomes heightened)

Allodynia

- A condition in which an otherwise innocuous stimulus causes pain

Autonomic Dysfunction

- Autonomic dysfunction may result in vasomotor nerve disturbances.
- Vasomotor nerves are nerves that have muscular control over the blood vessel walls. They control *vasodilation* and *constriction of blood vessels*.
- *Autonomic trophic changes* (also called nutritional changes) affect the skin, hair, and nails. The skin may become smooth and glossy, the nails can become thickened, and the hair on the denervated skin area thins or falls out.

Brachial Plexus

- The brachial plexus is a network of peripheral spinal nerves including C5, C6, C7, C8, and T1.
- The nerves that supply the upper extremities emerge from the brachial plexus.
- The brachial plexus is a site of common compression syndromes:
  - Radial Nerve Compression: wrist and finger drop, tennis elbow, and Saturday night palsy
  - Medial Nerve Compression: carpal tunnel
  - Ulnar Nerve Compression: clawhand deformity and deformity of the fourth and fifth digits

Lumbar Plexus

- The lumbar plexus is a network of peripheral spinal nerves formed by L1 through L4.
- It is also a common site of compression syndromes:
  - Sciatic Nerve Compression: sciatica
  - Peroneal Nerve Compression: foot drop

Types of Peripheral Neuropathy

Mononeuropathy

- Mononeuropathy involves *damage to a single nerve*.
- It is usually due to compression or entrapment.
  - Wrist drop: radial nerve entrapment
  - Foot drop: peroneal nerve entrapment

Radiculopathy

- *Nerve root impingement*
- Results from a lesion affecting the dorsal or ventral roots
- Can result from herniated vertebral discs

Plexopathy

- Caused by *damage* to one of the *plexus*—brachial or lumbar
- Involves multiple peripheral nerve damage

Polyneuropathy

- Usually involves *bilateral damage to more than one peripheral nerve*
- Example: Stocking and glove polyneuropathy. Usually caused by a disease process such as diabetes.
- With mild disease, only the distal lower extremities are involved.
- With more severe disease, the distal upper extremities are also involved.

Deep Tendon Reflexes

- When a peripheral neuropathy occurs, the individual often presents with an absence of distal deep tendon reflexes.
- An asymmetric decrease in deep tendon reflexes occurs in the following:
  - Radiculopathy (nerve root impingement on one side)
  - Plexopathy (brachial or lumbar plexus involvement)
  - Mononeuropathy (unilateral spinal nerve injury)

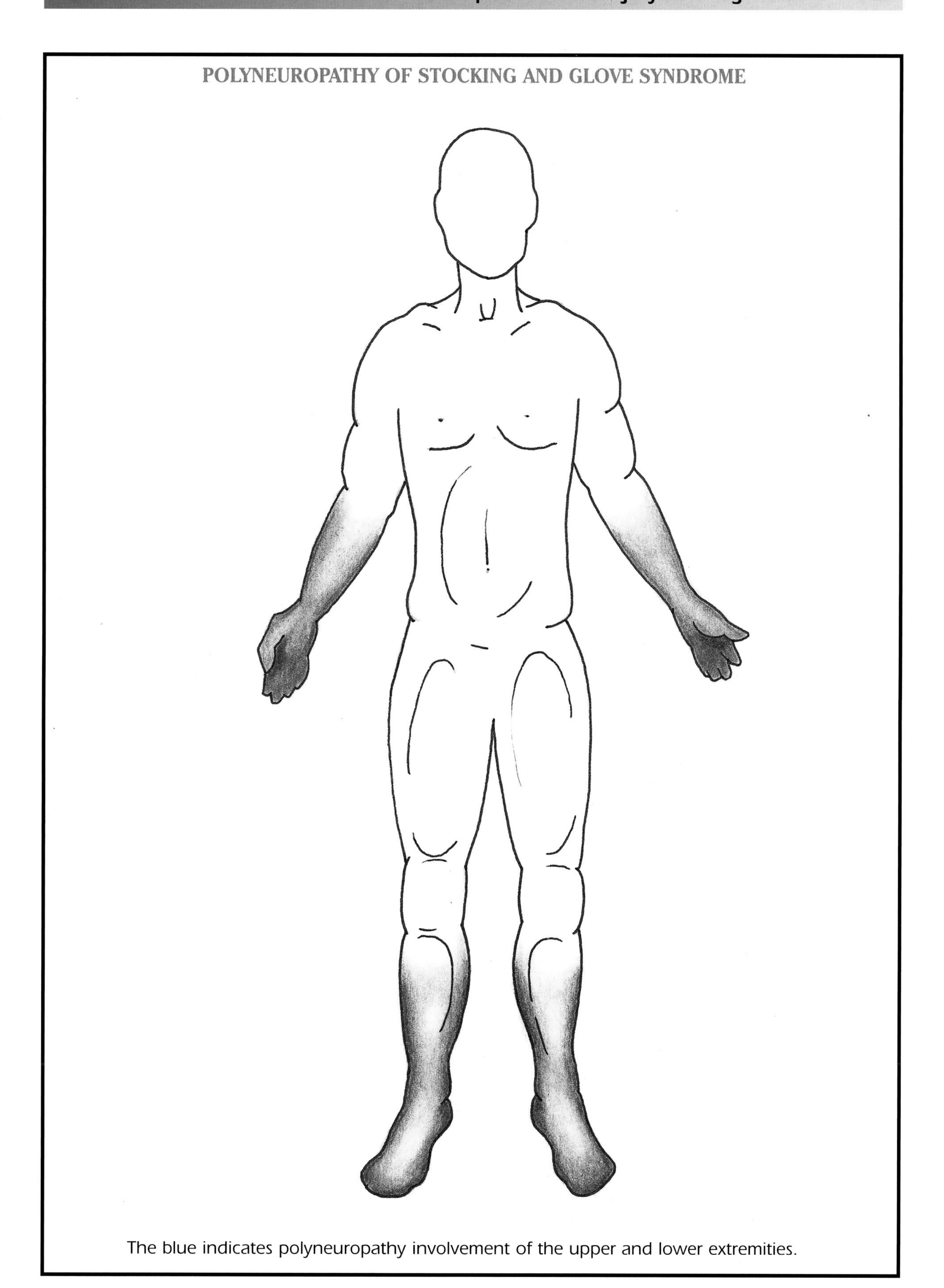

POLYNEUROPATHY OF STOCKING AND GLOVE SYNDROME

The blue indicates polyneuropathy involvement of the upper and lower extremities.

Neuropathies Caused by a Disease Process

Guillain-Barré

- Guillain-Barré is an acute inflammatory polyradiculopathy.
- It is preceded by an infectious illness that usually resolves before the neurologic dysfunction becomes evident.
- The disease is characterized by edema and demyelination of peripheral spinal roots.
- Guillain-Barré involves progressive ascending muscular weakness of the limbs (as in stocking and glove neuropathy). This produces flaccid paralysis that has a symmetric pattern.
- Paresthesia and numbness often accompany motor function loss.
- It also involves complete tendon areflexia.
- Autonomic nervous system (ANS) involvement can result in postural hypotension (when one rises from a horizontal position, blood pressure may drop to precariously low levels), arrhythmias, facial flushing, diarrhea, impotence, urinary retention, and increased sweating.
- Although progress is commonly slow, most patients experience complete recovery.

Diabetes Mellitus

- Diabetes mellitus is an endocrine disorder involving insulin deficiency or intolerance.
- Diabetic neuropathy is one of the most common complications of the disease process and affects at least 50% of patients with both type I and type II diabetes.
- While the exact mechanisms underlying diabetic neuropathy are unknown, damage to the peripheral nerves is thought to be a primary cause. This damage is mediated by inflammation and demyelination of large peripheral nerves, leaving smaller nerves intact. This results in lost inhibitory pain control from the spinal cord centers.
- Common symptoms include pain, numbness and tingling, mild weakness, and diminished proprioception, vibration, and discriminative touch. Patients commonly report burning pain in the distal lower extremities (bilaterally).
- Polyneuropathies, mononeuropathies, plexopathies, and autonomic neuropathies can all occur in diabetes.
- Diabetes can be associated with abnormalities at any level of the PNS.
- Autonomic neuropathies involve the gastrointestinal, cardiovascular, and genitourinary systems.
- ANS involvement can result in postural hypotension, arrhythmias, facial flushing, diarrhea, impotence, urinary retention, and increased sweating.
- In addition to pain and the loss of sensory and motor function, diabetic neuropathies can severely impair everyday function:
  - Patients are at an increased risk for falls due to the loss of sensation and position sense.
  - Burns and injuries to the feet are problematic due to the loss of temperature and pain sensation.
  - Impaired vasomotor reflexes may result in dizziness and syncope (fainting) when the patient moves from a supine to standing position.
  - Urinary retention can increase the risk for bladder infection and renal complications.

Severed Axon

- When an axon is severed, the portion connected to the cell body is referred to as the proximal segment.
- The portion that is now disconnected from the cell body is called the distal segment.

Leakage of Protoplasm

- Immediately after injury, protoplasm leaks out of each severed end, and the two segments retract away from each other.

Wallerian Degeneration

- The distal segment undergoes a process called wallerian degeneration.
- The myelin sheath pulls away from the distal segment.

- The distal segment swells and breaks into smaller segments.
- The axon terminals degenerate rapidly.
- The entire distal segment dies.
- Glial cells scavenge the area and clean up the debris from the degenerated distal segment.

Central Chromatolysis

- Sometimes the cell body of the proximal segment undergoes degenerative changes called central chromatolysis. This may lead to cell death.
- Simultaneously, the postsynaptic neuron—no longer innervated by the presynaptic neuron—may also degenerate and die.

Sprouting

- The regrowth of damaged axons is called sprouting.
- Sprouting takes two forms:
  - Collateral
  - Regenerative

Collateral Sprouting

- Collateral sprouting occurs when a denervated postsynaptic neuron is re-innervated by branches of an intact axon located near the damaged axon.
- In other words, another neuron projects a collateral axon branch to the cell body of the postsynaptic neuron.

Regenerative Sprouting

- Regenerative sprouting occurs when an axon and its postsynaptic neuron have both been damaged.
- The proximal segment of the damaged neuron projects side sprouts to form new synapses with other undamaged postsynaptic neurons.

Functional Regeneration Occurs in the Peripheral Nervous System

- Regeneration of axons occurs primarily in the PNS.
- This is partly due to the production of NGF by Schwann cells.
- Schwann cells are the myelin around the PNS axons; Schwann cells do not exist in the CNS.
- Oligodendrites are the myelin cells that wrap around the CNS axons; oligodendrites do not produce NGF.
- There is little or no recovery in most areas of the CNS.

Recovery Speed in the Periphery

- Regeneration of damaged axons in the PNS is slow—1 mm of growth per day or 1 inch of recovery per month.
- A severed nerve in the upper arm—such as C6—will take at least a year to recover.

Problems that Can Occur in the Nerve Regeneration Process

- Sometimes an axon will innervate a new postsynaptic neuron that is inappropriate.
- For example, after injury of a peripheral nerve, motor axons may innervate different muscles than before the injury.
- When the neuron fires, this results in unintended movements called *synkineses*.
- Synkineses are usually short-lived, as the person relearns muscle control.
- Similarly, in the sensory systems, innervation of a sensory receptor by axons that previously innervated a different type of sensory receptor can cause confusion of sensory modalities.

Synaptic Changes After Injury

Synaptic Effectiveness

- Local edema in the injured area causes compression of the presynaptic neuron's cell body or axon.
- When the edema is resolved, synaptic effectiveness returns.

Denervation Hypersensitivity

- Denervation hypersensitivity occurs when the postsynaptic neuron—of an injured presynaptic neuron—becomes hypersensitized to its own neurotransmitter.
- This occurs because the postsynaptic neuron develops new receptor sites that can respond to neurotransmitters released by other nearby neurons.
- This occurs only for a short time.
- It produces temporary muscle twitches and pain.

Synaptic Hypereffectiveness

- Synaptic hypereffectiveness occurs when only some branches of a presynaptic axon are destroyed, leaving the majority of branches intact.
- The remaining axon branches receive all of the neurotransmitter substance that would normally be shared among only the terminal end branches.
- This results in a larger than normal amount of neurotransmitter being released onto postsynaptic receptors.

Unmasking of Silent Synapses

- In the normal PNS, many synapses seem to be unused unless other pathways become injured.
- Unmasking of silent synapses occurs when previously nonused synapses become active after other pathways have been damaged.

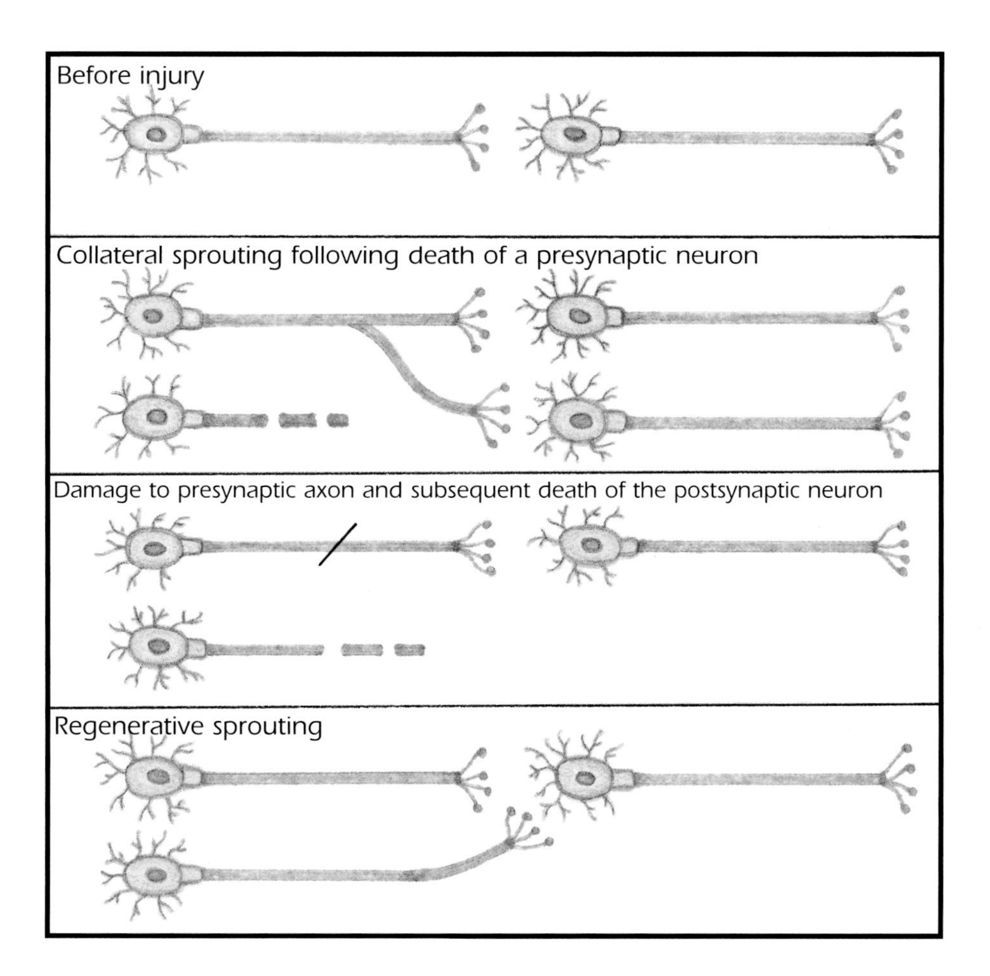
Before injury
Collateral sprouting following death of a presynaptic neuron
Damage to presynaptic axon and subsequent death of the postsynaptic neuron
Regenerative sprouting

SECTION 17

Phantom Limb Phenomenon

Phantom Limb Phenomenon

- Phantom limb phenomenon is the sensation that an amputated body part still remains.
- If the sensation is painful, it is called phantom pain.

Phantom Limb

- Any nonpainful sensation in the amputated limb

Phantom Pain

- Painful sensations that seem to occur in the lost limb

Stump Pain

- Painful sensations localized to the stump of an amputated body part

Phantom Pain

- Phantom pain is *cortically perceived*—there is *no peripheral component.*
- Phantom pain usually cannot be ended by nerve blocks.
- Phantom pain is often described as excruciating, sticking, cramping, burning, and squeezing.

Phantom Pain and Phantom Limb are Central Nervous System Phenomena

- Phantom pain is not mediated by peripheral nerve signals, as is stump pain.
- Phantom pain and phantom limb sensations are largely *due to central nervous system (CNS) phenomena.*
- The cortical map of the body (the sensory and motor homunculi) still retain the anatomical image of the amputated body part.
- The brain believes that the amputated body part still remains.
- On transcranial magnetic stimulation (TMS), no change can be observed in the cortical map area after the amputation.

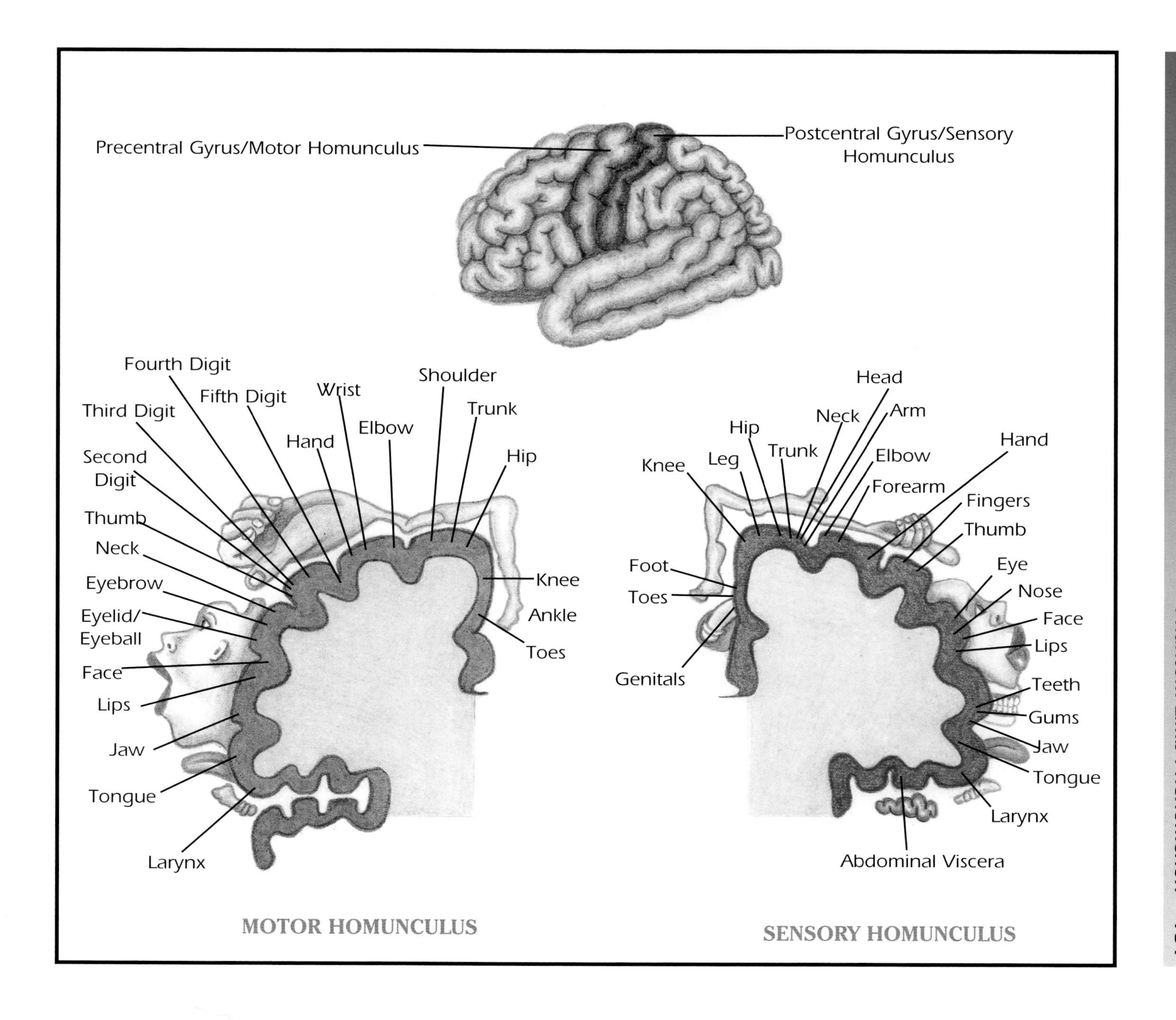
Precentral Gyrus/Motor Homunculus
Postcentral Gyrus/Sensory Homunculus
Fourth Digit
Fifth Digit
Wrist
Shoulder
Trunk
Third Digit
Hand
Elbow
Hip
Second Digit
Thumb
Neck
Eyebrow
Eyelid/ Eyeball
Face
Lips
Jaw
Tongue
Larynx
Knee
Ankle
Toes
MOTOR HOMUNCULUS
Head
Neck
Arm
Hip
Trunk
Knee
Leg
Elbow
Hand
Forearm
Fingers
Thumb
Foot
Toes
Eye
Nose
Face
Lips
Genitals
Teeth
Gums
Jaw
Tongue
Larynx
Abdominal Viscera
SENSORY HOMUNCULUS

Phantom Sensations Lessen Over Time

- Other areas of the cortical map eventually appropriate the cortical region that once mediated sensation/movement of the amputated limb.
- At this time, TMS will show change in the cortical map.

Phantom Limb Sensation Can Aid in Gait Training

- Patients with phantom sensations—that are not painful—report that they can use the phantom image to propel their prosthesis.
- Phantom pain interferes with gait training.

Treatment of Phantom Pain

- Treatment for phantom pain is often ineffective.
- Transcutaneous electrical nerve stimulation (TENS) works to a moderate degree.
- Vibration to the stump may work temporarily.
- Analgesics or painkillers work to a moderate degree and only temporarily.

Phantom Limb Sensation Can Be Experienced in Spinal Cord Injuries

- The phantom limb sensation that people with spinal cord injury (SCI) experience tends to resolve quickly.
- Other areas of the cortical map appropriate the areas once used to mediate the paralyzed limbs.
- This may relate to the fact that in SCI, no peripheral stimulation can continue, as it might in an amputee who still retains some portion of the limb.

Mirror Box

Ramachandran and Rogers-Ramachandran have shown that by using a mirrored box, the brain can be tricked into believing that the amputated limb is present again. The patient places his intact limb in a box with a mirror. The mirrored reflection of the intact limb appears as the amputated limb. The patient believes that he is looking down upon both his right and left hands and forearms—even though his left hand is actually amputated. When the patient is instructed to open and stretch his intact hand, it appears that the amputated hand is present and also stretching. The brain is tricked into perceiving that the amputated hand has responded to neural signals to open. Patients report that this phenomenon alleviates phantom limb pain and cramping in the amputated hand.

Phantom Limb Phenomena of the Hands and Feet

- Some patients with hand amputations report that if their amputated limb is touched, they experience the sensation of their face being touched.
- This phenomenon likely results from the organization of the sensory homunculus (ie, somatosensory cortex or primary somatosensory area [SS1]).
- On the somatosensory cortex, areas representing the face and hands are mapped next to each other.
- This organization may have formed in utero when the fetus' face and hands, and genitals and feet, lay next to each other in a fetal position.
- The topographical organization of the sensory homunculus then formed with these adjacent body parts mapped next to each other on SS1.
- When a hand is amputated, the contralateral area of the sensory homunculus representing the amputated hand becomes appropriated by adjacent areas of SS1 (ie, the face). This may account for the sensation of the face being touched when the amputated upper extremity limb is stimulated.
- Researchers have shown that when a prosthetic hand is grafted onto the amputated limb, the phenomenon often disappears.
- Therapists have also noted that techniques that enhance a patient's fine motor hand skills can often stimulate oral motor function. This, again, may result from the organization of the sensory homunculus.

SECTION 18

Spinal Cord Tracts

Review of Spinal Cord Anatomy

- 31 pairs of spinal nerves:
  - 8 cervical, 12 thoracic, 5 lumbar, 5 sacral, and 1 coccygeal

Ascending Spinal Nerves

- Ascending spinal nerves carry *sensory* information from the periphery (beginning with a sensory receptor) to the spinal cord (SC).
- The sensory spinal nerves enter the SC through the *dorsal root* and *rootlets* (in the peripheral nervous system [PNS]).
- Once in the *dorsal horn* (in the central nervous system [CNS]) the spinal nerves synapse with an interneuron.
- The *interneuron* synapses with an ascending sensory SC tract (in the CNS).

Ascending Spinal Cord Tracts

- Ascending SC tracts carry the sensory information up the SC, to the *brainstem*, *thalamus*, and *cortex*.
- Some sensory SC tracts carry sensory information from the *brainstem* to the *cerebellum*.

Descending Spinal Cord Tracts

- Descending SC tracts carry *motor* information from the *cortex*, through the *internal capsule*, *thalamus*, *brainstem*, and to the SC (in the CNS).
- Some motor tracts originate in the *cerebellum* and *brainstem*.
- When descending motor SC tracts are ready to *exit the cord*, they synapse on an *interneuron* in the *ventral horn* of the SC.
- The interneuron then synapses on a *motor neuron* in the ventral horn (in the PNS).
- The motor neuron in the ventral horn synapses on a *descending motor spinal nerve* and exits the cord through the *ventral rootlets* and *root* (in the PNS).
- The motor spinal nerve then travels to a *target muscle* in the PNS.

Spinal Cord Tract Names

- SC tract names indicate the tract's place of origin and destination.

Ascending Sensory SC Tracts

- Dorsal Columns
- Lateral Spinothalamic
- Anterior Spinothalamic
- Posterior Spinocerebellar

- Anterior Spinocerebellar
- Cuneocerebellar
- Rostral Spinocerebellar

Cortically Originated Descending Motor Tracts

- Lateral Corticospinal
- Anterior Corticospinal
- Corticobulbar

Mixed Pathways Originating from the Brainstem

- Medial Longitudinal Fasciculus

Descending Motor Tracts Originating in the Brainstem

- Vestibulospinal
- Rubrospinal
- Medullary Reticulospinal
- Pontine Reticulospinal

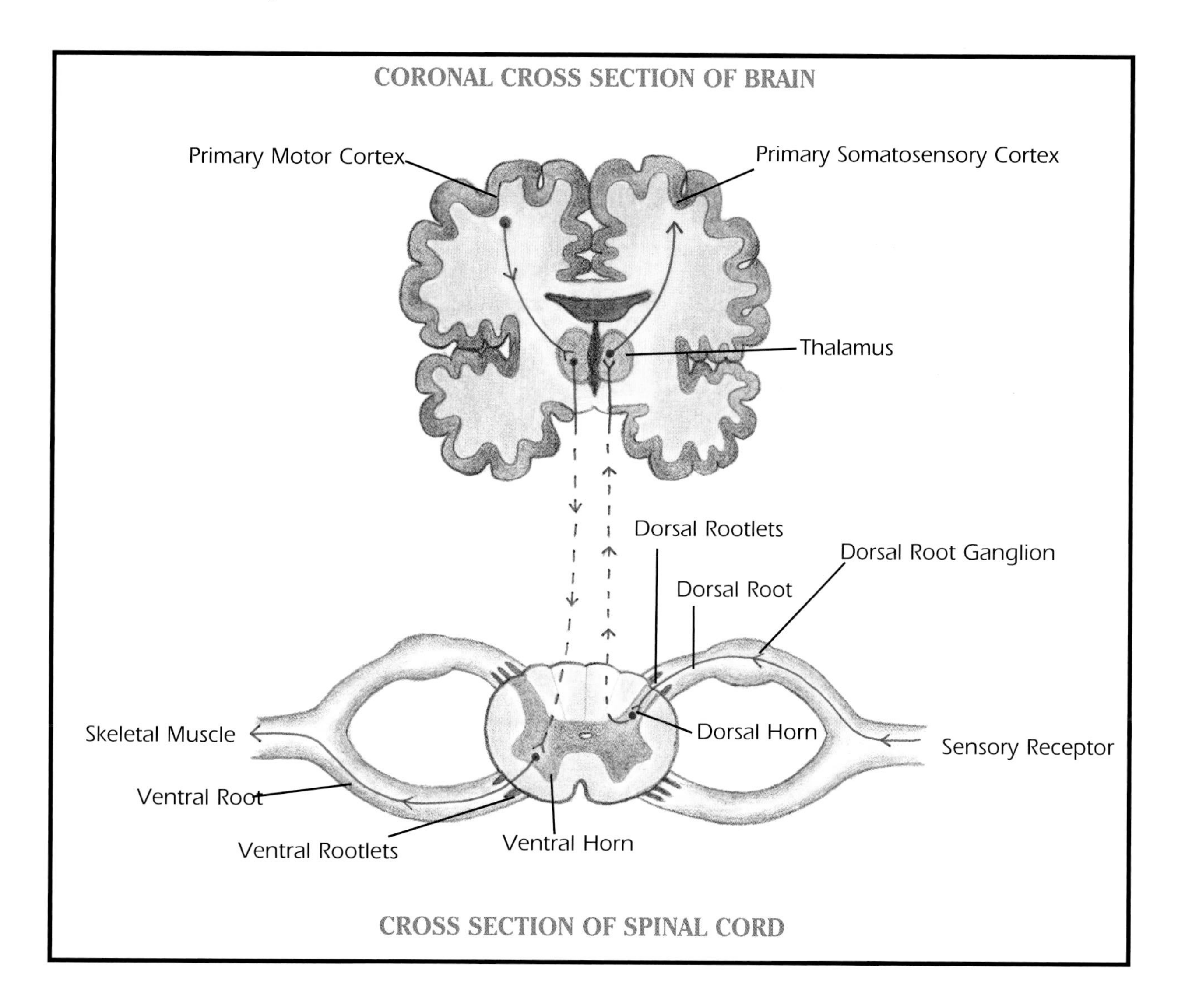

Ascending Sensory SC Tracts

Dorsal Columns (also called Medial Lemniscus or Posterior Columns)

- Believed to be a newer tract phylogenetically.

Function

- Carries conscious sensory information to the cortex regarding:
  - Discriminative touch
  - Pressure
  - Vibration
  - Proprioception
  - Kinesthesia

Origin

- Nucleus Gracilis and Nucleus Cuneatus (in the dorsal horn)

Decussation (where the tracts cross)

- Caudal Medulla Level

Destination

- Postcentral gyrus (primary somatosensory area [SS1])

Pathway

- Skin receptors in the PNS send sensory information along the peripheral spinal nerves.
- The sensory spinal nerves travel through the dorsal root and rootlets (still in the PNS).
- The spinal nerves then synapse on an interneuron in the dorsal horn of the SC (in the CNS).
- The interneuron then synapses on the cell bodies of the dorsal horn and ascends through the fasciculus gracilis and cuneatus in the dorsal columns.
- The dorsal column tract is considered to originate when the interneuron reaches the caudal medulla and synapses with the nucleus gracilis and cuneatus. At this point, the dorsal column tract decussates across the midline of the caudal medulla to the medial lemniscus.
- The tract ascends through the SC, brainstem, thalamus, and internal capsule and synapses in the postcentral gyrus.

Lesions

Complete Severance of the SC

- Bilateral loss of sensation: discriminative touch, etc (below the severed cord level)

A Hemi-Lesion (on one side of the SC) Below the Decussation

- Ipsilateral loss
- Example: If the right SC is severed at T1, it causes right-sided sensory loss.

A Hemi-Lesion in the Brainstem (above the medulla level)

- Contralateral sensory loss

A Lesion in the Cortex

- Contralateral sensory loss

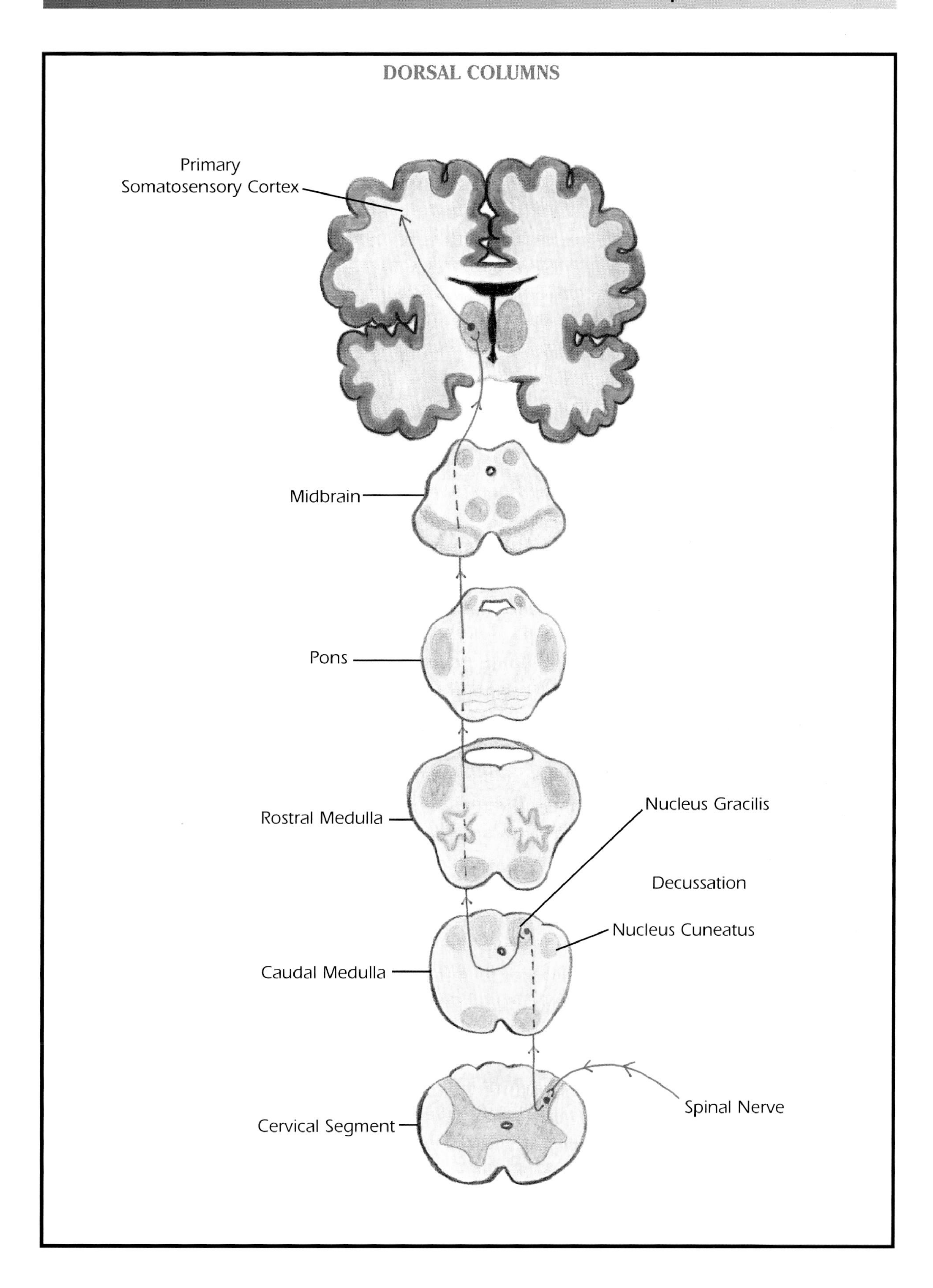
DORSAL COLUMNS
Primary
Somatosensory Cortex
Midbrain
Pons
Rostral Medulla
Nucleus Gracilis
Decussation
Nucleus Cuneatus
Caudal Medulla
Cervical Segment
Spinal Nerve

Lateral Spinothalamic Tract

- Joins the *dorsal column* at the level of the *pons*.

Function

- Carries conscious sensory information to the cortex regarding:
  - Pain and temperature

Origin

- Nucleus Proprius (in the dorsal horn)

Decussation

- SC Level (crosses as soon as the spinal nerve enters the cord and synapses on the tract).

Destination

- Postcentral Gyrus (SS1)

Pathway

- Skin receptors in the PNS send sensory information along peripheral sensory spinal nerves.
- The sensory spinal nerves travel through the dorsal root and rootlets (in the PNS) and enter the dorsal horn (in the CNS).
- Once in the dorsal horn, the spinal nerves synapse on an interneuron that joins to the cell bodies of the lateral spinothalamic tract. This occurs in the nucleus proprius of the dorsal horn.
- The lateral spinothalamic tract crosses the midline as soon as it synapses in the dorsal horn. The tract then travels to the anterior white funiculus.
- From the anterior white funiculus, the tract travels to the lateral white funiculus and ascends up the SC.
- The tract joins the dorsal column in the pons and continues to ascend to the thalamus, internal capsule, and finally synapses in the postcentral gyrus in the cortex.

Lesions

Complete Severance of the SC

- Bilateral loss of sensation: pain and temperature (below the severed cord level)

A Hemi-Lesion (on one side of the cord) in the SC

- At the lesion level: bilateral sensory loss
- Below the lesion level: contralateral sensory loss

A Hemi-Lesion (on one side) of the Brainstem

- Contralateral sensory loss

A Unilateral Lesion in the Postcentral Gyrus

- Contralateral sensory loss

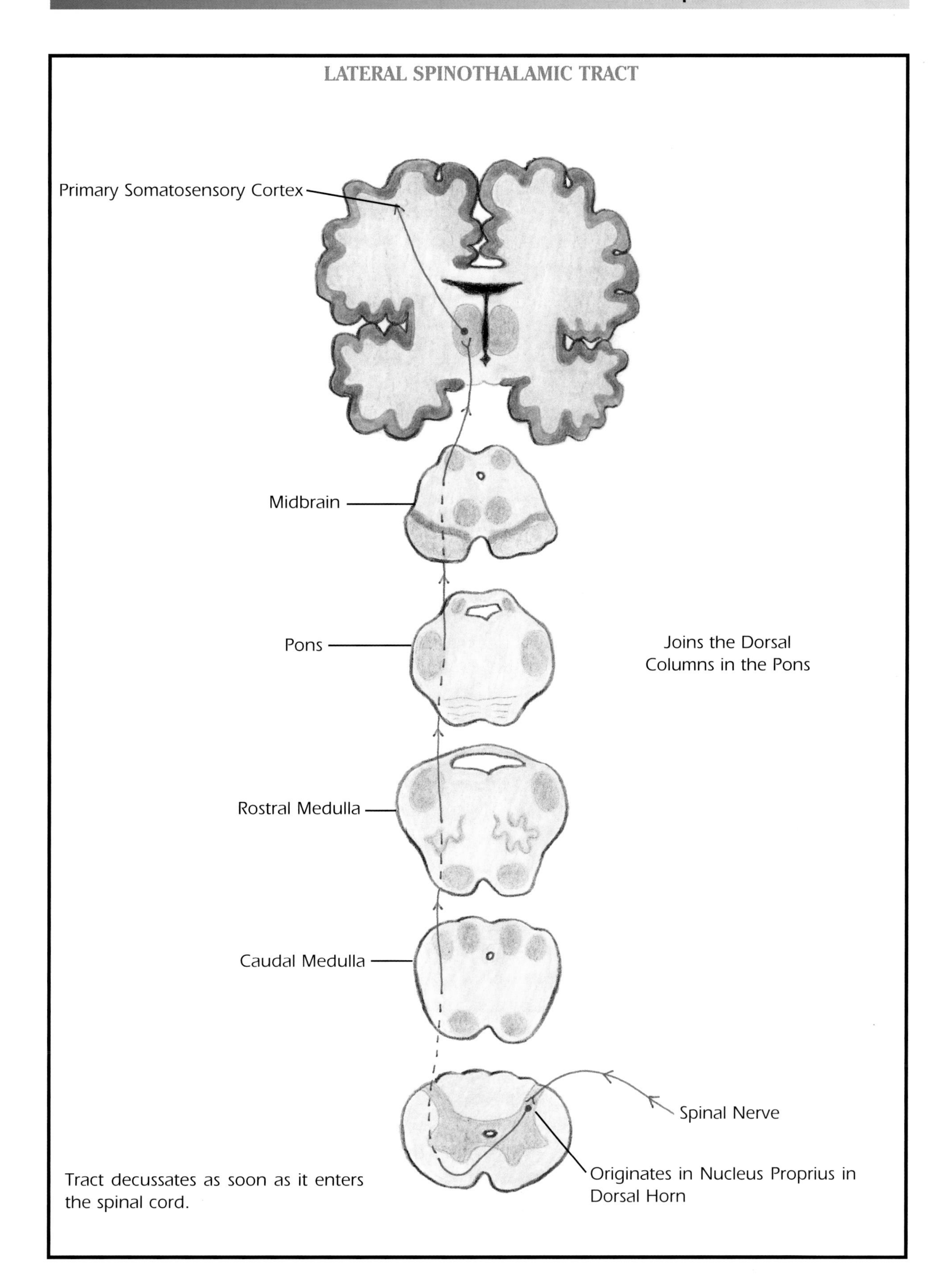
LATERAL SPINOTHALAMIC TRACT
Primary Somatosensory Cortex
Midbrain
Pons
Joins the Dorsal Columns in the Pons
Rostral Medulla
Caudal Medulla
Spinal Nerve
Originates in Nucleus Proprius in Dorsal Horn
Tract decussates as soon as it enters the spinal cord.

Anterior Spinothalamic Tract

Function

- Carries conscious sensory information to the cortex regarding:
  - Crude touch, light touch
  - Some researchers believe that this tract may carry pain and temperature if the lateral spinothalamic is damaged.

Origin

- Nucleus Proprius (in the dorsal horn)

Decussation

- SC Level
- The tract crosses the midline as soon as spinal nerves have synapsed on its cell bodies in the dorsal horn.

Destination

- Postcentral Gyrus (SS1)

Pathway

- Skin receptors in the PNS send sensory messages along the ascending sensory spinal nerves to the dorsal root and rootlets.
- The sensory spinal nerves synapse on an interneuron in the dorsal horn (specifically in the nucleus proprius).
- The interneuron then synapses on the cell bodies of the ascending spinothalamic tract.
- At this time, the tract crosses the midline of the SC and travels to the ventral white funiculus and begins to ascend.
- As it ascends, the tract travels through the posterolateral funiculus and eventually joins the lateral spinothalamic tract in the medulla.
- The tract ascends through the brainstem, thalamus, and internal capsule to the SS1.

Lesions

Complete Severance of the SC

- Bilateral loss of sensation: crude and light touch (below the severed cord level)

A Hemi-Lesion (on one side) of the SC

- At the lesion level: bilateral sensory loss
- Below the lesion level: contralateral sensory loss

A Hemi Lesion (on one side) of the Brainstem

- Contralateral sensory loss

A Unilateral Lesion in the Postcentral Gyrus

- Contralateral sensory loss

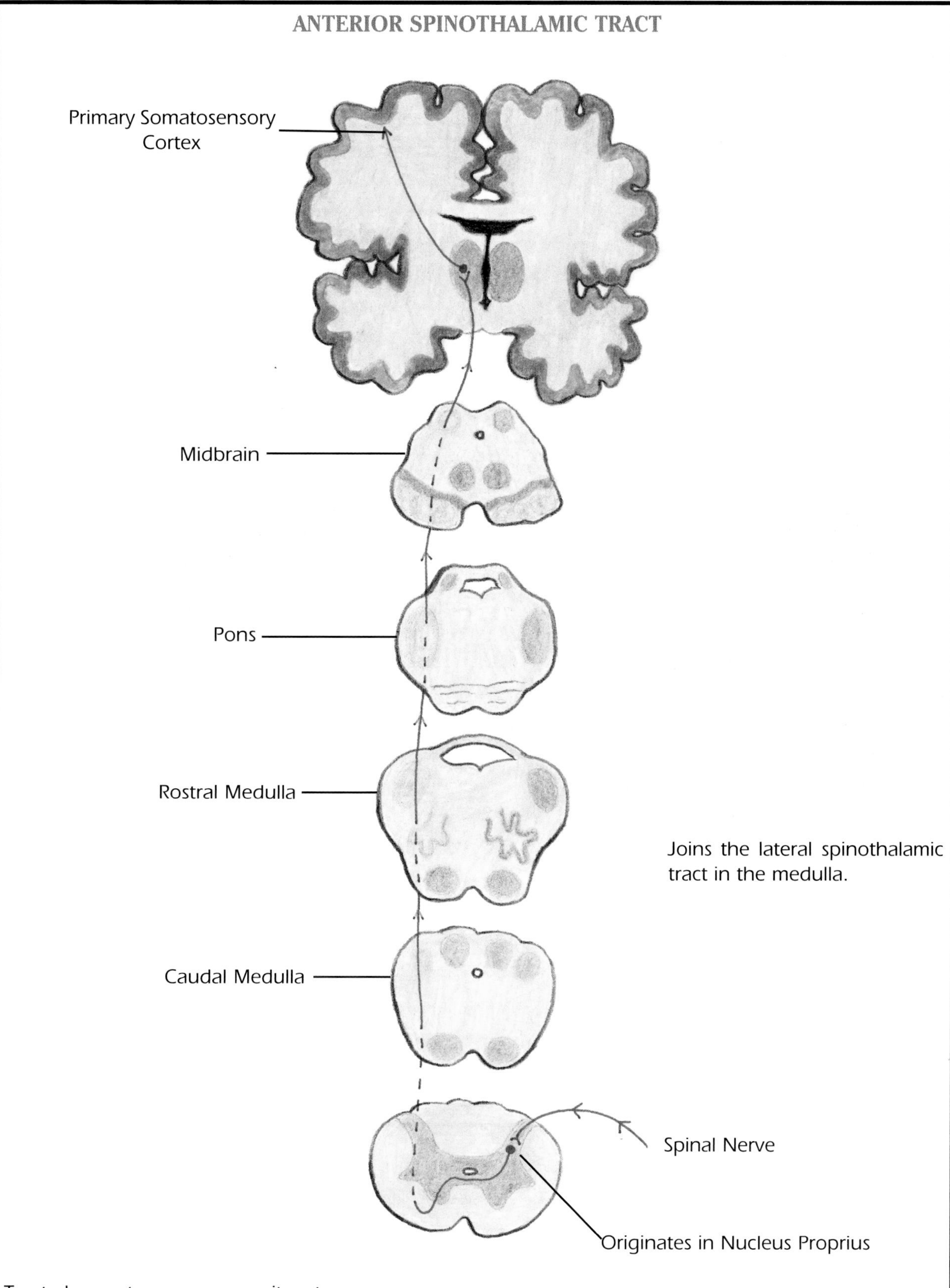
ANTERIOR SPINOTHALAMIC TRACT
Primary Somatosensory Cortex
Midbrain
Pons
Rostral Medulla
Joins the lateral spinothalamic tract in the medulla.
Caudal Medulla
Spinal Nerve
Originates in Nucleus Proprius
Tract decussates as soon as it enters the spinal cord.

Posterior Spinocerebellar Tract

Function

- Carries unconscious sensory information from the lower extremities to the cerebellum regarding:
  - Proprioception:
    - Pressure and tension of skeletal muscles
    - Coordination of motoric movement of individual muscles
- Carries sensory information from the muscle spindles (MS), Golgi tendon organs (GTO), and the joint receptors in the PNS.
- This information never reaches the cortex for conscious detection and interpretation.

Origin

- Clark's column in the dorsal horn in SC levels T6 and below (because this tract serves the *lower extremities*)
- Clark's column contains the lower extremity proprioceptive and kinesthetic inputs.

Decussation

- None. This is an *ipsilateral* tract that does *not* cross.

Destination

- Cerebellum

Pathway

- The MSs, GTOs, and joint receptors in the PNS send proprioceptive information along the ascending sensory spinal nerves to the dorsal root and rootlets.
- The sensory spinal nerves then synapse on an interneuron in the dorsal horn (specifically in Clark's column).
- The interneuron joins to the cell bodies of the posterior spinocerebellar tract in the dorsal horn, and the tract begins to ascend up the posterolateral funiculus on the ipsilateral side.
- The tract ascends through the medulla to the inferior cerebellar peduncle.
- From the inferior cerebellar peduncle, the tract synapses in the cerebellum.

Lesions

- All lesions are *ipsilateral* because the tract does not decussate.

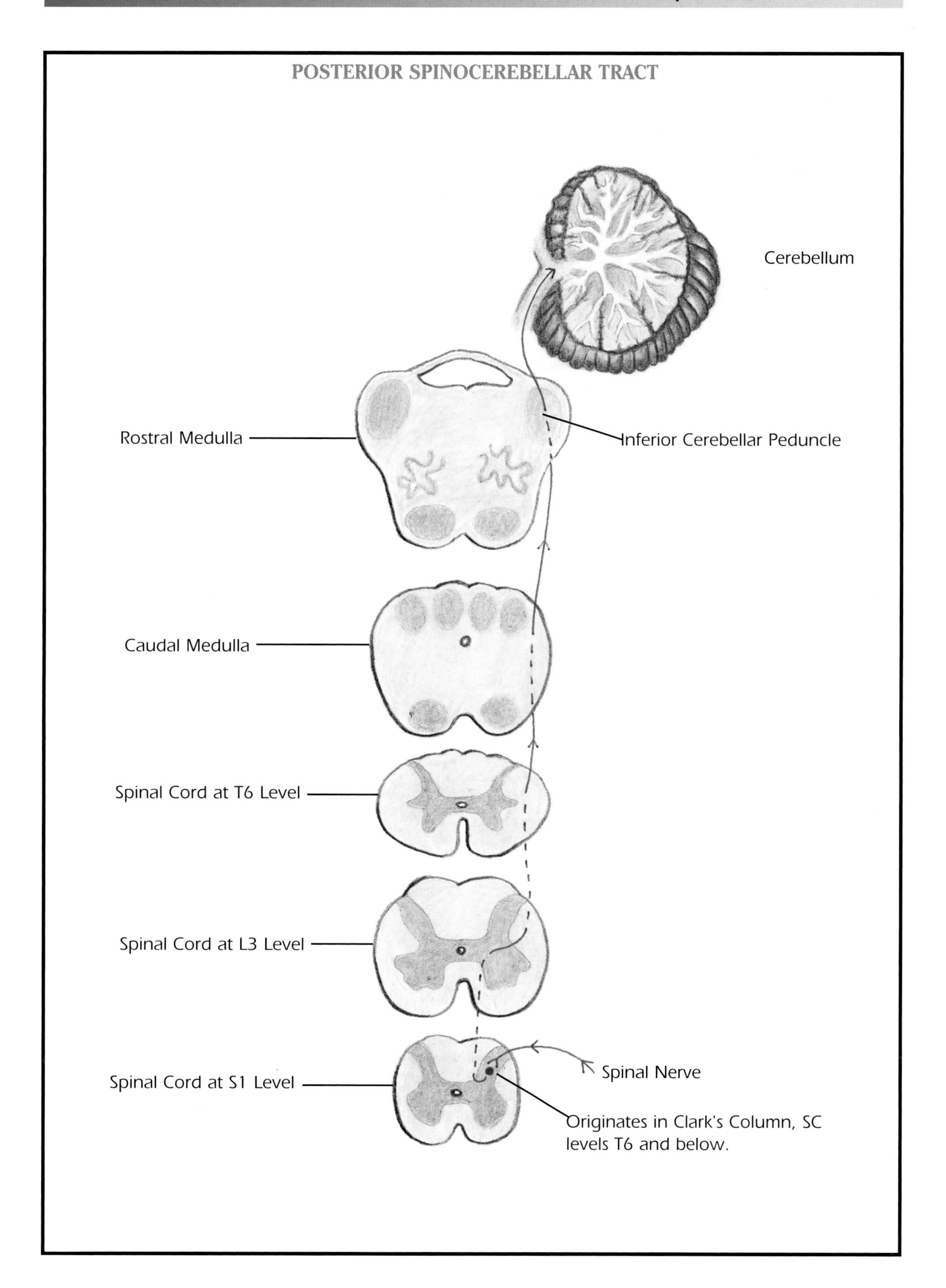
POSTERIOR SPINOCEREBELLAR TRACT
Cerebellum
Rostral Medulla
Inferior Cerebellar Peduncle
Caudal Medulla
Spinal Cord at T6 Level
Spinal Cord at L3 Level
Spinal Nerve
Spinal Cord at S1 Level
Originates in Clark's Column, SC levels T6 and below.

Anterior Spinocerebellar Tract

Function

- Carries unconscious sensory information from the *lower extremities* to the cerebellum regarding:
  - Proprioception:
    - Pressure and tension of skeletal muscles
    - Coordination of posture and movement of limbs (not individual muscles as in the posterior spinocerebellar tract)
- Carries sensory information from the MSs, GTOs, and joint receptors to the cerebellum

Origin

- Nucleus Proprius in the dorsal horn of the lumbar sections (because the tract serves the *lower extremities*)

Decussation

- SC Level in the Lumbar Sections
- The tract decussates as soon as the spinal nerves synapse on the cell bodies of the anterior spinocerebellar tract.

Destination

- Cerebellum

Pathway

- The MSs, GTOs, and joint receptors in the PNS send proprioceptive information along the ascending sensory spinal nerves to the dorsal root and rootlets.
- The sensory spinal nerves synapse on an interneuron in the dorsal horn (specifically the nucleus proprius).
- The interneuron then synapses on the cell bodies of the anterior spinocerebellar tract in the dorsal horn and the tract then decussates across the midline to the anterior white funiculus.
- The tract begins to ascend in the lateral white funiculus and travels to the superior cerebellar peduncle in the pons.
- From the superior cerebellar peduncle the tract synapses in the cerebellum.

Lesions

Complete Severance of the SC

- Bilateral loss of proprioception from the lower extremities

A Hemi-Lesion (on one side) of the SC

- At the lesion level: bilateral proprioceptive loss
- Below the lesion level: contralateral proprioceptive loss

A Lesion in the Superior Cerebellar Peduncle

- Contralateral proprioceptive loss

A Lesion in One Hemisphere of the Cerebellum

- Contralateral proprioceptive loss

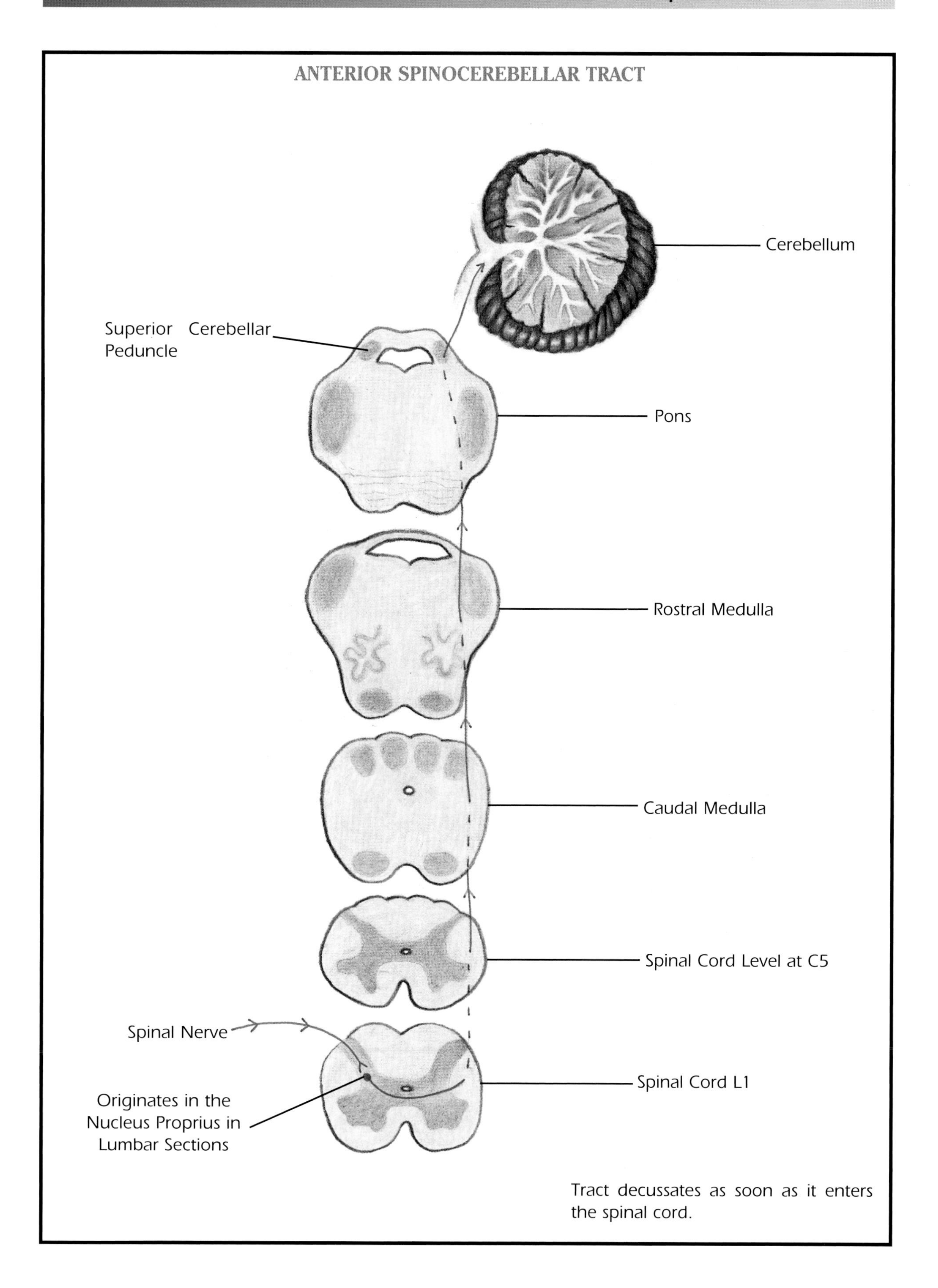
ANTERIOR SPINOCEREBELLAR TRACT
Cerebellum
Superior Cerebellar Peduncle
Pons
Rostral Medulla
Caudal Medulla
Spinal Cord Level at C5
Spinal Nerve
Spinal Cord L1
Originates in the Nucleus Proprius in Lumbar Sections
Tract decussates as soon as it enters the spinal cord.

Cuneocerebellar Tract

Function

- Carries unconscious sensory information from the *trunk* and *upper extremities* to the cerebellum regarding the following:
  - Proprioception:
    - Pressure and tension of skeletal muscles
    - Coordination of motoric movement of individual muscles
- Carries sensory information from the MSs, GTOs, and the joint receptors to the cerebellum.

Origin

- Nucleus Cuneatus in the dorsal horn of the SC of T6 and above (because the tract serves the *upper extremities*)

Decussation

- None. This is an *ipsilateral* tract that does *not* cross.

Destination

- Cerebellum

Pathway

- The MSs, GTOs, and joint receptors in the PNS send proprioceptive information along the ascending sensory spinal nerves to the dorsal root and rootlets of T6 and above.
- The sensory spinal nerves synapse on an interneuron in the dorsal horn (specifically in the nucleus cuneatus).
- The interneuron then synapses with the cell bodies of the cuneocerebellar tract in the dorsal horn.
- The tract begins to ascend in the posterolateral white funiculus and travels to the inferior cerebellar peduncle in the medulla.
- From the inferior cerebellar peduncle, the tract synapses in the cerebellum.

Lesions

- All lesions are *ipsilateral* because the tract does *not* decussate.

Bilateral Representation of Unconscious Proprioception

- The below two sets of tracts provide bilateral representation of unconscious proprioceptive information to the cerebellum.
  - Proprioception from the lower extremities:
    - Posterior Spinocerebellar
    - Anterior Spinocerebellar
  - Proprioception from the trunk and upper extremities:
    - Cuneocerebellar
    - Rostral Cerebellar

- If one tract of each set is lost, the other will still send proprioceptive information to the cerebellum. An individual will not lose unconscious proprioceptive information from the upper or lower extremities unless both tracts within a set are lost.

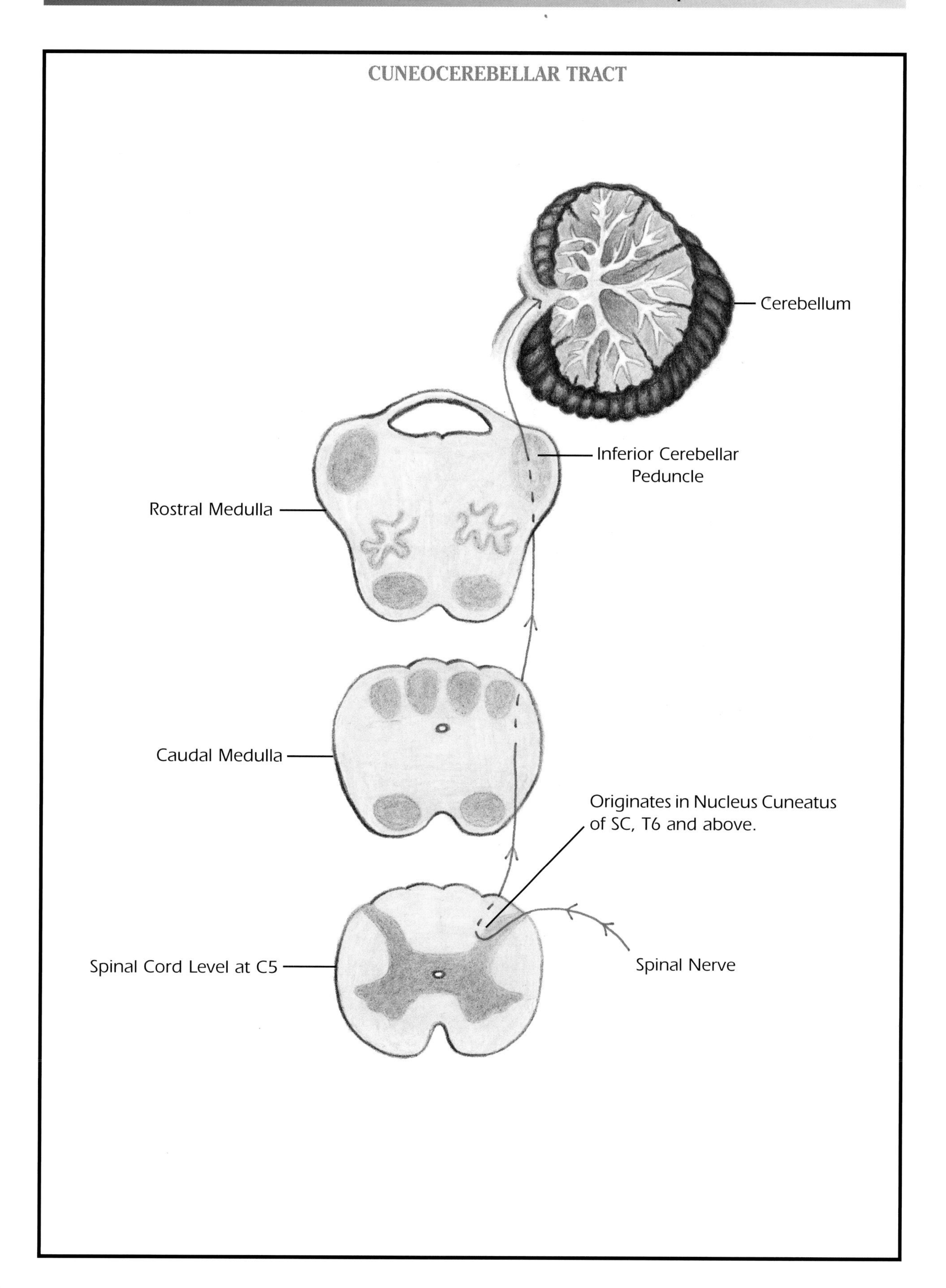
CUNEOCEREBELLAR TRACT
Cerebellum
Inferior Cerebellar Peduncle
Rostral Medulla
Caudal Medulla
Originates in Nucleus Cuneatus of SC, T6 and above.
Spinal Cord Level at C5
Spinal Nerve

Rostral Spinocerebellar Tract

Function

- Carries unconscious sensory information from the *trunk* and *upper extremities* to the cerebellum regarding:
  - Proprioception:
    - Pressure and tension of skeletal muscles
    - Coordination of posture and movements of limbs
- Carries sensory information from the MSs, GTOs, and joint receptors to the cerebellum

Origin

- Ventrolateral Gray of the SC in the Cervical Levels (because the tract serves the *upper extremities*)

Decussation

- None. This is an *ipsilateral* tract that does *not* cross.

Destination

- Cerebellum

Pathway

- The MSs, GTOs, and joint receptors in the PNS send proprioceptive information along the ascending sensory spinal nerves to the dorsal root and rootlets of the cervical levels.
- The sensory spinal nerves enter the dorsal horn of the SC and synapse in the ventrolateral gray on an interneuron.
- The interneuron then synapses on the cell bodies of the rostral spinocerebellar tract, and the tract begins to ascend up the lateral white funiculus.
- The tract joins the anterior spinocerebellar tract in the lateral white funiculus.
- The tract then ascends to either the inferior cerebellar peduncle or the superior cerebellar peduncle.
- From one of these peduncles, the tract then synapses in the cerebellum.

Lesions

- All lesions are *ipsilateral.*

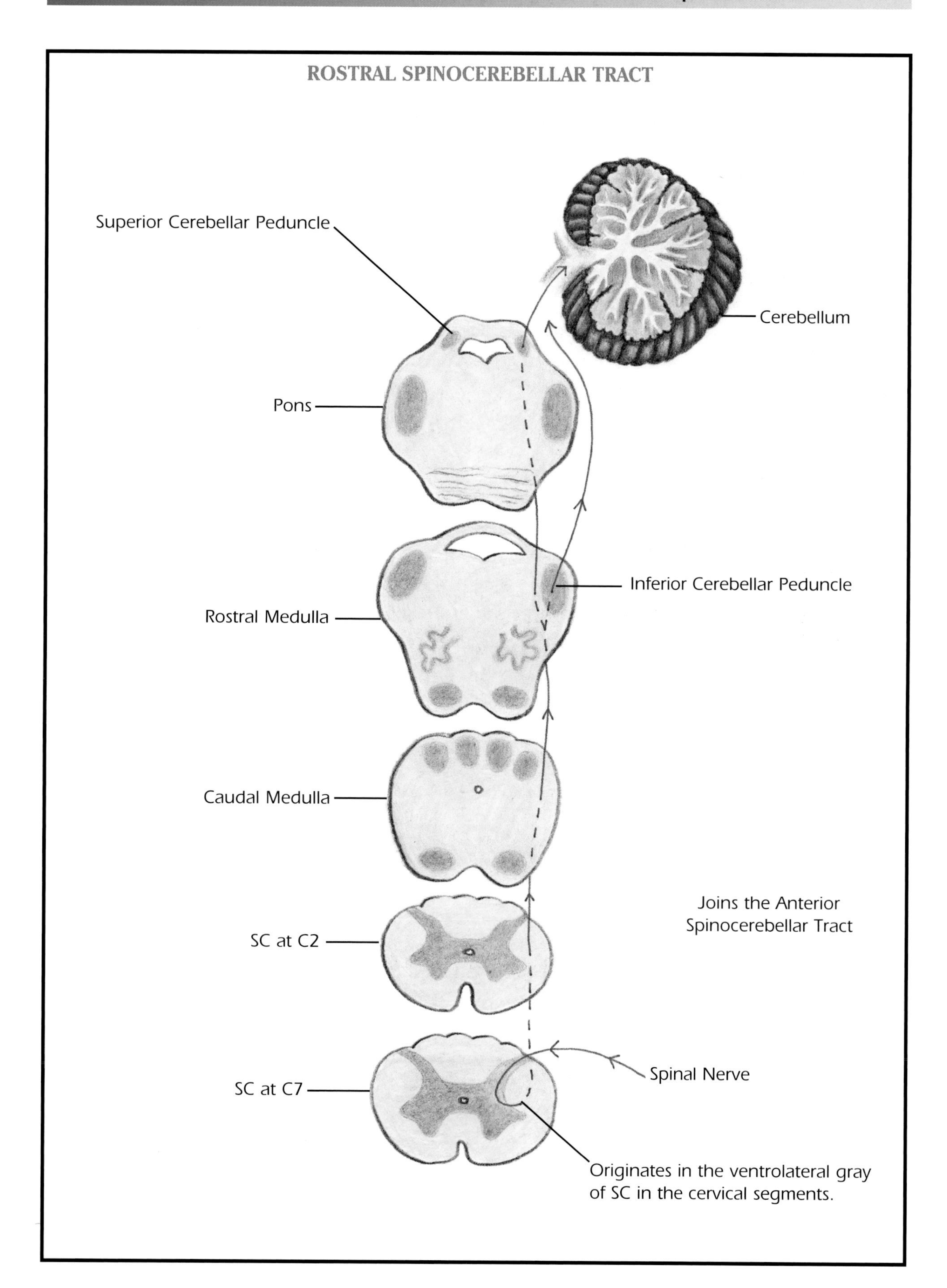
ROSTRAL SPINOCEREBELLAR TRACT
Superior Cerebellar Peduncle
Cerebellum
Pons
Inferior Cerebellar Peduncle
Rostral Medulla
Caudal Medulla
Joins the Anterior Spinocerebellar Tract
SC at C2
Spinal Nerve
SC at C7
Originates in the ventrolateral gray of SC in the cervical segments.

Cortically Originated Descending Motor Tracts

Lateral and Anterior Corticospinal Tracts

Part of the Pyramidal System

- The lateral and anterior corticospinal tracts are *descending motor tracts* that travel through the *pyramids* of the medulla.
- The lateral corticospinal tract decussates at the pyramidal decussation; the anterior corticospinal tract remains ipsilateral and descends through the pyramids.
- These tracts are considered to be *upper motor neurons* (UMNs) in the CNS.

Function

- Carry *conscious/voluntary motor information* from the precentral gyrus (primary motor area [M1]) up to, but not including the ventral horn.
- The tracts then synapse on motor spinal nerves (in the ventral horn) that innervate skeletal muscles.

Origin

- Precentral Gyrus (M1)

Decussation

- Lateral Corticospinal Tract:
    - Decussates in the Pyramidal Decussation of the Medulla
- Anterior Corticospinal Tract:
    - Does *not* decussate

Destination

- Synapses on an *interneuron* in the *ventral horn*.
- This interneuron then synapses on the *motor neurons* of the motor spinal nerves that innervate skeletal muscles in the PNS.

Pathway

- The cell bodies of the tract are located in the precentral gyrus. They send descending motor signals through the internal capsule, thalamus, and brainstem.
- In the brainstem (at the medulla level), the tract separates into the lateral and anterior corticospinal tracts.
- The lateral corticospinal tract (90% of the original tract before separation) decussates in the pyramidal decussation in the medulla.
- The anterior corticospinal tract (10% of the original tract before separation) remains ipsilateral and continues to descend through the pyramids.
- When the tract is ready to exit the cord, it synapses on an interneuron in the ventrolateral gray of the SC.
- The interneuron then synapses on a motor neuron (in the ventral horn) that innervates a descending motor spinal nerve.
- The motor spinal nerve then exits through the ventral rootlets and root and travels to a target skeletal muscle in the PNS.

Lesions

A Lesion in M1 (in one hemisphere)

- Contralateral loss of voluntary muscle movement (because 90% of the tract decussates)
- Spasticity of the distal musculature below the lesion level (UMN damage)
- Hyperactive reflexes

A Lesion in the Internal Capsule

- Contralateral spastic paralysis
- Hyperactive reflexes

A Unilateral Lesion in the Brainstem above the Decussation

- Contralateral spastic paralysis

LATERAL AND ANTERIOR CORTICOSPINAL TRACTS

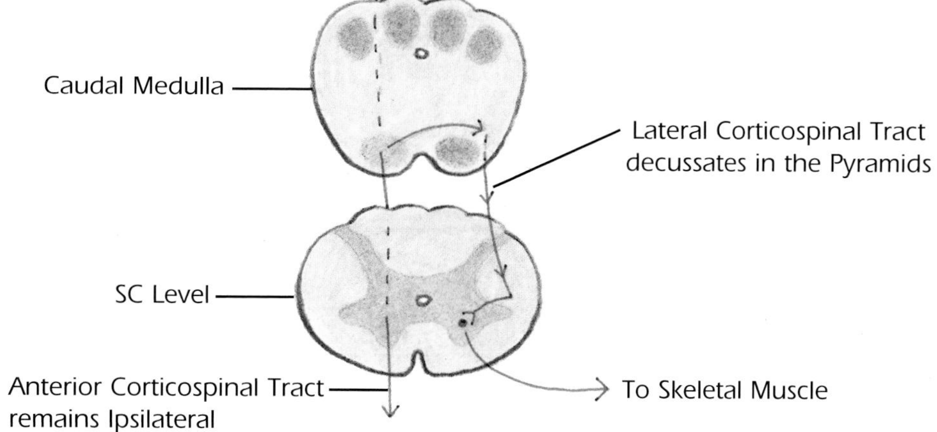

A Unilateral Lesion in the SC below the Decussation

- Ipsilateral loss of voluntary motor control
- Spasticity in the distal musculature (below the lesion level)
- Flaccidity (at the lesion level)
- Hyperactive reflexes

A Complete Severance of the SC below the Decussation

- Bilateral loss of voluntary motor control
- Spasticity below the lesion level
- Flaccidity at the lesion level

Decorticate Rigidity

- Damage to the corticospinal tracts will result in decorticate rigidity.
- In decorticate rigidity:
  - The upper extremities are in a spastic flexed position.
  - The lower extremities are in a spastic extended position.
- Decorticate rigidity occurs because the corticospinal tracts fire without modification from the cortex.
- The rigidity that presents is more accurately described as spasticity, as it is derived from dysfunction of the pyramidal system and is characterized by increased tone on only one side of the joint rather than both.

Corticobulbar Tract

Pathway

- The corticobulbar tract descends from the corticospinal tract and projects to certain *cranial nerve (CN) nuclei* that have a *motor component*:
  - In the midbrain, the corticobulbar tract projects to CN *nuclei* 3, 4, 6 (the *extraocular* CNs).
  - In the pons, the corticobulbar tract projects to CN *nuclei* 5, 7 (for *facial muscle* innervation).
  - In the medulla, the corticobulbar tract projects to CN *nuclei* 9, 10, 11, 12 (muscles for *swallowing, eating, speaking*).

Function

- The corticobulbar tracts control CN lower motor neurons.

Lesions

- Lesions involve the above listed CN nuclei.
- Lesions result in *flaccidity* of the anatomy innervated by the above CNs.

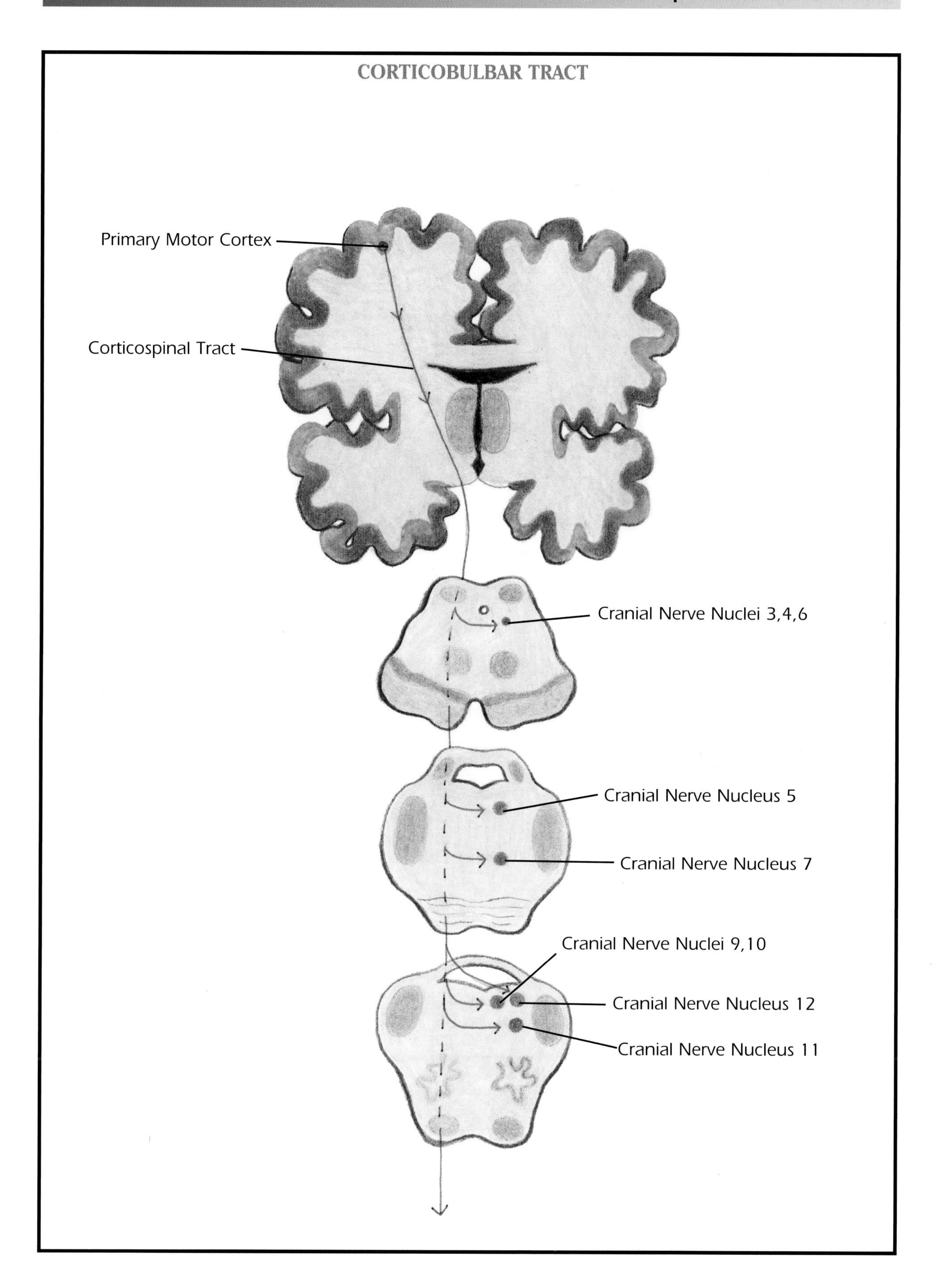
CORTICOBULBAR TRACT
Primary Motor Cortex
Corticospinal Tract
Cranial Nerve Nuclei 3,4,6
Cranial Nerve Nucleus 5
Cranial Nerve Nucleus 7
Cranial Nerve Nuclei 9,10
Cranial Nerve Nucleus 12
Cranial Nerve Nucleus 11

Mixed Pathways Originating from the Brainstem

Medial Longitudinal Fasciculus

- A fiber bundle situated near the midline of the brainstem
- Composed of ascending and descending fibers
- Part of the UMN system

Descending Fibers of the Medial Longitudinal Fasciculus

Function

- The medial longitudinal fasciculus (MLF) exerts an inhibitory effect on the motor neurons of the ventral horn in the cervical SC.
- This inhibitory effect allows for the *coordination* of *head* and *neck movements.*

Origin

- Vestibular Nuclei (in the medulla)

Decussation

- None. This is an *ipsilateral* tract that does *not* cross.

Destination

- From the cervical segments of the SC, the tract projects to spinal nerves in the PNS that innervate head and neck musculature.

Pathway

- The *vestibular nuclei in the medulla* (the origin of the tract) receive information from the following:
    - Superior colliculi in the midbrain
    - Oculomotor nuclei in the midbrain
    - Pontine reticular formation
- The tract then descends through the anterior white funiculus of the SC.
- When the tract is ready to exit the cord, it synapses on an interneuron in the ventral horn of the cervical cord segments.
- The interneuron then synapses on motor neurons (in the ventral horn) of descending motor spinal nerves that innervate head and neck musculature in the periphery.

Ascending Fibers of the Medial Longitudinal Fasciculus

Function

- The ascending fibers of the MLF are responsible for the *visual tracking of a moving object* through the coordinated movements of the eyes, head, and neck.
- The ascending fibers of the MLF are responsible for the refinement of extraocular eye movements.

Origin

- Vestibular Nuclei (in the medulla)

Decussation

- None. This is an *ipsilateral* tract that does *not* cross.

Destination

- Oculomotor Nerve Nuclei
- Trochlear Nerve Nuclei
- Abducens Nerve Nuclei

Pathway

- The tract begins in the vestibular nuclei in the medulla and terminates on CN nuclei 3, 4, 6.

MEDIAL LONGITUDINAL FASCICULUS: ASCENDING AND DESCENDING FIBERS

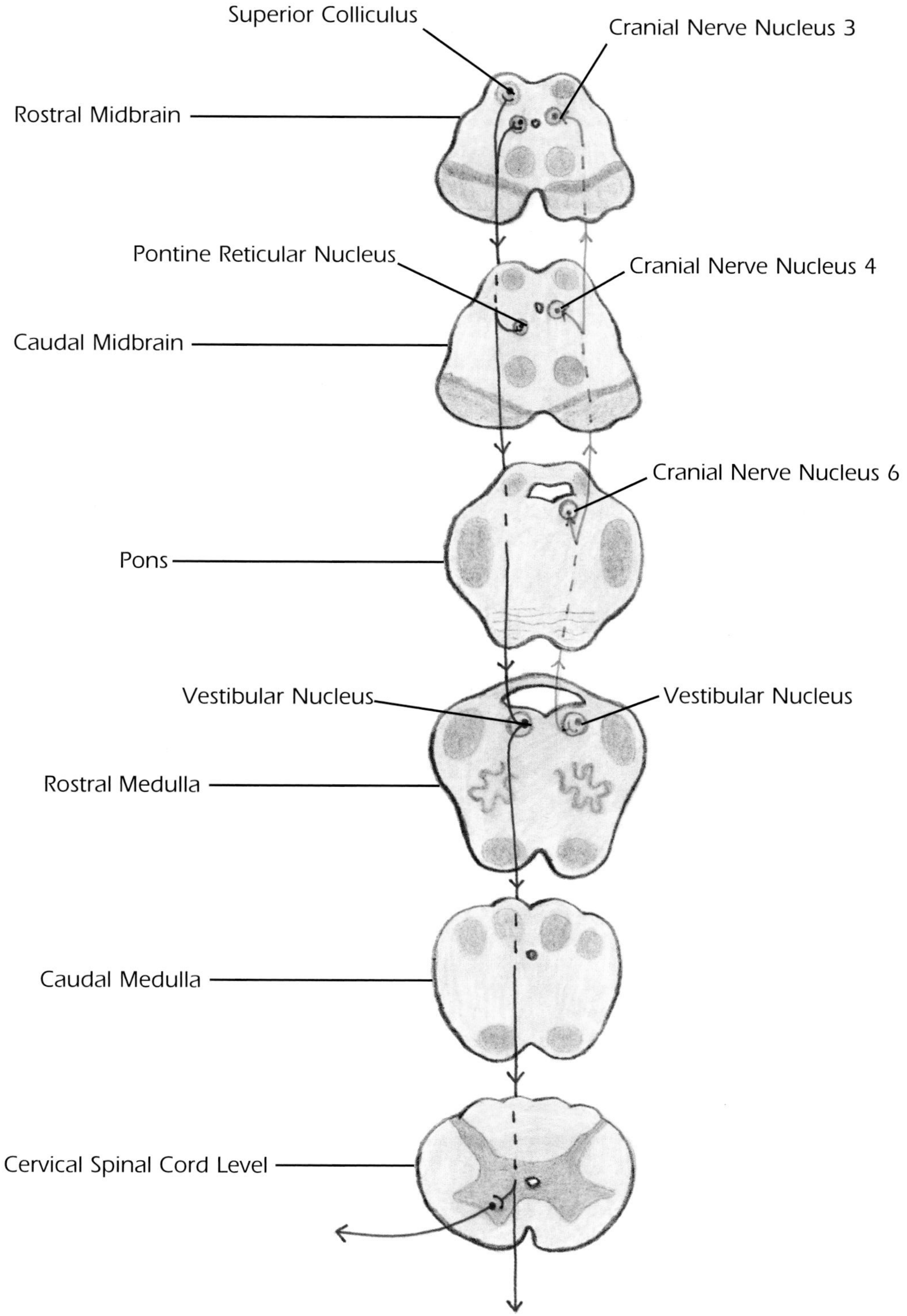

Superior Colliculus
Cranial Nerve Nucleus 3
Rostral Midbrain
Pontine Reticular Nucleus
Cranial Nerve Nucleus 4
Caudal Midbrain
Cranial Nerve Nucleus 6
Pons
Vestibular Nucleus
Vestibular Nucleus
Rostral Medulla
Caudal Medulla
Cervical Spinal Cord Level

Descending Motor Pathways Originating in the Brainstem

Vestibulospinal Tract

Extrapyramidal Tract

- The vestibular tract does *not* travel through the pyramids and is thus considered to be an *extrapyramidal tract.*

Function

- Facilitation of antigravity (extensor) muscles
- Facilitation of muscles responsible for posture and stance

Origin

- Vestibular Nuclei (in the medulla)

Decussation

- None. The tract remains *ipsilateral.*

Destination

- The vestibulospinal tract *innervates extensor muscle groups* in the PNS.

Pathway

- The vestibular nuclei in the medulla (the tract's origin) receive messages from the following:
  - Vestibular apparatus in the inner ear via the vestibulocochlear nerve
  - Cerebellum
- From the vestibular nuclei, the tract then travels to the anterolateral white funiculus of the SC.
- When the tract is ready to exit the cord, it synapses on an interneuron that joins with a motor neuron in the ventral horn.
- The motor neuron synapses on a descending motor spinal nerve that innervates a target extensor muscle.

Decerebrate Rigidity

- Damage to any of the tracts that originate in the brainstem—including the vestibulospinal tract—result in *decerebrate rigidity.*
- When damage occurs to the brainstem, the vestibulospinal tract fires without modification from the brainstem and cortex, causing decerebrate rigidity.
- Decerebrate rigidity involves *spastic extension of both the upper and lower extremities.*
- The rigidity in decerebrate rigidity is more accurately described as spasticity (as it emerges from the extrapyramidal system and involves increased muscle tone on one side of the joint rather than both).
- The occurrence of decerebrate rigidity indicates a much poorer prognosis than does decorticate rigidity (because it signifies severe damage of the brainstem).

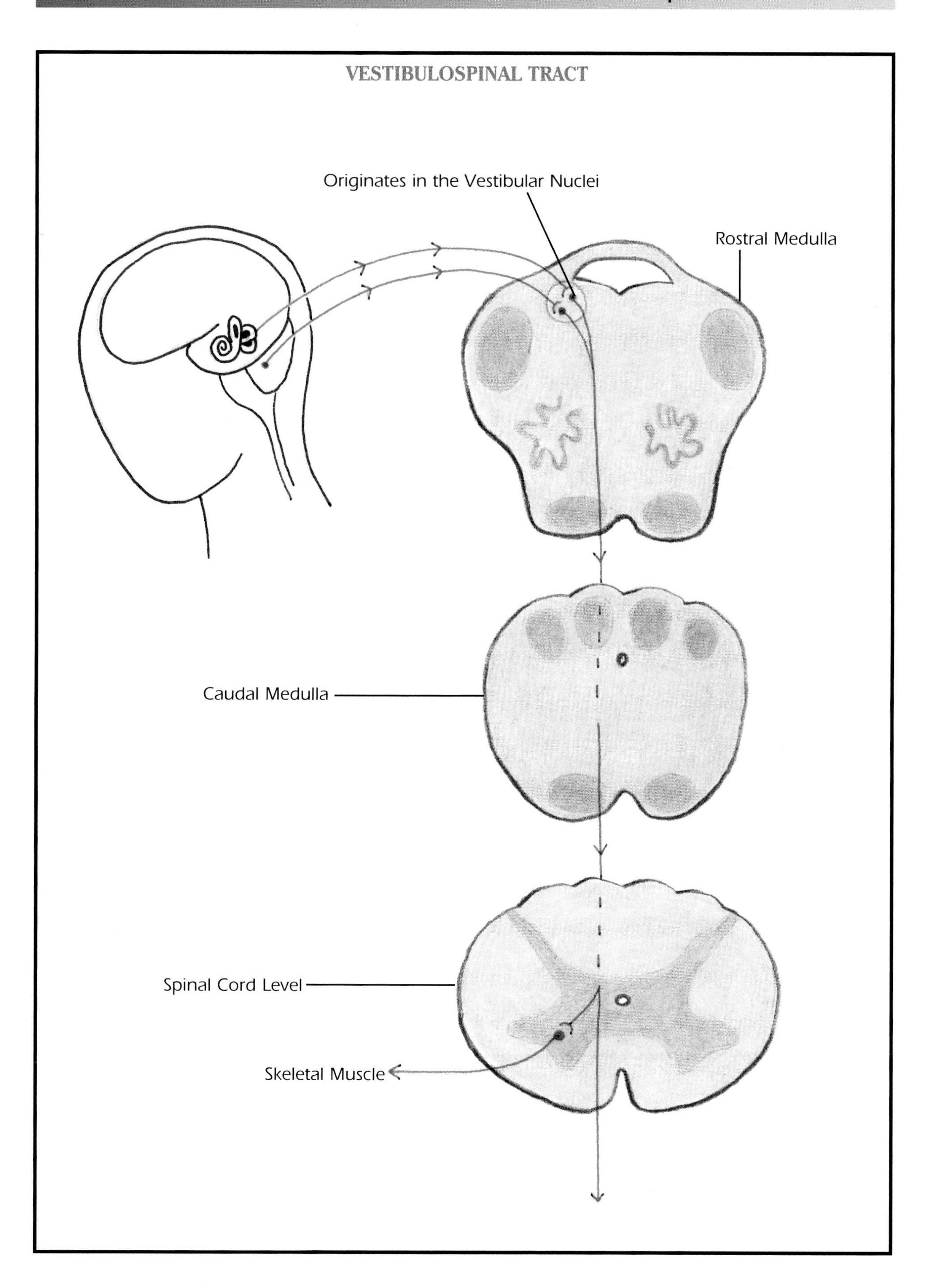
VESTIBULOSPINAL TRACT
Originates in the Vestibular Nuclei
Rostral Medulla
Caudal Medulla
Spinal Cord Level
Skeletal Muscle

Rubrospinal Tract

Extrapyramidal Tract

- The rubrospinal tract does *not* travel through the pyramids and is thus considered to be an *extrapyramidal tract.*

Function

- The rubrospinal tract facilitates the *antagonist* of antigravity muscles (mostly in the limbs) causing *facilitation* of *flexor muscle groups*.
- Activity of the rubrospinal tract is modified by the cerebellum and the cortex.
- The cerebellum and cortex modify the activity of the red nucleus in the midbrain—the origin of the rubrospinal tract.

Origin

- Red Nucleus (in the midbrain)

Decussation

- Occurs at the Midbrain Level

Destination

- The rubrospinal tract innervates flexor muscles in the limbs.

Pathway

- The tract begins in the red nucleus of the midbrain.
- In the midbrain, the tract crosses the midline and enters the crus cerebri.
- Once the tract decussates, it begins to descend to the SC.
- The rubrospinal tract travels very close to the corticospinal tract as it descends in the SC.
- When the tract is ready to exit the cord, it synapses with an interneuron in the ventral horn.
- The interneuron joins with a descending motor spinal nerve that innervates flexor muscles in the PNS.

Decerebrate Rigidity

- A lesion to the rubrospinal tract results in decerebrate rigidity (involves *spastic extension of the upper and lower extremities*).
- If the rubrospinal tract is lost, the antagonists of the antigravity muscles are gone—in other words, the flexors are lost.
- The upper and lower extremities present with *increased extensor tone*.

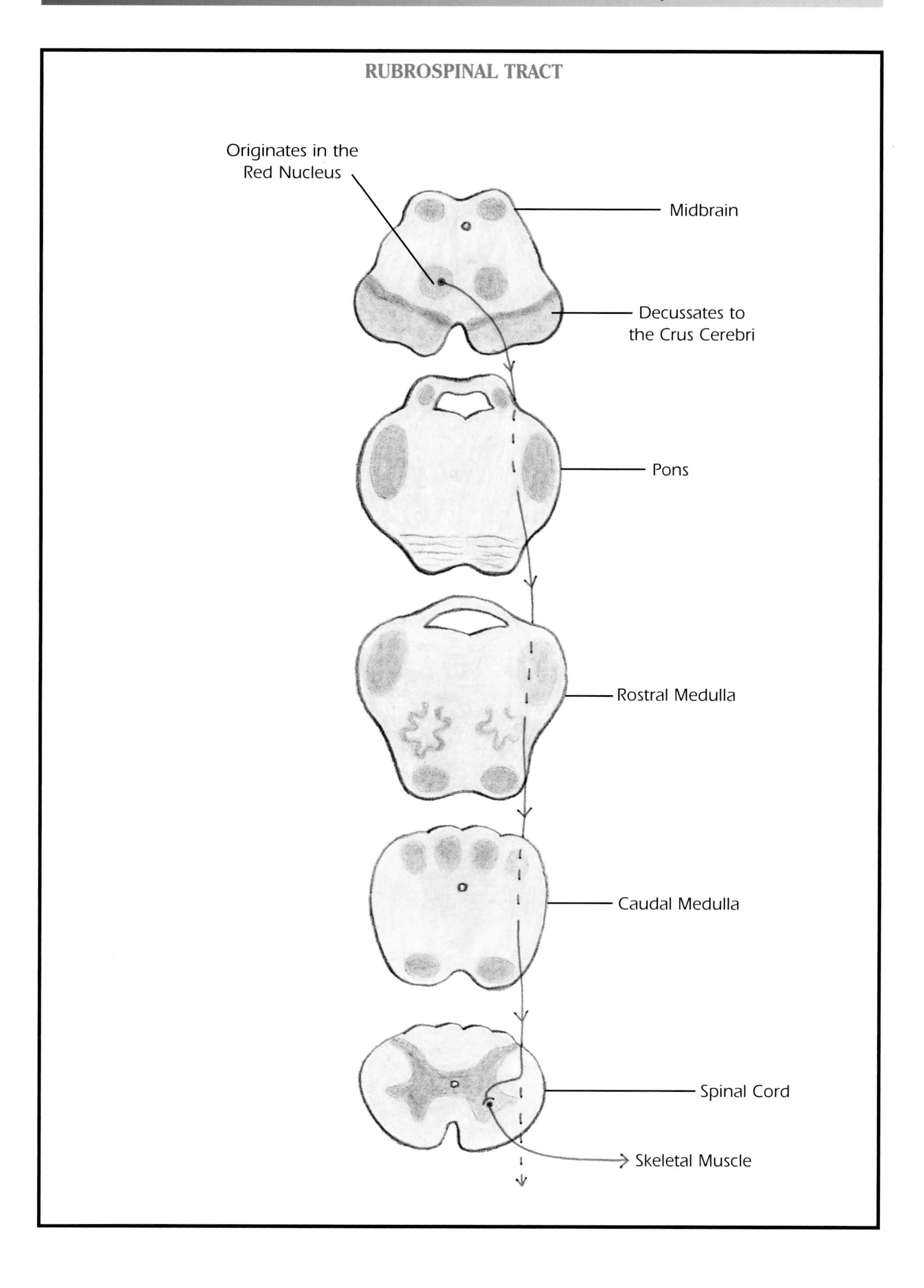
RUBROSPINAL TRACT
Originates in the Red Nucleus
Midbrain
Decussates to the Crus Cerebri
Pons
Rostral Medulla
Caudal Medulla
Spinal Cord
Skeletal Muscle

Medullary Reticulospinal Tract

Extrapyramidal Tract

- The medullary reticulospinal tract does *not* travel through the pyramids and is thus considered to be an *extrapyramidal tract.*

Function

- Inhibits antigravity muscles—*inhibits extensor tone.*
- This tract also depresses cardiovascular responses (blood pressure, heart rate) and the inspiratory phase of respiration.
- The tract receives substantial modification from the corticospinal tracts.

Origin

- Medullary Reticular Formation (in the medulla)

Decussation

- Most of the fibers of this tract remain uncrossed.

Destination

- The medullary reticulospinal tract modifies antigravity or extensor muscles in the PNS.

Pathway

- This tract begins in the medullary reticular formation in the medulla.
- It descends through the anterior white funiculus.
- When this tract is ready to exit the cord, it synapses on an interneuron in the ventral horn.
- The interneuron then synapses on a motor spinal nerve that travels to an extensor muscle in the PNS.

Decerebrate Rigidity

- When the medullary reticulospinal tract is damaged, the tract that inhibits extensor tone is lost.
- This causes an *increase in extensor tone* or decerebrate rigidity (*spastic extension of the upper and lower extremities*).

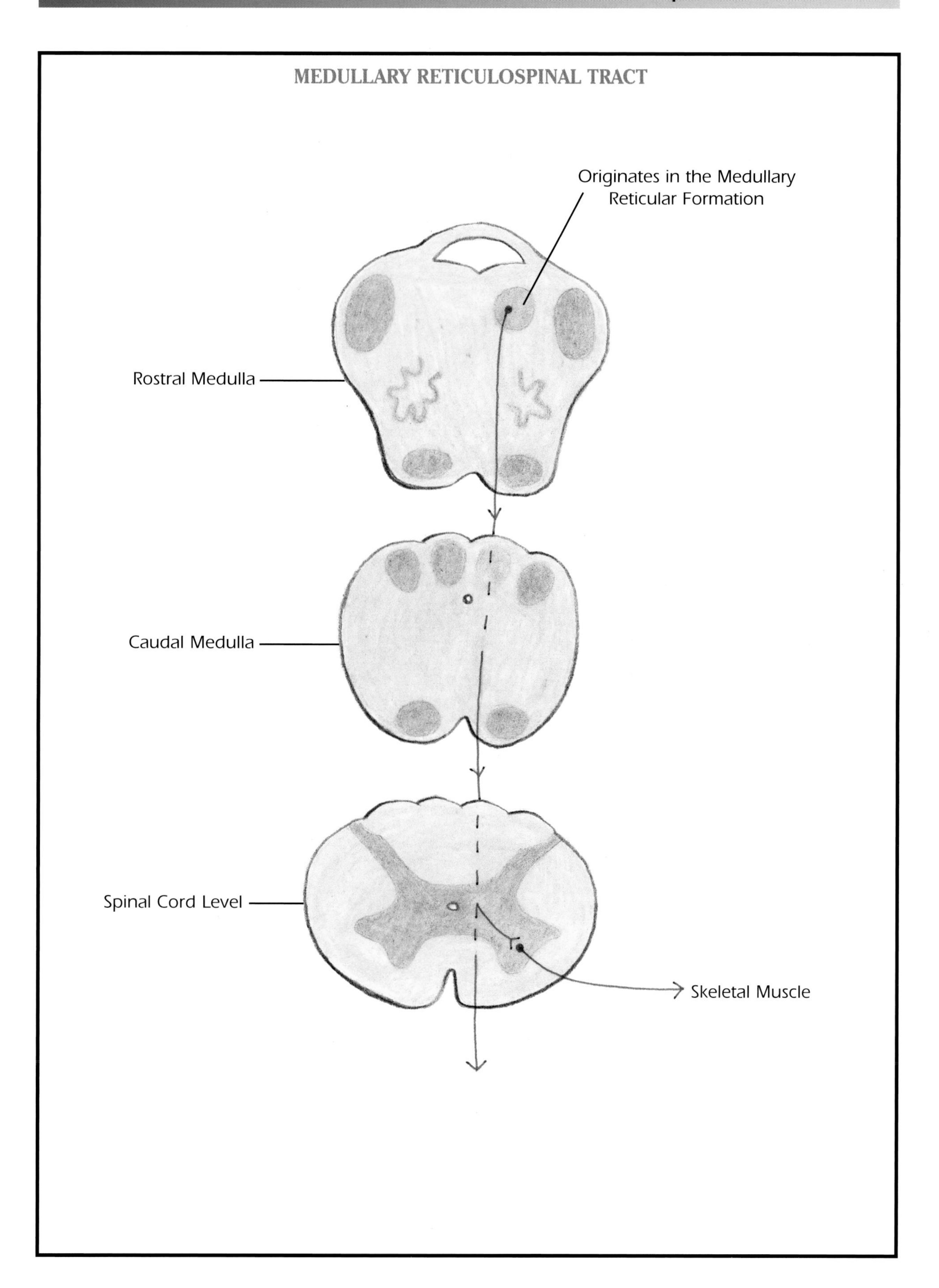
MEDULLARY RETICULOSPINAL TRACT
Originates in the Medullary Reticular Formation
Rostral Medulla
Caudal Medulla
Spinal Cord Level
Skeletal Muscle

Pontine Reticulospinal Tract

Extrapyramidal Tract

- The pontine reticulospinal tract does not travel through the pyramids and is thus considered to be an *extrapyramidal tract.*

Function

- Facilitates antigravity muscles—*facilitates extensor tone*
- This tract receives substantial modification from the corticospinal tracts.

Origin

- Pontine Reticular Formation (in the pons)

Decussation

- None. This is an *ipsilateral* tract.

Destination

- The pontine reticulospinal tract innervates extensor muscles.

Pathway

- The tract begins in the pontine reticular formation.
- It descends through the anterior white funiculus of the SC.
- When this tract is ready to exit the cord, it synapses on an interneuron in the ventral horn.
- The interneuron then synapses with a motor spinal nerve that innervates an extensor muscle in the PNS.

Decerebrate Rigidity

- Damage to this tract results in decerebrate rigidity (*spastic extension of the upper and lower extremities*).
- Usually, damage to this tract is less important than damage to the tracts and neural regions surrounding it.
- If cortical or brainstem damage has occurred, then nothing modifies this tract's firing rate. Consequently, the tract fires without higher center modification. This leads to *increased extensor tone.*

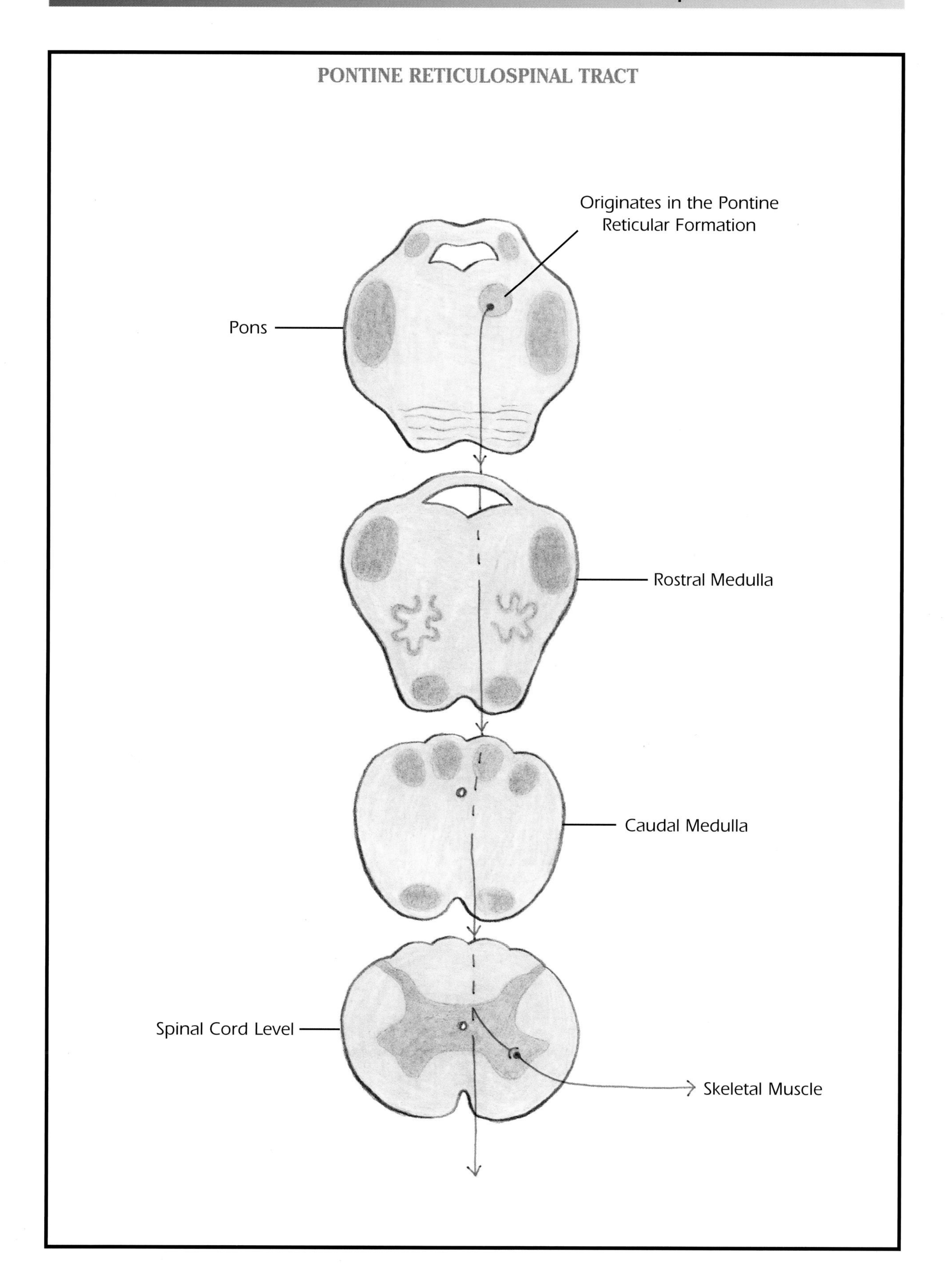
PONTINE RETICULOSPINAL TRACT
Originates in the Pontine Reticular Formation
Pons
Rostral Medulla
Caudal Medulla
Spinal Cord Level
Skeletal Muscle

Decorticate Rigidity. Results from a lesion to the corticospinal tracts. The upper extremities are in spastic flexion: head and neck are flexed; scapulae are elevated and retracted; shoulders are internally rotated and adducted; forearms are pronated; elbows, wrists, and fingers are flexed; wrists are ulnarly deviated. The lower extremities are in spastic extension: hips are rotated and adducted; feet are plantar flexed and inverted.

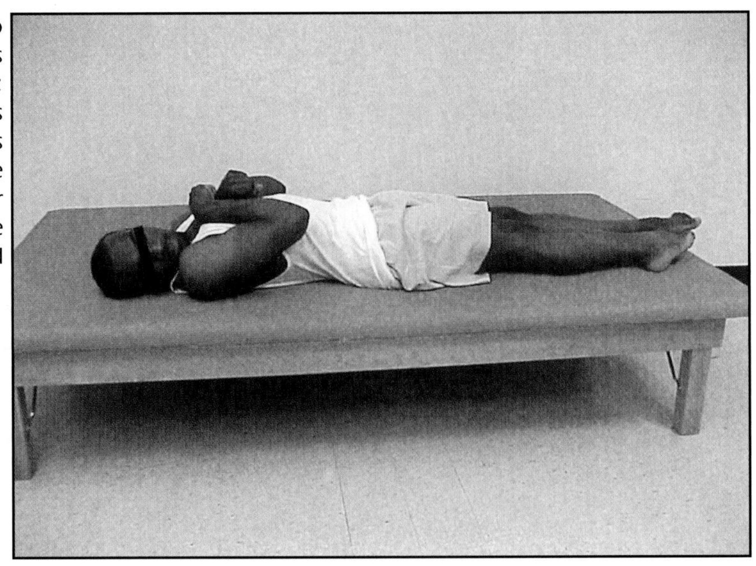

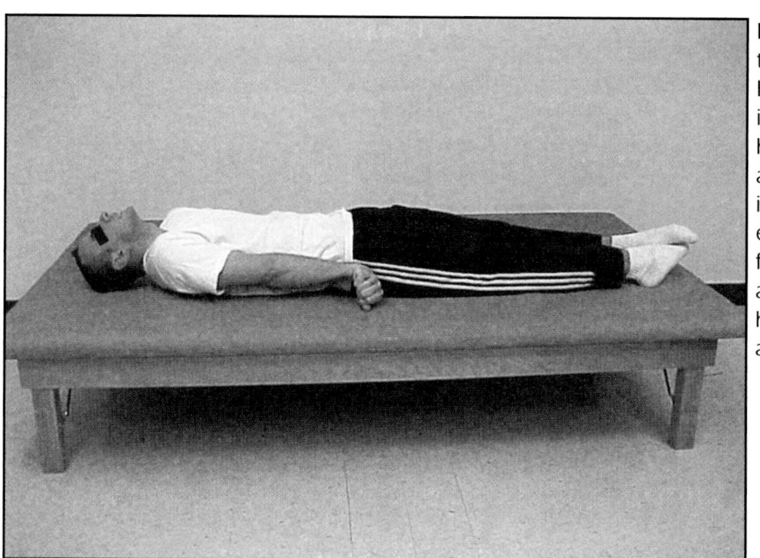

Decerebrate Rigidity. Results from a lesion to the extrapyramidal spinal cord tracts. Both the upper and lower extremities are in spastic extension. Head and neck are in hyperextension; jaws are clenched; scapulae are elevated and retracted; shoulders are internally rotated and adducted; elbows are extended; forearms are pronated; wrists are flexed and ulnarly deviated; fingers are flexed and adducted; hips and knees are extended; hips are internally rotated and adducted; feet are plantar flexed and inverted.

Spinal Cord Injury and Disease

INCOMPLETE VS COMPLETE SPINAL CORD INJURY

Complete Spinal Cord Injury

- Involves the *absence* of *sensory* and *motor* functions *below* the *lesion level.*
- Return of function of the last preserved spinal cord (SC) level is enhanced by the administration of the corticosteroid *methylprednisone* (stabilizes cell membranes, decreases inflammation, increases nerve impulse generation, and improves blood flow to the damaged area). Methylprednisone must be administered in the first 3 to 8 hours after injury for it to be effective.

Incomplete Spinal Cord Injury

- Involves *partial preservation* of *sensory* and *motor* functions *below* the *lesion level.*
- *Prognosis* is *better* than in a complete SC injury (SCI) as a result of preserved axon function.
- Return of some function is enhanced by the administration of methylprednisone.
- Incomplete SCIs occur more frequently than complete severances.

COMMON CAUSES OF SPINAL CORD INJURY/DISEASE

Transection

- A transection is a *complete severance* of the cord.
- It involves interruption of all sensory and motor information *at* and *below* the lesion level.
- Transections can result from traumatic injury including auto accidents, knife wounds, gun shot wounds, and diving accidents.

Compression

- Compression involves *impingement* of the cord.
- Symptoms depend on the severity of the injury.
- Compressions can result from trauma, tumor, or vertebral degenerative joint disease.

Infection

- Infection may compromise the integrity of the cord.
- An example is *polio*—involves damage to the cell bodies in the ventral horn causing lower motor neuron (LMN) loss.

Degenerative Disorders

- Degenerative diseases can damage the motor SC tracts.
- One example is *amyotrophic lateral sclerosis* (ALS)—results in bilateral degeneration of the ventral horn and pyramidal tracts.
- Involves both LMN and upper motor neuron (UMN) damage.

Five Most Important Tracts to Clinically Evaluate

1. **Lateral Corticospinal Tracts**	Voluntary motor control on the contralateral side
2. **Dorsal Columns**	Conscious discriminative touch, pressure, vibration, and proprioception on the contralateral side
3. **Lateral Spinothalamic Tracts**	Conscious pain and temperature on the contralateral side
4. **Spinocerebellar Tracts**	Unconscious proprioception
5. **Vestibulospinal Tracts**	Facilitation of extensor tone (important to assess in neurologic injury)

Upper Motor Neuron vs Lower Motor Neuron Lesions

Definition of Upper Motor Neuron

- An UMN is a motor neuron (MN) that carries motor information from the cortex or subcortical regions to either of the following:
  - The *cranial nerve* (CN) *nuclei* in the *brainstem*. The CN *nuclei* are considered to be part of the UMN system. The CN fibers that travel to target muscles are considered to be within the LMN system.
  - *Interneurons* that synapse with *motor cell bodies* in the *ventral horn*. An UMN travels *up to but does not enter* the ventral horn. The ventral horn is considered to be part of the LMN System.
- An UMN lesion includes all SC injuries and diseases that affect the cord between the levels of C1 – T12.

Definition of Lower Motor Neuron

- A MN that carries information from the *motor cell bodies* in the *ventral horn* to *skeletal muscles*.
- LMNs include the following:
  - CNs
  - Conus medullaris (at L1 – L2 vertebrae)
  - Cauda equina
- Thus, cord lesions at the L1 vertebra area and lower are considered LMN injuries.
- All lesions to the peripheral nerves are also considered to be LMN conditions.

Signs and Symptoms of Upper and Lower Motor Neuron Lesions

Upper Motor Neuron Lesion Signs

Below the Lesion Level

- Spasticity:
  - An *increase* in *muscle tone* with an associated inability to voluntarily control the muscle
  - Difficulty actively and passively moving the muscles on *one side of the joint*—but *not* both sides
  - Either the *flexors* or the *extensors* are spastic, but *not* both. (If both the flexors and the extensors display increased tone, *rigidity* is occurring. Rigidity usually results from basal ganglia dysfunction rather than SCI.)
- Hyperactive reflexes
- Clonus:
  - A sustained series of rhythmic jerks in a muscle.
  - Usually caused by a *quick stretch* of the *spastic muscle group*.

At the Lesion Level

- Flaccidity—loss of muscle tone.

Lower Motor Neuron Lesion Signs

- Flaccidity
- Hyporeflexia
- Within a few weeks of LMN injury, muscles begin to *atrophy*.

- Muscles undergoing the early stages of atrophy may display the following:
  - *Fibrillations*—fine twitches of single muscle fibers that usually cannot be detected on clinical exam but can be identified on an electromyogram.
  - *Fasciculations*—brief contractions of motor units, which can be observed in skeletal muscle and detected on clinical exam.

Spinal Cord Disease

Dorsal Column Disease (also called Posterior Cord Syndrome or Tabes Dorsales)

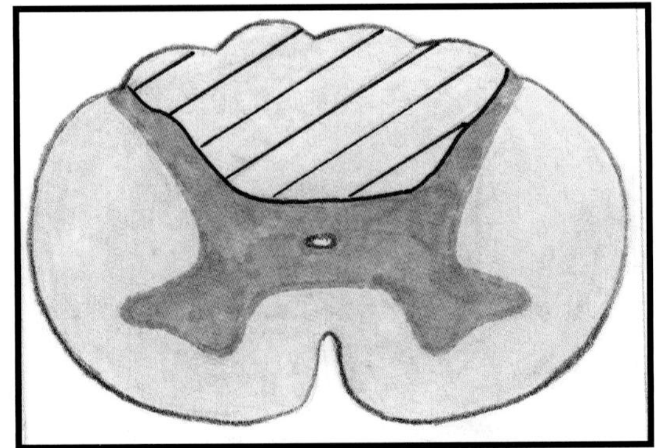

Etiology

- Seen in patients with neurosyphilis.

Pathology

- Dorsal columns are lost bilaterally.

Symptomatology

- Causes a bilateral loss of the following:
  - Tactile discrimination
  - Vibration
  - Pressure
  - Proprioception (often accompanied by ataxia)

Brown-Séquard Syndrome

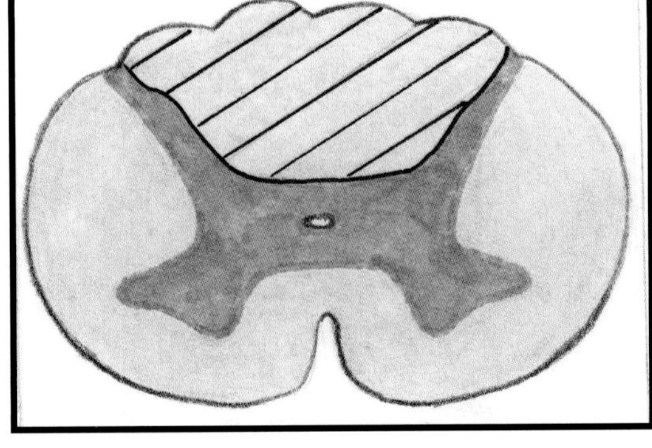

Etiology

- Multiple sclerosis
- Stab wound
- Tumor

Pathology

- Brown-Séquard syndrome is a SC hemisection.

Symptomatology

The Lateral Corticospinal Tract Is Lost Ipsilaterally

- Because the injury occurs in the SC, the patient presents with an ipsilateral loss of motor control and *spasticity below* the lesion level.
- The patient presents with *ipsilateral flaccidity* at the lesion level.

The Dorsal Column Is Lost Ipsilaterally

- Because the decussation of this tract occurs at the caudal medulla level, the patient presents with an *ipsilateral loss of discriminative touch*, *pressure*, *vibration*, and *proprioception.*

The Spinothalamic Tract Is Lost Contralaterally

- Because the spinothalamic tracts decussate as soon as they enter the cord, *pain* and *temperature* will be lost on the *contralateral side* (below the lesion level).
- At the lesion level, the patient experiences *bilateral loss of pain* and *temperature*.

Anterior Cord Syndrome

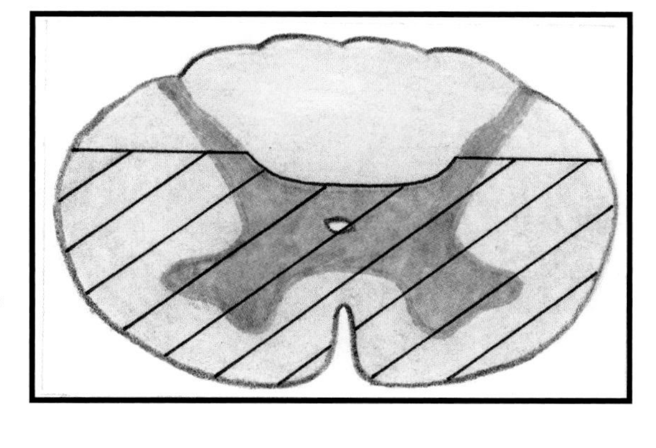

Etiology

- Infarct
- Ischemia
- Trauma

Pathology

- Anterior cord syndrome occurs when two-thirds of the anterior cord is lost.

Symptomatology

The Dorsal Columns Are Spared

- Discriminative touch, vibration, pressure, and proprioception are spared.

The Lateral Corticospinal Tracts Are Lost

- Because the lateral corticospinal tracts descend down the lateral white funiculus, they are lost.
- This results in *bilateral spastic paralysis*.
- The patient loses *bilateral voluntary motor control* below the level of the lesion.

The Ventral Horn Is Lost

- The MNs in the ventral horn are part of the LMN system.
- Because the ventral horn MNs are lost, the patient presents with *flaccidity at* and *below* the *lesion* level.

The Spinothalamic Tracts Are Lost

- Because the spinothalamic tracts synapse in the dorsal horn and decussate across the anterior white funiculus as soon as they enter the cord, the patient presents with *bilateral loss* of *pain* and *temperature*.

Central Cord Syndrome (also called Syringomyelia)

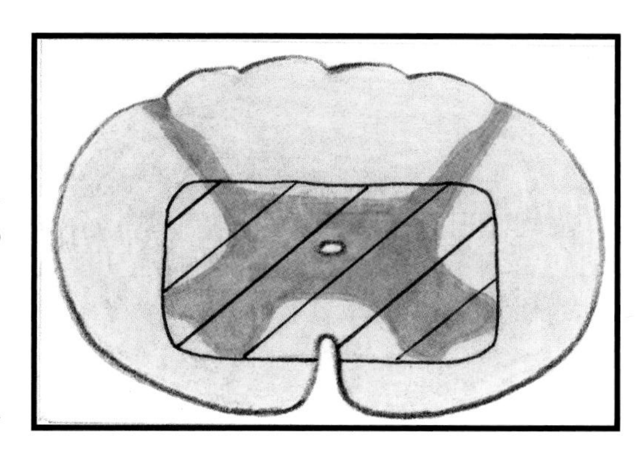

Etiology

- Unknown
- More commonly occurs in elderly persons who have narrowing, or stenotic changes, in the spinal canal related to arthritis.
- Damage may also occur in people with congenital stenosis.

Pathology

- Central cord syndrome involves a *cavitation* of the central cord in the *cervical segments*.

Symptomatology

The Spinothalamic Tracts Are the First to Be Lost

- Because the spinothalamic tracts synapse in the dorsal horn and decussate to the anterior white funiculus as soon as they enter the cord, the spinothalamic tracts are lost.
- This results in *bilateral loss* of *pain* and *temperature*.

The Ventral Horn Is Lost

- This results in *flaccidity* of the *upper extremities* because the disease occurs in the cervical regions of the SC.

Posterolateral Cord Syndrome

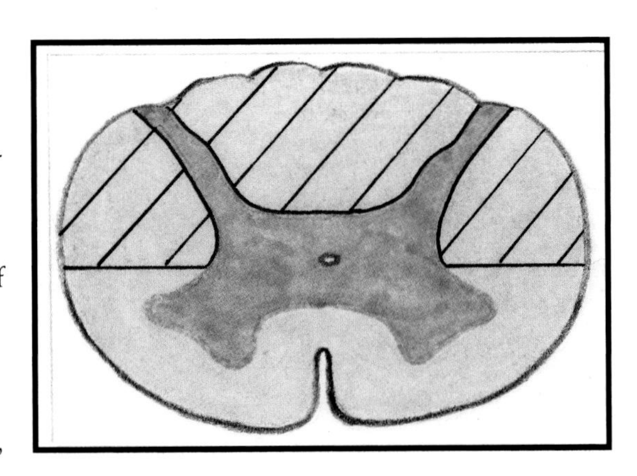

Etiology

- Degeneration of the SC from severe vitamin B12 deficiency, pernicious anemia, or AIDS

Pathology

- Affects the posterior and posterolateral white funiculi of the SC

Symptomatology

The Dorsal Columns Are Lost Bilaterally

- Results in *bilateral loss* of *discriminative touch, pressure, vibration*, and *proprioception.*

The Lateral Corticospinal Tracts Are Lost Bilaterally

- Because the lateral corticospinal tracts descend in the lateral white funiculus, they are lost.
- Results in *bilateral spastic paralysis.*

The Spinocerebellar Tracts Are Lost

- This results in *bilateral ataxia.*

Anterior Horn Cell Syndrome (also called Ventral Horn Syndrome)

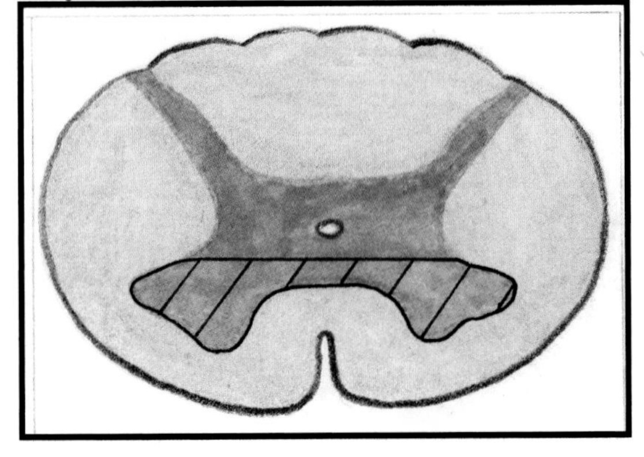

Etiology

- Disease process that destroys the MNs in the ventral horn

Pathology

- Anterior horn cell syndrome involves LMN damage.

Symptomatology

- Results in *bilateral flaccidity* in the muscles innervated by the affected SC levels
- An example is *poliomyelitis*—an acute viral disease affecting the ventral horn motor cell bodies.

Spinal Shock (also called Neurogenic Shock)

- A state of *areflexia* that occurs immediately after SCI
- Involves the *loss of all spinal reflexes* below the lesion level
- Because the motor pathways below the lesion level are also lost, the patient experiences the following:
  - Flaccid paralysis (because no spinal reflex arcs are firing)
  - Loss of tendon reflexes
  - Loss of autonomic function
- Spinal shock can last for hours, days, or weeks.
- Upon resolution, spinal reflex activity returns.
- Frequently, the severity of spinal shock increases as the level of injury increases.

Autonomic Dysreflexia in Spinal Cord Injury

- Involves an *acute episode of exaggerated sympathetic reflex* responses in SCI that occurs because higher center reflex regulation is lost
- Autonomic dysreflexia occurs only after spinal shock has resolved and autonomic reflexes return.
- Characterized by the following:
  - Severe hypertension
  - Bradycardia
  - Severe headache
  - Vasodilation
  - Flushed skin
  - Profuse sweating above the lesion level
- Patients can experience one or several of the above symptoms in one episode. Autonomic dysreflexia can initially occur or recur at any time during the patient's lifespan.
- Usually associated with SCIs at T6 and above
- Common causes include the following:
  - Full bladder or rectum (frequent cause—catheter should be checked for overfill)
  - Stimulation of pain receptors (as occurs with pressure ulcers)
  - Ingrown toenails
  - Dressing changes
  - Visceral contractions (eg, ejaculation, bladder spasms, uterine contractions)
- Autonomic dysreflexia is a *clinical emergency* requiring immediate treatment.
- *Convulsions, loss of consciousness,* and even *death* may occur if autonomic dysreflexia goes unattended.

Orthostatic Hypotension in Spinal Cord Injury

- This is also called postural hypotension.
- This usually occurs in patients with injuries at T6 and above.
- This occurs when sympathetic outflow to blood vessels in the extremities and abdomen is interrupted.
- There is a decrease in cardiac output when the *patient is placed in an upright position*. Venous return to the heart is impaired, and pooling of blood in the feet occurs.
- Blood pressure can drop precipitously.
- Signs of orthostatic hypotension include the following:
  - Dizziness
  - Pallor in the face (blood has shifted away from the head to the feet)
  - Excessive sweating above the lesion level
  - Blurred vision
  - Possible fainting (syncope)
- Orthostatic hypotension can often be prevented by helping the patient to assume an upright position in a gradual and slow manner.

Body Temperature Dysregulation in Spinal Cord Injury

- Body temperature is regulated by the *sympathetic nervous system*.
- The mechanisms for body temperature regulation are located largely in the *hypothalamus*.
- After SCI, communication between the hypothalamic temperature regulators and sympathetic function below the lesion level become disrupted.
- In other words, the body's ability to control blood vessel responses that conserve or dissipate heat is lost. Lost also are the abilities to sweat and shiver.
- Higher level injuries produce greater disturbances in temperature control.
- *Poikilothermy* is the condition in which the body takes on the temperature of its external environment. This commonly occurs in patients with injuries at T6 and above.

Edema, Skin Breakdown, and Deep Vein Thrombosis in Spinal Cord Injury

Edema

- Edema is the presence of an *abnormal accumulation of fluid in interstitial tissue.*
- It can frequently occur in SCI as a result of immobility causing increased venous pressure and abnormal pooling of blood in the abdomen, lower limbs, and upper extremities.

Skin Breakdown

- The sympathetic nervous system influences skin integrity through the control of vasomotor and sweat gland activity.
- These functions ensure that the skin has adequate circulation, excretion of body fluids, and temperature regulation.
- *SCI compromises skin integrity* as a result of lost sensation and circulatory changes. Skin breakdown is a common but frequently preventable complication of SCI.

Deep Vein Thrombosis

- Deep vein thrombosis is a *clot in the venous system that may produce infarction.* Such clots commonly originate in the legs and can travel to the lungs, causing pulmonary embolism.
- In SCI, deep vein thrombosis occurs as a result of impaired vasomotor tone, loss of muscle tone, trauma to the vein wall, immobility, and hypercoagulation.

Disorders of Bowel and Bladder Function in Spinal Cord Injury

Bladder Function

- Bladder function is controlled by the sympathetic nerve fibers from T1 – L3. These allow for relaxation of the detrusor muscle during bladder filling, awareness of signals indicating bladder distention (S2 – S4), and relaxation of the external sphincter for bladder voiding.
- Patients with UMN lesions have a *spastic bladder.*
    - They lack awareness of bladder distension and have no voluntary control of voiding.
    - Involuntary voiding reflexes may be elicited during bladder filling, causing incontinence and an inability to completely empty the bladder.
- Patients with a LMN lesion have a *flaccid bladder.*
    - These patients lack an awareness of bladder distention and cannot void voluntarily or involuntarily.
- Because there is no bladder function other than storage, retention (with overfill) and urine leakage occur.

Bowel Function

- Bowel function is controlled by parasympathetic fibers from S2 – S4 that innervate the colon, rectum, and internal anal sphincter.
- Somatic innervation from S2 – S4 allow for voluntary control of the external anal sphincter.
- Defecation involves a reflex that increases peristaltic movements of the colon, rectum, and anus.
- Patients with SCI *above* S2 – S4 have a *spastic defecation reflex* and *lose voluntary control* of the *external anal sphincter.*
- Patients with SCI directly at the S2 – S4 level have a *flaccid defecation reflex* and *lose anal sphincter tone.* Without the defecation reflex, peristaltic movements cannot evacuate the stool.

Sexual Function in Spinal Cord Injury

- Sexual function is mediated by the S2 – S4 segments of the SC.
- The T11 – L2 cord segments are responsible for sexual arousal due to *mental stimulation* (referred to as the *psychogenic response*). This is the area where autonomic nerve pathways, in communication with the cortex, leave the cord and innervate the genitalia.
- The S2 – S4 cord segments are responsible for sexual arousal due to touch (referred to as the *reflexogenic response*).
- In patients with a *T12/L1 or higher SCI* (UMN injury):
    - Sexual arousal from touch remains intact—although the patient cannot feel the erection or vaginal lubrication.
    - Sexual arousal from mental stimulation is lost.
- In patients with a *cord injury from L2 – S1* (LMN injury):
    - Sexual arousal from touch is possible, but the patient cannot feel the erection or vaginal lubrication.
    - Sexual arousal from mental stimulation is possible.

Level of Spinal Cord Injury and Functional Ability

SCI Level	*Preserved Sensorimotor Function*	*Activities of Daily Living Function*
C1 – C3	Head and neck sensation Some neck control Respirator dependent	Dependent in activities of daily living (ADL)
C4	Good head and neck sensation and motor control Scapular elevation Diaphragmatic movement (respiration)	Requires maximum assistance with all ADL
C5	Full head and neck control and sensation Some shoulder strength Shoulder external rotation Shoulder abduction to 90 degrees Elbow flexion and supination	Self-feeding with adaptive equipment Limited upper extremity dressing with adaptive equipment Limited self-care with adaptive equipment (brushing teeth, grooming)
C6	Fully innervated shoulder movement Forearm pronation Wrist extension Tenodesis	Self-feeding with adaptive equipment Upper and lower extremity dressing with adaptive equipment Requires greater assistance with lower extremity dressing Self-care with adaptive equipment
C7	Elbow extension Wrist flexion Finger extension	Independent in self-feeding, dressing, and grooming with adaptive equipment Independent bed mobility and transfers Meal prep with adaptive equipment Can drive with hand controls
C8 – T1	All upper extremity muscles are innervated Fine motor coordination present Full grasp available	Independent in all self-care, grooming, and meal prep with adaptive equipment Can drive with hand controls
T1 – T6	Top half of intercostal muscles are innervated allowing increased respiratory reserve Long muscles of back are innervated allowing for improved trunk control	Standing is possible with assistance but is not practical for dynamic ADL Independent self-catheterization
T6 – T12	All intercostal muscles and lower abdominals are innervated, providing improved trunk control and endurance	Limited ambulation is possible with lower extremity orthotics and assistive devices
T12 – L4	Hip flexion Hip adduction Knee extension	Functional ambulation is possible with bilateral lower extremity orthotics and assistive devices Wheelchair is used for energy conservation
L4 – L5	Knee extension is present but weak. Ankle dorsiflexion	Functional ambulation is possible with bilateral lower extremity orthotics Wheelchair is used for energy conservation

SECTION 20

Proprioception

- Sherrington named the term proprioception and defined it as the sixth sense: "The sense by which the body knows itself, judges with perfect, automatic, instantaneous precision the position and motion of all of its movable parts, their relations to one another, and their alignment in space."
- Proprioception is the ability to *sense one's body position in space.*
- The Latin root proprius means to "own oneself" or to feel one's body as one's own.
- Proprioception occurs mostly on an unconscious level because it is primarily mediated by the cerebellum.
- The term *kinesthesia* refers to the ability to sense one's body *movement* in space.

The Ability to Sense One's Body in Space Is Based on Three Systems

Visual System

- Humans use visual cues to negotiate the environment.

Vestibular System

- The *labyrinthine system in the inner ear*—the semicircular canals—contains continuously moving liquid that is constantly monitored by the vestibular system to provide feedback regarding the *head's position in space.*
- The vestibular system also has neural connections to the cerebellum to provide feedback about the *head's position in space.*

Proprioceptive System

- The proprioceptive system is a feedback/feedforward loop between the *muscle spindles (MSs), Golgi tendon organs, joint receptors,* and the *cerebellum.*
- Together, these provide constant information about the *body's position in space.*

Muscle Spindle

- Muscle spindles (MSs) are proprioceptors located in skeletal muscle.
- MSs are sensory receptors that provide a constant flow of information regarding length, tension, and load on the muscles.
- MSs detect when a muscle has been stretched and initiate a reflex that resists the stretch.

Extrafusal Muscle Fibers

- Extrafusal muscle fibers are the *bulk* of the muscle.

Intrafusal Muscle Fibers

- Intrafusal muscle fibers are the MSs that sit *within the bulk* of the muscle.
- The intrafusal muscle fibers are attached to the extrafusal muscle fibers.

Density of Muscle Spindles in an Extrafusal Muscle Fiber

- The more MSs in an extrafusal muscle fiber, the *greater the muscle's precision control.*
- Muscles with the *highest density* of MSs are *small muscles* designed for *fine motor control.*

Nuclear Chain Fibers and Nuclear Bag Fibers

- The nuclear chain and nuclear bag are structures within the MS.

Nuclear Bag

- The nuclear bag is responsive to *changes in muscle length.*
- When the length of the muscle changes, the nuclear bag fires.
- The nuclear bag also detects the velocity of the muscular stretch, or how *quickly* the *muscle stretched.*

Nuclear Chain

- The nuclear chain only fires in response to a *new muscle length*—it does not respond to velocity, like the nuclear bag.

Properties of the Equatorial Regions of the Nuclear Bag and Chain

- The equatorial region of the bag is *elastic* and *phasic* (*quick responding*).
- The equatorial region of the chain is *elastic* and *tonic* (*slow responding*).

Properties of the Polar Regions of the Nuclear Bag and Chain

- The polar regions of the bag are *contractile* and *tonic* (*slow responding*).
- The polar regions of the chain are *contractile* and *phasic* (*quick responding*).

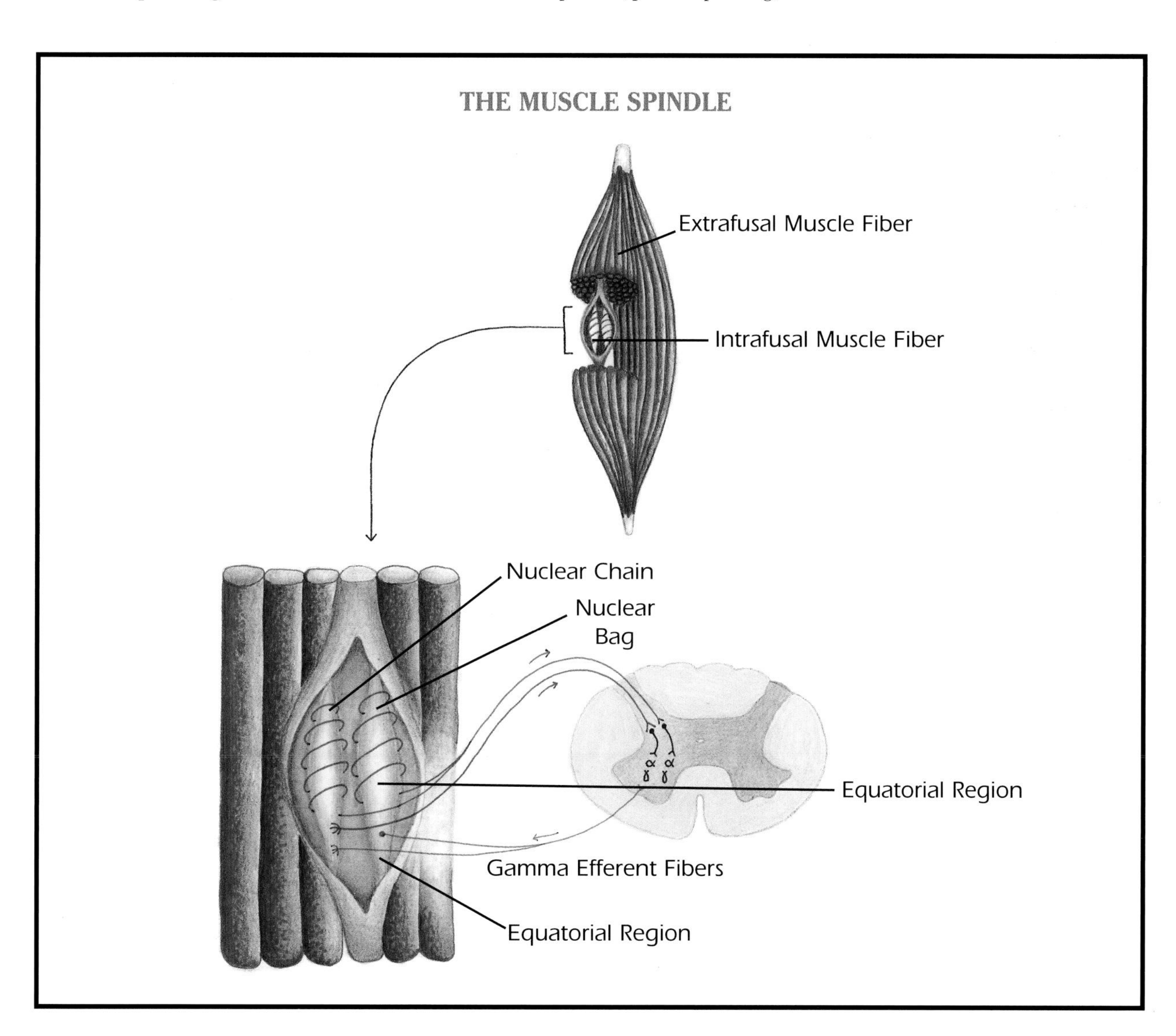

The Muscle Spindles Use Two Types of Sensory Fibers to Send Information to the Ventral Horn

Ia (Primary Ending)

- The Ia sensory fibers are *large* and *heavily myelinated.*
- They are *fast conducting.*
- The Ia fibers wrap around the *equatorial region* of both the *bag* and *chain.*
- The Ia fibers respond to the *rate of muscle stretch* (velocity) and to *changes in muscle length.*
- The Ia fibers are fast adapting (phasic).

II (Secondary Ending)

- II fibers are *medium* size fibers.
- They terminate on the *equatorial region* of the *chain* only.
- II fibers are predominantly located on the chain.
- II fibers respond to *changes* in the *length* of the MS (the rate of the stretch is not involved).
- II fibers are slow adapting (tonic).

Muscle Spindle Sequence of Events

- The extrafusal muscle fiber stretches.
- This causes the MS (the intrafusal muscle fibers) to stretch.
- The equatorial region of the bag stretches right away—because the equatorial region of the bag is elastic and responds more to stretch than does the polar region. (If the stretch is a sustained stretch, the equatorial region of the chain will also stretch.)
- Stretching of the MS causes the Ia fibers to fire.
- The Ia fibers will fire in response to a quick or phasic response—because the Ia fibers from the bag are phasic.
- The Ia fibers will also fire in response to a tonic or sustained and slow response—because the Ia fibers from the chain are tonic.
- The secondary fibers, the II fibers, then fire.
- The II is attached only to the chain and the chain only detects length and position changes.
- The II fibers are tonic and respond to a slow, sustained stretch.

Proprioceptive Information from the Muscle Spindles Travels to Four Places

- Proprioceptive information from the MS travels along the Ia and II fibers to the dorsal horn.
- In the dorsal horn, the fibers synapse on an interneuron and connect to alpha motor neurons (MNs) in the ventral horn.
- Information from the MS travels to the following:
  - **An Alpha MN of the Same Muscle** (the agonist)—for facilitation of that muscle. Referred to as autogenic excitation
  - **An Alpha MN of the Antagonist Muscle**—for inhibition of the antagonist. Referred to as reciprocal inhibition: every time an alpha turns on an agonist, an alpha turns off the antagonist.
  - **Renshaw Cells** (or special interneurons that modify the action of synergy muscles)—a Renshaw cell is a short axon that connects motor nerve fibers with each other to produce refined motor movement. Referred to as recurrent inhibition: every time an alpha turns on an agonist, an alpha also modifies (usually inhibits) the action of the synergy muscles.
  - **The Cerebellum**—MSs send proprioceptive signals to inform the cerebellum of all changes in muscle position and length.

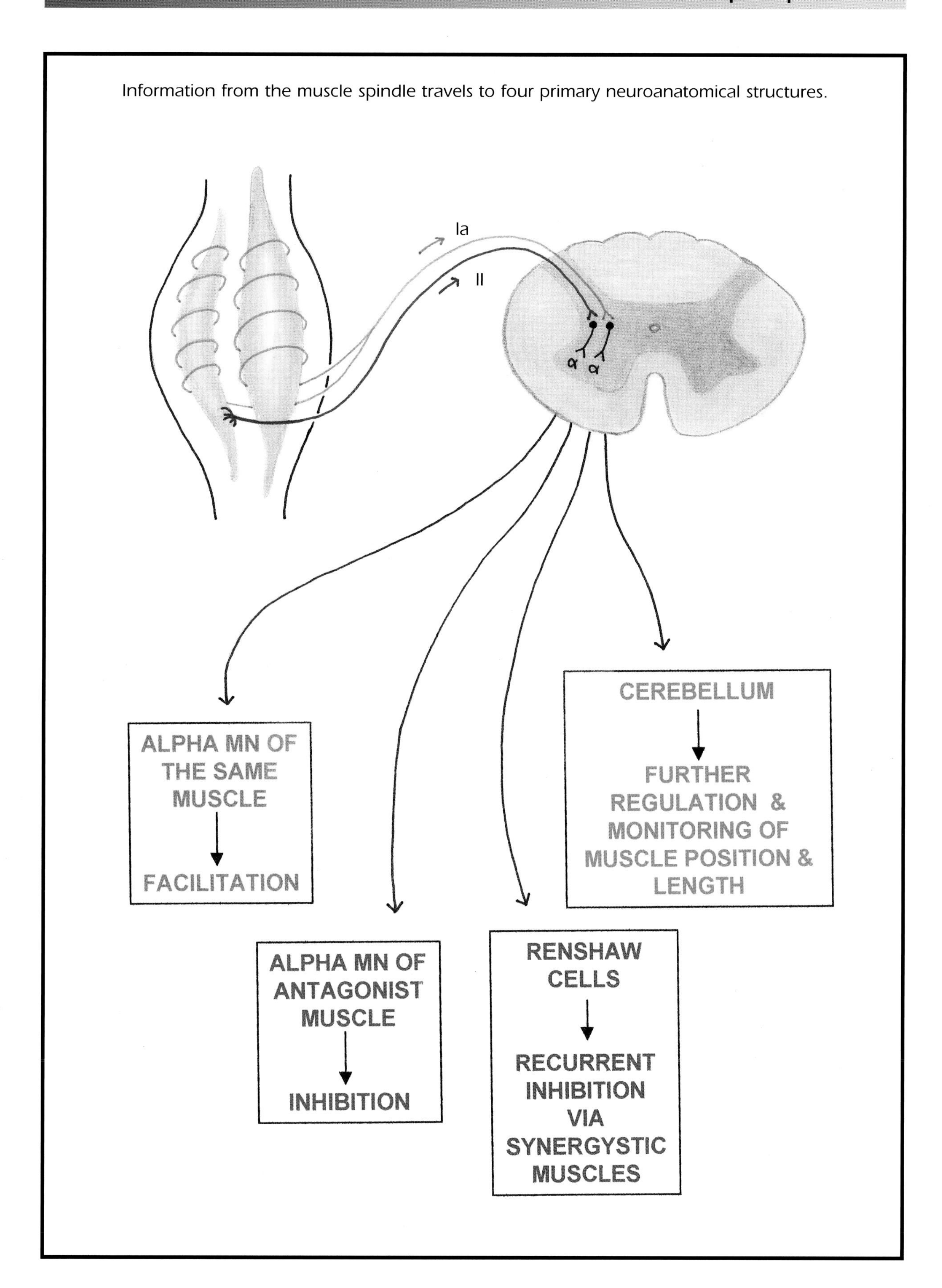
Information from the muscle spindle travels to four primary neuroanatomical structures.
Ia
II
α α
ALPHA MN OF THE SAME MUSCLE
FACILITATION
ALPHA MN OF ANTAGONIST MUSCLE
INHIBITION
RENSHAW CELLS
RECURRENT INHIBITION VIA SYNERGYSTIC MUSCLES
CEREBELLUM
FURTHER REGULATION & MONITORING OF MUSCLE POSITION & LENGTH

The Gamma Motor Neuron System

- The gamma motor neurons (MNs) are also located in the *ventral horn*, along with the alpha MNs.
- The gamma MNs send proprioceptive information from the ventral horn back to the MS.
- Gamma MNs are *fast conducting* because they are *heavily myelinated* and large.
- Gamma MNs *innervate* the MSs.
- Alpha MNs *innervate* the *extrafusal muscle fibers* (the muscle bulk).

Two Types of Gamma Motor Neurons

Gamma 1 Fibers

- Gamma 1 fibers have plate endings and terminate on the polar regions of the nuclear bag.

Gamma 2 Fibers

- Gamma 2 fibers have trail (or multi-branching) endings and terminate predominately on the nuclear chain, adjacent to the equatorial region.

Gamma Motor Neuron Stimulation

- When the gamma MNs in the ventral horn are stimulated, they fire causing the polar regions of the bag and chain to contract.
- This causes the equatorial regions to stretch (or distort).
- The Ia fibers then fire, causing the alpha MNs in the ventral horn to fire.
- An alpha MN in the ventral horn innervates the agonist, causing the extrafusal muscle fiber to contract.

Golgi Tendon Organs

Location

- The Golgi tendon organs (GTOs) are embedded in the tendons, close to the skeletal muscle insertions.

Function

- The GTOs are proprioceptors that detect tension in the tendon of a contracting muscle.

Golgi Tendon Organs Use Ib Afferent Neurons

- The GTOs use Ib afferent (or sensory) neurons to send proprioceptive information to the dorsal horn.
- The Ib fibers synapse on interneurons in the ventral horn.
- In the ventral horn, the interneurons synapse on alpha MNs.

Autogenic Inhibition

- Activation of the GTOs causes a contracting muscle to be inhibited; in other words, it relaxes.
- This is a protective function. If the GTOs did not become activated in response to a muscle's stretch, an individual could easily tear his or her muscles.

Sequence of Golgi Tendon Organs Events

1. The *agonist* muscle contracts.
2. This activates the GTOs (which are embedded in the contracting muscle).
3. The GTOs send proprioceptive information along the Ib sensory fibers to the dorsal horn.
4. In the dorsal horn, the Ib fibers synapse with an interneuron.
5. The interneuron connects with an alpha MN in the ventral horn.
6. Information from the GTOs travels to three places:
   - **An Alpha MN** to inhibit the contracting agonist muscle. Referred to as autogenic inhibition
   - **An Alpha MN** to facilitate the antagonist of the contracting muscle
   - **The Cerebellum** for further proprioceptive feedback

Golgi Tendon Organ Functional Implications

- The GTOs help to control the speed of a contraction for *coordinated, fine, precision movements.*
- Protective mechanism—humans need the action of the GTOs to protect against muscle tears and pulls.
- The GTOs help to reduce muscle cramps:
  - When an individual has a cramp in a muscle group (eg, when the *plantar flexors* are cramping), placing a *sustained stretch* on that muscle group will kick in the GTOs to reduce the cramp. In other words, position the plantar flexors in a sustained stretch (ie, pull the plantar flexors into *dorsiflexion* for a sustained stretch).
  - The GTOs in the *agonist* become activated and *inhibit the cramping agonist* (the plantar flexors).
- The GTOs can help to reduce *clasp knife syndrome*:
  - Clasp knife syndrome occurs when a muscle group is hypertonic (involves severe spasticity at a joint).
  - To reduce spasticity, perform a sustained stretch on the contracting spastic muscle.
  - This will activate the GTOs and relax the spastic muscle.

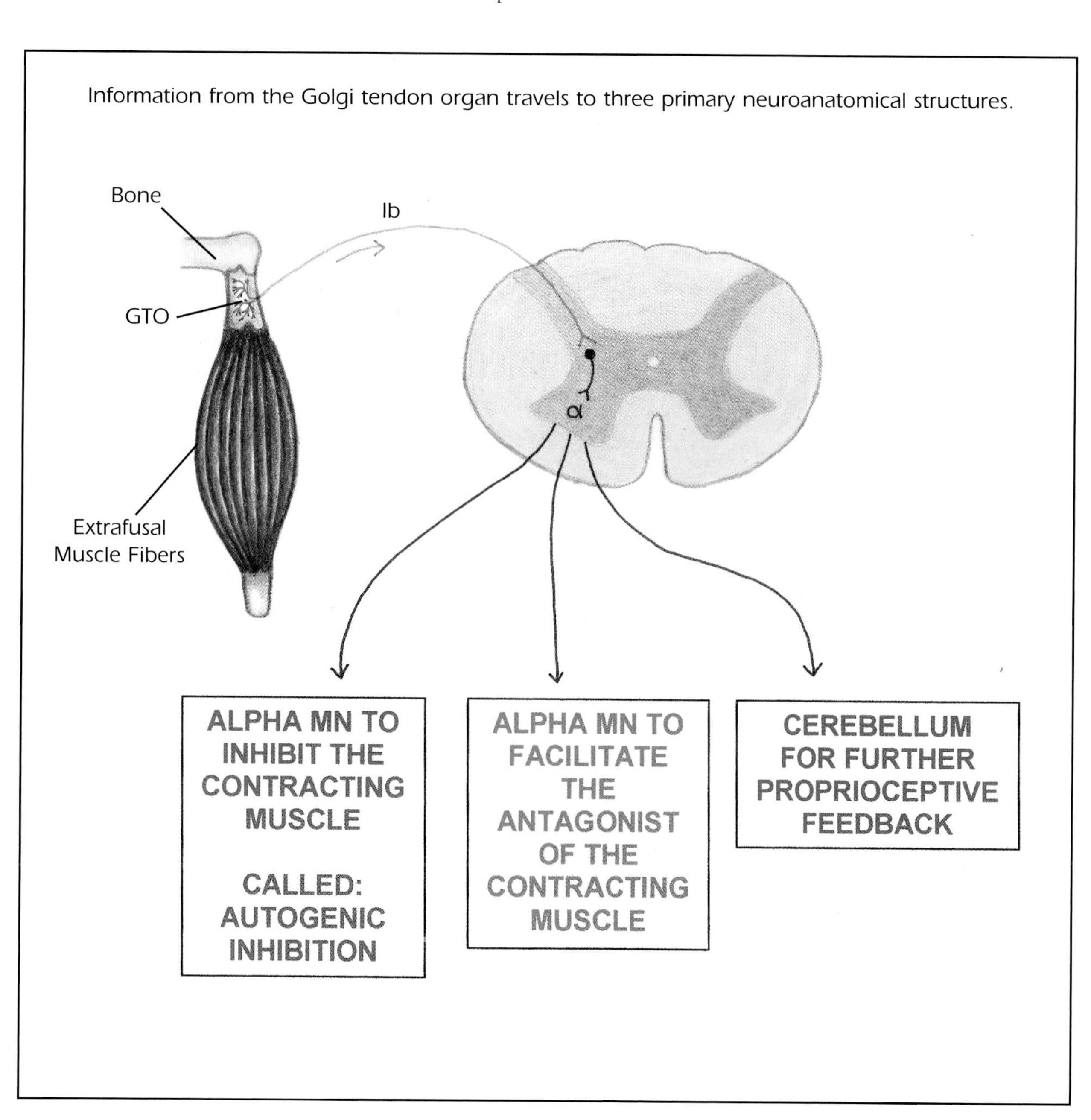

Neurologic Concepts Underlying the Basis of Many Therapeutic Techniques

- **Slow Sustained Stretch** will reduce spasticity in a muscle group by activating the GTOs. The GTOs will inhibit the contracting spastic agonist and facilitate the antagonist.
- **Splinting** a spastic muscle group essentially positions the spastic limb in a sustained stretch. The GTOs become activated and inhibit the spastic contracting muscle groups (agonists) while also facilitating the antagonists.
- **Serial Casting** also places a spastic contracting muscle group (agonists) on a sustained stretch. This activates the GTOs, causing inhibition of the spastic muscle group and facilitation of the antagonists.
- **Seating and Positioning Techniques** are also used to place a spastic contracting muscle group on a sustained stretch. Again, the GTOs will become activated, causing a reduction in the spastic muscle groups and facilitating the antagonist muscle groups.

Muscle Spindles and Gamma Motor Neurons Work Together in a Feedback/Feedforward Loop

Sequence of Events

- Some change occurs in the extrafusal muscle fiber (eg, a muscle stretch).
- The intrafusal muscle fibers (the nuclear bag and chain of the MS) then stretch.
- The bag and chain send proprioceptive information to the alpha MNs in the ventral horn via the Ia and II fibers.
- This proprioceptive information from the MS travels to four places:
  - **An Alpha MN of the Same Muscle** (the agonist)—for facilitation of that muscle. Referred to as autogenic excitation
  - **An Alpha MN of the Antagonist Muscle**—for inhibition of the antagonist. Referred to as reciprocal inhibition
  - **Renshaw Cells**—for modification (usually inhibition) of synergy muscles. Referred to as recurrent inhibition
  - **The Cerebellum**—for further monitoring and modification of that muscle group's movement. The cerebellum uses this proprioceptive information for the fine tuning of precision, coordinated movement.
- Alpha excitation also stimulates the gamma MNs in the ventral horn.
- Gamma MNs fire and send proprioceptive information along gamma 1 and 2 fibers to the MSs.
- This causes the intrafusal muscle fibers (the muscle spindles) to contract.
- The MSs send their proprioceptive information to the ventral horn via the Ia and II fibers.
- The feedback/feedforward system begins over again, sending information to the four places described previously.

Gamma Biasing

- The function of the gamma MNs is to make the MS more sensitive and keep the *muscles primed*, or ready for action.
- Gamma biasing *raises* the *level of firing* of the MS.
- When activation of the gamma MNs raises the level of MS firing, it causes the muscle to prepare, or set, for an *anticipated activity*.

Examples of Gamma Biasing

- A person begins walking up a set of steps in which the step height is not uniform. She raises her foot upward to meet the next higher step level. However, because the step is lower than anticipated, she places her foot down too heavily.
- A person lifts a box of supplies that he believes is much heavier than it actually is. He lifts the box too quickly and too high as a result of anticipating the box to be heavier. His muscles were primed, by gamma biasing, to lift a heavier box.

Three Events Cause the Gammas to Fire

- **Stimulation of the Supraspinal Motor Centers**: The supraspinal motor centers include the basal ganglia, vestibular nuclei, and the reticular formation. These motor centers send projections to the gammas in the ventral horn through descending pathways—that is, the vestibulospinal and reticulospinal tracts.
- **Cutaneous Stimulation**: Sensory information from the skin is sent to the dorsal horn. This information synapses on interneurons and excites the gamma MNs.
- **Alpha and Gamma Coactivation**: Every time an alpha MN is activated, a gamma MN is coactivated.

JOINT RECEPTORS

Location

- Joint receptors are located in the connective tissue of a joint capsule.

Function

- Joint receptors respond to mechanical deformation occurring in the joint capsule and ligaments.
- Joint receptors send this proprioceptive information to the cerebellum and to the ventral horn.

What Stimulates the Joint Receptors?

Ruffini's Endings

- Located in the joint capsule
- Signal the extremes of joint range
- Respond more to passive than active movement

Paciniform Corpuscles

- Located in the joint capsule
- Respond only to dynamic movement—that is, when the joint is moving

Ligament Receptors

- Similar in function to GTOs
- Located in the ligaments of a joint capsule
- Respond to tension in the joint capsule

Free Nerve Endings

- Located in the joint capsule
- Most often stimulated by inflammation
- Signal the detection of pain in a joint

Joint Receptors Use Several Different Sensory Fibers to Send Proprioceptive Information to the Cerebellum and Ventral Horn

- Ligament receptors use Ib sensory fibers.
- Ruffini's endings and paciniform corpuscles use II sensory fibers.
- Free nerve endings use delta A and C sensory fibers.

Joint Receptor Pathway

- The joint receptors use sensory fibers to send proprioceptive information to the dorsal horn.
- Here, the messages synapse on an *interneuron*.
- The interneuron connects with a MN in the *ventral horn* and sends proprioceptive messages back to the *muscles* surrounding the joint.
- An *alpha* MN in the ventral horn also sends joint receptor information to the *cerebellum* for constant feedback about joint position and movement.

Joint Receptors and Proprioception

- While MSs, GTOs, and joint receptors are critical for normal proprioception, joint receptors *alone* may not be essential for proprioception.
- This is suggested because patients with *joint replacements still retain good proprioception* in joint midrange.

SECTION 21

Disorders of Muscle Tone

Muscle Tone

- Muscle tone is a *continuous state of muscle contraction at rest.*
- Tone is an *unconscious phenomenon*—humans cannot consciously will muscles to increase or decrease in tone.

Primary Mechanisms that Mediate Tone

Extrapyramidal Structures (Part of the Upper Motor Neuron System)

- The extrapyramidal structures are motor centers and pathways located outside of the pyramidal system.
- These include brainstem centers such as the vestibular nuclei and the reticular nuclei.
- The extrapyramidal structures also include extrapyramidal motor tracts such as the vestibulospinal, rubrospinal, and reticulospinal pathways.

Basal Ganglia

- Includes the caudate, putamen, and globus pallidus (of the upper motor neuron [UMN] system).
- Also includes the substantia nigra and the subthalamic nuclei.

Pyramidal Structures

- Includes the corticospinal tracts (of the UMN system).

Cerebellum

- The cerebellum works in a feedback/feedforward loop with the (a) brainstem, basal ganglia, extrapyramidal structures; and (b) the muscle spindles (MSs), Golgi tendon organs (GTOs), and joint receptors.
- The cerebellum also mediates tone through the afferent sensory tracts that travel to the cerebellum: posterior and anterior spinocerebellar tracts, cuneocerebellar tracts, and rostral spinocerebellar tracts.

Motor Neurons of the Ventral Horn (of the Lower Motor Neuron System)

- The alpha and gamma motor neurons (MNs) of the ventral horn mediate tone.

Peripheral Nerves that Innervate Skeletal Muscle (of the Lower Motor Neuron System)

- The peripheral nerves that innervate the skeletal muscles also mediate tone.

Classifications of Tone

Hypotonicity

- Hypotonicity is an *abnormal decrease in muscle tone* (eg, floppy babies and individuals with spinal cord injury [SCI] below L1).
- Hypotonicity is caused by lower motor neuron (LMN) lesions:
  - *Damage* to the MNs in the *ventral horn*

- *Damage* to the *spinal nerves* in the *periphery*
- Lesions to the *posterior cerebellar lobe* (neocerebellar lobe) produce hypotonicity and hyporeflexia.

Hypertonicity

- Hypertonicity is an *abnormal increase in muscle tone* accompanied by resistance to active and passive movement.
- Hypertonicity is caused by UMN damage.
- Lesions to the anterior cerebellar lobe (paleocerebellar lobe) produce hypertonicity and hyperactive reflexes.

Spasticity

- Spasticity is a form of hypertonicity.
- Spasticity involves difficulty actively and passively moving the spastic muscle groups on one side of a joint.
- Either the flexors or extensors are spastic but not both.
- Spasticity is associated with such disorders as SCI (T12 and above), head injury, cerebrovascular accident, and cerebral palsy.

Rigidity

- Rigidity is a form of hypertonicity.
- Rigidity involves difficulty actively and passively moving the muscle groups on *both* sides of a joint.
- Rigidity is associated with such disorders as Parkinson's disease.

Clasp Knife Syndrome

- Clasp knife syndrome involves severe spasticity at a joint.
- A sustained stretch will relax the muscle group, and the spasticity will suddenly give way.

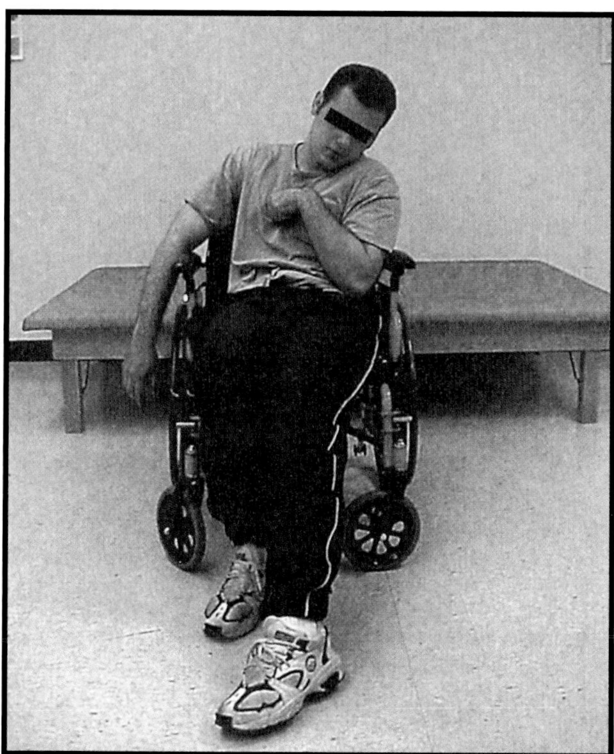

Typical posture of an individual with left hemiplegia. Increased tone on the contralateral side of the body. Head and neck are flexed; scapula is retracted and depressed; shoulder is adducted and internally rotated; elbow, wrist, and fingers are flexed; forearm is pronated; wrist is ulnarly deviated; fingers are adducted; hip, leg, knee, and ankle are extended; pelvis is posteriorly tilted; thigh is internally rotated; foot is inverted.

Cogwheel Rigidity

- In cogwheel rigidity, the resistance is jerky and characterized by a pattern of release/resistance in a quick jerky movement.
- This is often seen in Parkinson's disease.

Lead Pipe Rigidity

- Lead pipe rigidity is characterized by a uniform and continuous resistance to passive movement as the extremity is moved through its range of motion (in all planes).

Clonus

- Clonus is an uncontrolled oscillation of a muscle that occurs in a spastic muscle group (results from UMN lesions).

Theories of Spasticity (Etiology)

Hyperactive Reflex Arcs

- The reflex arc is firing without modification from the cortex.
- This occurs in an UMN SCI when the corticospinal tracts are lost.
- Below the lesion level, the corticospinal tracts are lost; however, the reflex arc remains intact and fires without modification from the higher centers.
- This results in spasticity.

Reduced Reciprocal Inhibition of the Antagonist and Synergy Muscles

- When lesions cause a reduction in the ability of the alpha MNs to inhibit the activity of antagonist and synergy muscle groups, the agonist fires without modification.
- This can result in spasticity.

Loss of the Cortical Modification of the Alpha Motor Neurons in the Ventral Horn

- When lesions to the brainstem and subcortical motor center occur, cortical inhibition of these motor centers is lost.
- This causes the alpha and gamma MNs to fire without modification from the cortex.
- This can result in increased tone.

Damage to the Primary Motor Area

- When damage to the primary motor area occurs, the corticospinal tracts fire without cortical modification, causing spasticity.

Damage to the Brainstem Regions that Contain the Supraspinal Motor Centers

- The supraspinal motor centers include the vestibular nuclei, reticular nuclei, and pontine nuclei.
- When these are damaged, severe spasticity can occur in the form of extensor tone.

Therapeutic Techniques to Influence Tone

Sustained Stretch on the Agonist

- Any sustained stretch on an agonist (of a spastic contracting muscle group) will activate the GTOs, located in the spastic agonist.
- The GTOs serve to inhibit the spasticity of the agonist and facilitate the antagonist muscle group.
- Example: If the biceps are spastic, place the elbow joint on a sustained stretch (in extension). This will activate the GTOs; the GTOs will inhibit the spastic contracting agonist and facilitate the antagonist muscle groups.

Quick Stretch on the Agonist

- A quick stretch on the agonist activates the MS.
- The MS sends messages to the alpha MNs in the ventral horn to continue innervating the agonist.
- Example: In a floppy baby, quick stretches on the biceps may increase tone in the biceps for functional upper extremity use.

Placing Pressure on the Tendon of the Agonist

- Placing pressure on the tendon of the agonist of a spastic muscle group will activate the GTOs.
- The GTOs will inhibit the spastic contracting agonist and facilitate the antagonist.
- Example: If the biceps are spastic, place sustained pressure on the tendons of the biceps to activate the GTOs. This will relax the spasticity in the biceps and facilitate activity in the triceps.
- *Never* place pressure on the *muscle belly* of the contracting spastic muscle. This will activate the MS and continue to increase spasticity in the spastic muscle.

Splinting

- Splinting works on the premise of positioning the spastic muscle group on a sustained stretch to reduce tone in the agonist.
- Example: Spasticity in the wrist flexors
- Use a resting pan splint to place the wrist flexors on a sustained stretch in extension.
- The GTOs will become activated and will serve to reduce the spasticity in the agonist and facilitate activity in the antagonist (the wrist extensors).
- It is very important to make sure that the splint is well-fitting or the cutaneous receptors will activate the gamma MNs. The gamma MNs will send signals to the MS to fire, thus facilitating spasticity in the agonist (the wrist flexors).

Serial Casting

- Serial casting involves placing the spastic muscle groups around a joint in a cast to increase range of motion (ROM) over time.
- Example: Elbow flexors are spastic.

- In the initial cast, the therapist pulls the elbow flexors into slightly greater extension and places a cast on the elbow joint in this position.
- Essentially, the cast places the elbow flexors on a prolonged sustained stretch.
- The GTOs become activated and reduce the spasticity in the elbow flexors.
- Every few days or once a week, the therapist places a new cast on the spastic joint.
- With each successive cast, the elbow joint is positioned in greater extension to gradually increase ROM.
- A common problem with serial casting is skin breakdown. If skin breakdown occurs, the cast must be removed and the skin must heal until any further casting can be done. Unfortunately, while the skin is healing, the patient's spastic muscle group will often resume its initial degree of spasticity. Any ROM that has been gained as a result of serial casting is often lost.

Reducing Clonus Through Sustained Stretching

- Clonus is an uncontrolled oscillation of a muscle that occurs in a spastic muscle group.
- Clonus can frequently be observed in the ankle joint of individuals with quadriplegia.
- When the therapist places the patient's foot on the wheelchair footplate, the therapist is essentially giving a quick stretch to the plantar flexors.
- This quick stretch activates the MS and causes the plantar flexors to contract uncontrollably.
- To reduce clonus:
  - Position the plantar flexors on a sustained stretch; in other words, pull the foot into dorsiflexion on a sustained stretch.
  - This will activate the GTOs. The spastic muscle group (the plantar flexors) will relax.

Modified Ashworth Scale: Used to Evaluate Tone

0 No increase in muscle tone

1 Slight increase in tone manifested by a catch and release or by minimal resistance at the end of the range of motion (ROM) when the affected joint is flexed or extended

–1 Slight increase in muscle tone manifested by a catch followed by minimal resistance throughout the remainder (less than half) of the ROM.

2 More marked increase in muscle tone through most of the ROM, but the affected joint is easily moved

3 Considerable increase in muscle tone, making passive movement difficult

4 Affected joint is rigid in flexion or extension

Synergy Patterns of the Upper and Lower Extremities

- A synergy pattern is a stereotyped set of movements that occur in response to a stimulus or voluntary movement.
- Synergy patterns involve pathology of muscle tone that affects joint position after neurologic damage—such as traumatic brain injury (TBI) and cerebrovascular accident (CVA).
- Synergies are described as patterns because the involved joint positions occur consistently as a result of specific neurologic damage.
- Specific flexor and extensor synergies can be observed in the upper and lower extremities.
- Synergy patterns can change as the patient experiences stages of recovery, or they may continue if recovery of damaged brain structures cannot occur.

Joint	*Flexor Synergy Pattern*	*Extensor Synergy Pattern*
Scapula	Elevation and retraction	Protraction and depression
Shoulder	Abduction and external rotation	Horizontal adduction and internal rotation
Elbow	Flexion	Extension and pronation
Forearm	Supination	Pronation
Wrist	Flexion and ulnar deviation	Extension
Fingers	Flexion and adduction	Flexion and adduction
Thumb	Flexion and adduction	Flexion and adduction
Hip	Flexion, abduction, and external rotation	Extension, adduction, and internal rotation
Knee	Flexion	Extension
Ankle	Dorsiflexion and inversion	Plantar flexion and inversion
Toes	Dorsiflexion	Plantar flexion

Associated Reactions

- Associated reactions are stereotyped movements in which effortful use of one extremity influences the posture and tone of another extremity (usually the opposite extremity). In other words, voluntary movements of one extremity produce unintentional movements in another extremity.
- Associated reactions can occur in normal movement as a result of reflex stimulation. For example, yawning, sneezing, coughing, and stretching all involve unintentional movements.
- Associated reactions can also occur as a result of pathology. Example: A patient several days post-stroke has a right upper extremity flexor synergy pattern. When she uses her uninvolved left limb to brush her hair, her affected right upper extremity becomes more flexed and abducted.
- Associated reactions can also occur during voluntary strenuous movement. Example: A patient with a right upper extremity flexor synergy pattern is attempting to ambulate. As he ambulates, a marked increase in spasticity can be observed in his right upper extremity.
- Associated reactions result from an overflow of activity into the opposite limb. This occurs because of an inability to selectively inhibit the interneurons that synapse with the motor cell bodies of the opposite limb.

SECTION 22

Motor Functions and Dysfunctions of the Central Nervous System: Cortex, Basal Ganglia, Cerebellum

CEREBRAL CORTEX

Cortical Mapping of the Brain

- The first neuroanatomist to attempt brain mapping was Brodmann. He numbered each area of the cortex.
- In 1909, he mapped 52 brain areas.
- Brodmann believed that each numbered area corresponded to a precise brain function.
- The boundaries between each area were often imprecise.
- Today, we know that the correlation between Brodmann's areas and a specific function is not precise.

Positron Emission Tomography and Magnetic Resonance Imaging Scans

- Today positron emission tomography (PET) and magnetic resonance imaging (MRI) scans are used to more precisely map brain areas.

Functional Divisions of the Cerebrum

Archicortex

- The archicortex is the core of the brain. It is comprised of the hippocampal formation.
- The hippocampal formation is located deep in the brain and is involved in learning and memory.
- The archicortex is believed to be phylogenetically ancient.

Paleocortex

- The paleocortex is the outer layer that sits over the core. It includes the parahippocampal gyrus.
- The parahippocampal gyrus is a region of the limbic system that is adjacent to the hippocampus.
- The parahippocampal gyrus relays information between the hippocampus and other brain regions.

Neocortex

- The neocortex is considered to be the newest part of the brain phylogenetically.
- It is comprised of the most superficial layers of the brain.
- It includes the primary motor cortex, the primary sensory cortex, and the association cortices.
- The neocortex is found only in mammals. It makes up 50% to 80% of the brain's total volume.

THE CEREBRUM'S ROLE IN MOTOR CONTROL

Primary Motor Area

- The primary motor area (M1) is the *precentral gyrus*. This is where voluntary/conscious movement is initiated.
- The corticospinal tracts originate here.
- It is also the site of the *motor homunculus*.

Motor Homunculus

- The motor homunculus is the map that denotes each body part's cortical representation for voluntary movement.
- The face and mouth have a large amount of cortical representation. This is for the purpose of speech and eating.
- The hands also have a large cortical representation. This is for fine motor control necessary to explore the environment.
- The cortical representation for the lower extremities is located in the medial longitudinal fissure.

Organization of the Motor Homunculus

- The organization of the motor homunculus is *not permanent.*
- One needs to use each body part or the cortical representation for that body part disappears—as with amputees over time.

> **Motto of the Brain: "If you don't use it, you'll lose it."**
>
> - Cortical representations of each body part on the sensory and motor homunculi are use dependent.
> - In other words, each body part must be consistently used, or it will lose its place on the topographical map of the homunculi.
> - The cortical representation for a specific body region can be rapidly appropriated by other body regions if that specific body part loses its function due to pathology (eg, amputation, peripheral nerve injury, orthopedic injury causing prolonged inactivity of a limb).
> - Once a cortical area representing a specific body region is gone from the topographical map, the person must recreate it through new experiences. For example, if a patient has lost hand function from a peripheral nerve injury lasting several months, the patient must recreate the hand area on the cortical map through use of the hand as the muscles become reinnervated.

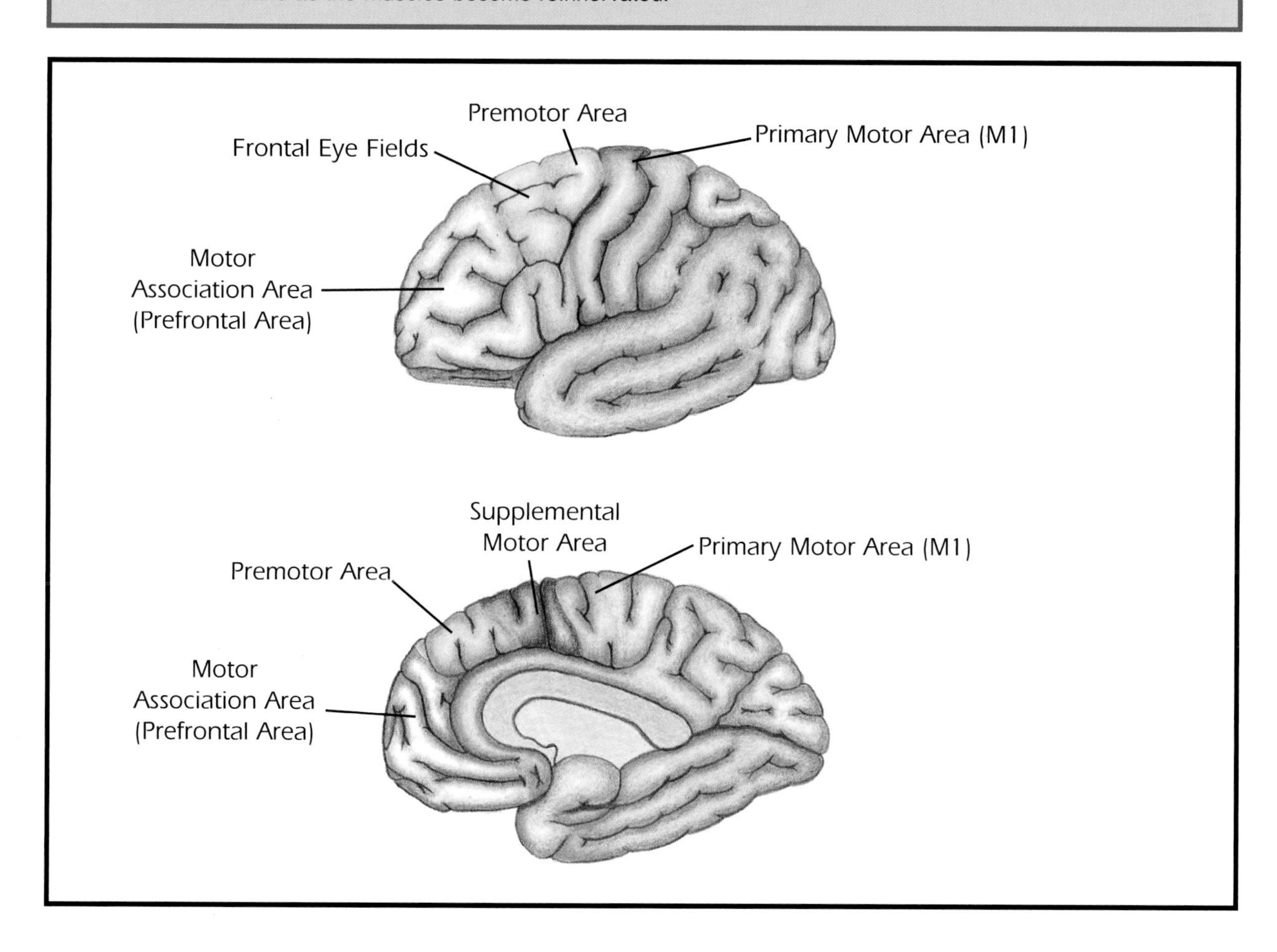

Lesion to the Primary Motor Area

- A lesion to M1 causes a loss of voluntary movement of the contralateral body part.
- The loss of voluntary movement in a body part corresponds to the area of the motor homunculus that was damaged.
- A lesion to M1 may also result in a loss of the ability to implement a specific motor plan.

Premotor Area

- The premotor area is located just anterior to M1.
- The premotor area has a role in *motor planning*, or *praxis*.

Lesion to the Premotor Area

- This causes *apraxia*, or motor planning difficulties.
- Apraxia involves either the inability to understand the demands of the task or the inability to access the appropriate motor plan.

Motor Association Area

- Also called the *prefrontal area*
- Located in the anterior frontal lobe
- The motor association area has a role in the *cognitive planning of movement.*

Lesion to the Motor Association Area

- Causes the storage of specific motor plans to be lost
- This involves an inability to cognitively understand how to carry out a motor task that was once known before injury/disease.
- The premotor area may be able to take over and compensate for damage to the motor association area.

Supplementary Motor Area

- The supplementary motor area is part of the premotor area.
- Located inside the medial longitudinal fissure
- Has a role in the *bilateral control of posture*

Lesion to the Supplementary Motor Area

- May result in loss of bilateral control of posture
- Other areas may take over and compensate for this loss—for example, the cerebellum or vestibular system.

Frontal Eye Field

- Located in the middle frontal gyrus, anterior to the premotor area
- Responsible for *visual saccades*

Lesion to the Frontal Eye Field

- Results in *deviation of the eyes* to the same side of the lesion

Praxis Involves Three Processes

Ideational Praxis

- Ideational praxis is the ability to cognitively understand the motor demands of the task.
- For example, when shown a shirt, the patient must be able to understand that this is an article of clothing that is worn on the trunk and upper extremities. The patient must also understand the motor plans needed to don a shirt.
- It is largely a function of the motor association area.

Ideomotor Planning I

- This is the ability to access the appropriate motor plan. For example, this would involve the ability to sort through all stored motor plans and identify the specific one for shirt donning.
- Such motor plans are commonly stored in the premotor area.

Ideomotor Planning II

- This is the ability to implement the appropriate motor plan—or put it into action. For example, after identifying the appropriate motor plan for donning a shirt, the patient would have to put that plan into action.
- Implementing motor plans commonly involves M1.

Pathology can occur in any of these three stages of motor planning.

Emergence of Primitive Reflexes as a Result of Neurologic Damage

- When serious neurologic damage occurs to the cortex, internal capsule, diencephalon, brainstem, and/or basal ganglia—often as a result of stroke or traumatic brain injury—it is common for primitive reflexes to reemerge.
- Primitive reflexes develop during gestation and infancy and become integrated by the central nervous system (CNS) in the first months or years of life.
- These reflexes facilitate normal movement.
- The reemergence of these reflexes in an adult with neurologic damage indicates severe CNS pathology. Volitional movement is compromised by the presence of such reflexes.
- There are two primary types of primitive reflexes:
  - Spinal Level (or elemental) Reflexes
  - Brainstem Level Reflexes

Spinal Level Reflexes

Flexor Withdrawal	When the sole of the foot is stimulated, the toes extend, the foot dorsiflexes, and the leg flexes. In an adult with neurologic damage, this often interferes with attempts to stand.	Develops at 28 weeks of gestation Integrated at 1 to 2 months
Crossed Extension	When an extended leg is passively flexed, the opposite leg extends. In an adult with neurologic damage, this can interfere with attempts to assume a seated position—such as returning to a seated position in a wheelchair. It can also interfere with transfer training.	Develops at 28 weeks of gestation Integrated at 1 to 2 months
Extensor Thrust	When the ball of the foot of a flexed leg is stimulated, that leg extends. In an adult with neurologic damage, this can interfere with attempts to assume a seated position—such as returning to a seated position in a wheelchair. It can also interfere with transfer training.	Develops between birth and 2 months Integrated at 1 to 2 months

(Continued)

Emergence of Primitive Reflexes as a Result of Neurologic Damage

(Continued)

Brainstem Level Reflexes

Asymmetrical Tonic Neck (ATNR)	When the head is rotated to one side, the opposite shoulder and elbow flex in a bow and arrow position. In an adult with neurologic damage, this interferes with volitional use of the upper extremities.	Develops at birth Integrated at 4 to 6 months
Symmetrical Tonic Neck (STNR)	When the head flexes, the arms flex and the legs extend. When the head extends, the arms extend and the legs flex. In an adult with neurologic damage, this interferes with volitional movement of the upper and lower extremities.	Develops at 4 to 6 months Integrated at 8 to 12 months
Symmetrical Tonic Labyrinthine (STLR)	When placed in a prone position, the patient's arms and legs flex. When placed in a supine position, the arms and legs extend. In an adult with neurologic damage, this interferes with learning to transfer from bed to wheelchair.	Develops at birth Integrated at 6 months
Positive Support Reaction	When the ball of the foot makes contact with the floor (in a standing position), the legs experience rigid extensor tone. In an adult with neurologic damage, this interferes with relearning to walk.	Develops at birth Integrated at 6 months
Associated Reactions	In response to effortful, voluntary movement (in any limb), the patient experiences involuntary movement usually in the opposite limb. In an adult with neurologic damage, this interferes with volitional movement.	Develops between birth and 3 months Integrated at 8 to 9 years

Babinski Sign

- The Babinski sign is a superficial cutaneous reflex that is elicited by stroking the outer border of the plantar surface of the foot.
- The reflex is caused by a brief contraction of muscles that are innervated by the same spinal segments that respond to the cutaneous stimulation.
- A positive Babinski sign is indicated by extension of the first toe accompanied by fanning of the other toes.
- A positive Babinski sign in an adult with neurologic damage is always indicative of damage to the corticospinal tracts (or to CNS damage involving the corticospinal tracts—such as damage to the internal capsule, thalamus, and brainstem).

Hoffman's Sign

- Hoffman's sign is another superficial cutaneous reflex.
- This reflex can be elicited by flicking the nail of the patient's third finger.
- A positive Hoffman's sign is indicated by prompt adduction of the thumb and flexion of the index finger.
- The presence of Hoffman's sign in an adult with neurologic damage is indicative of damage to the corticospinal tracts (or to CNS damage involving the corticospinal tracts—such as damage to the internal capsule, thalamus, and brainstem).

Absence of Higher Level Reactions as a Result of Neurologic Damage

- Higher level reactions are reflexes that develop during infancy or early childhood and remain throughout the lifespan.
- These types of reactions are controlled by centers in the midbrain, basal ganglia, and cortex:
  - Righting Reactions
  - Equilibrium Reactions
  - Protective Extension
- Higher-level reactions are important for normal postural control and movement.
- When neurologic damage occurs in an adult, higher-level reactions typically disappear.
- Their absence leads to postural instability, balance problems, and an inability to protect oneself when falling.

Midbrain Level Reactions

Righting Reactions

Neck Righting on Body (NOB)	When the head is passively rotated to one side (in supine), the body also rotates as a whole (log rolls) to align with the head.	Develops at 4 to 6 months Integrated at 5 years
Body Righting on Body (BOB)	When the upper or lower trunk is passively rotated (in supine) as an isolated segment, the other segment also rotates to become aligned.	Develops at 4 to 6 months Integrated at 5 years
Labyrinthine Head Righting (LR)	When the body is tipped in different positions (with vision occluded), the head orients to a vertical position. When this reaction is lost, the ability to maintain the head in a vertical position for orientation is lost.	Develops between birth and 2 years Persists throughout life

Basal Ganglial Level Reactions

Protective Extension (PE)	When the patient's center of gravity is displaced, the arms and legs extend outward to protect the body when falling. Patients who lose protective extension are at an increased risk for injuring themselves when they lose their balance.	PE of the arms develops at 4 to 6 months PE of the legs develops at 6 to 9 months Persists throughout life
Equilibrium Reactions (ER) (Tilting)	When the center of gravity is displaced by moving the support surface, the trunk will curve and the extremities will extend and abduct. The maintenance of balance and postural control is severely impaired when this reaction is lost.	Develops between 6 and 21 months Persists throughout life
Equilibrium Reactions (ER) (Postural Fixation)	When the body is pushed, altering the center of gravity, the trunk will curve toward the external force. The extremities will extend and abduct. The maintenance of balance and postural control is severely impaired when this reaction is lost.	Develops between 6 and 21 months Persists throughout life

Cortical Level Reactions

Optic Righting (OR)	When the body is repositioned by tipping it in different directions, the head orients to a vertical position.	Develops between birth to 2 months Persists throughout life

Basal Ganglia's Role in Motor Control

Basal Ganglia

- The basal ganglia have a role in *stereotypic* and *automated movement patterns* (eg, walking, riding a bike, and writing).
- The basal ganglia structures include the *caudate*, the *putamen*, and the *globus pallidus*. The *subthalamic nucleus* of the diencephalon and the *substantia nigra* of the midbrain are also considered to be part of the basal ganglia.
- The axons of the substantia nigra form the *nigrostriatal pathway*, which supplies dopamine to the striatum. Dopamine is produced by the substantia nigra.
- While the basal ganglia have traditionally been thought of as a subcortical (and involuntary) motor system, recent research suggests that the basal ganglia may also function in *cognitive* and *affective processes requiring precision timing*—such as the ability to know when it is appropriate to contribute to social interaction or the ability to inhibit one's desire to act upon personal impulses.
- Children with *attention deficit hyperactivity disorder (ADHD)* who display difficulty taking turns, refraining from calling out in class, and controlling the urge to physically move may experience such difficulty, in part, because of pathology in the basal ganglia's ability to balance inhibitory and excitatory actions.
- With regard to motor performance, the basal ganglia achieve precision control as a result of a balance between their inhibitory and excitatory effects.
- The basal ganglia are also modulated by two neurotransmitters—dopamine and acetylcholine.
- Generally, *dopamine* stimulates the basal ganglia pathways and, as a result, exerts an excitatory effect upon the cerebral cortex.
- Conversely, dopamine depletion—such as that found in Parkinson's disease—causes a severe inhibitory effect upon movement.
- *Acetylcholine* works in opposition to dopamine; however, both work collaboratively to achieve a balanced inhibition and excitation of movement—which is necessary for precision and timed action.
- Motor disorders of the basal ganglia result in a disequilibrium between inhibitory and excitatory movement.

Divisions of the Basal Ganglia

Neostriatum

- This is considered to be the newest region of the basal ganglia phylogenetically.
- It includes the caudate and the putamen.
- The putamen is an excitatory structure.
- The caudate has a role in the inhibitory control of movement. The caudate acts like a brake on certain motor activities.
- When that brake is gone, hyperkinetic movement disorders can occur—like tics or Tourette's Syndrome.

Paleostriatum

- This is considered to be the older part of the basal ganglia phylogenetically.
- It includes the globus pallidus—an excitatory structure.

Corpus Striatum

- The corpus striatum is the collective name for all three structures: the caudate, putamen, and globus pallidus.

Collective Functions of the Basal Ganglia

Stereotypic Movement

- Stereotypic movements are motor patterns that are "hard-wired" (ie, do not have to be learned on a conscious level) and develop normally as the individual's nervous system matures (eg, stretching the arms during yawning, using a reciprocal arm swing when walking).

Automated Movements

- Automated movements are movement patterns that used to be mediated by conscious cortical control but have been assumed by the basal ganglia structures once learned.

BASIL GANGLIA AFFERENT PATHWAY

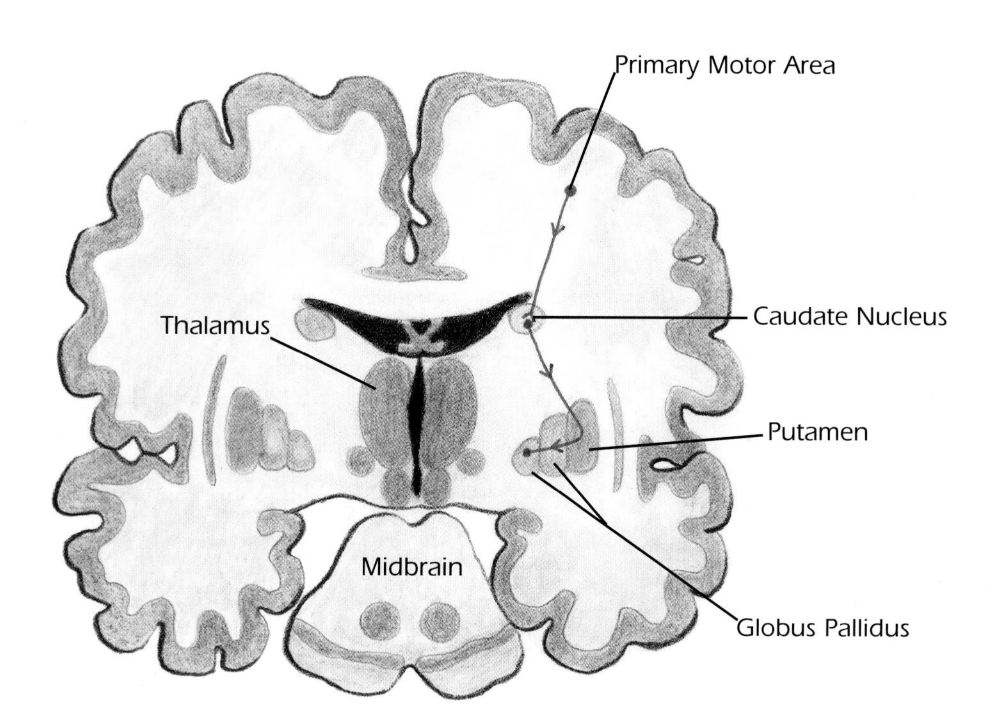

LENTICULAR FASCICULUS PATHWAY

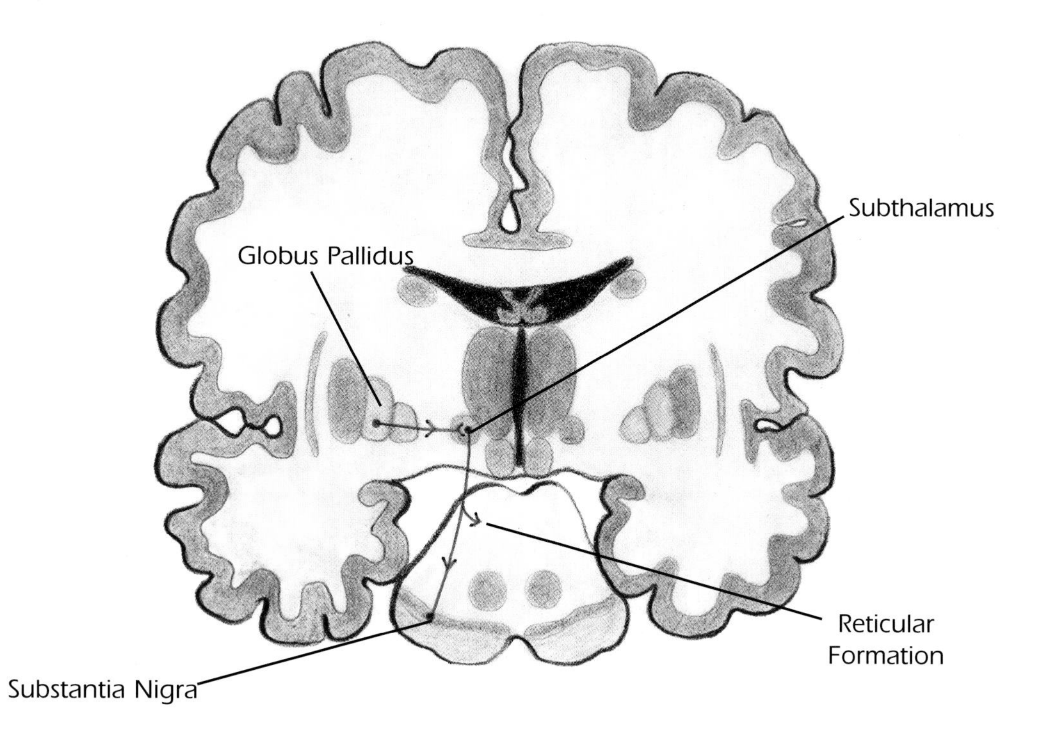

- This process underlies the saying, "Once you learn to ride a bike, you never forget," unless damage is sustained to the basal ganglia.
- Automated movements include driving, riding a bike, and writing.

The Basal Ganglia Also Influence Tone

- The basal ganglia send input to the alpha and gamma motor neurons (MNs) in the ventral horn to influence muscle tone.

Basal Ganglia Pathways

Afferent Basal Ganglia Pathways (Leading to the Basal Ganglia)

- The primary motor area and the premotor area send motor signals to the caudate, putamen, and globus pallidus.

Lenticular Formation (Traveling from the Basal Ganglia)

- This pathway travels from the globus pallidus to the subthalamus, to the reticular formation, and to the substantia nigra.

Ansa Lenticularis (Traveling from the Basal Ganglia)

- This pathway travels from the globus pallidus to the ventrolateral nucleus of the thalamus, to M1, and to the premotor area.

Nigrostriatal Pathway (Feedback Circuit of the Basal Ganglia)

- This pathway travels from the substantia nigra to the globus pallidus and subthalamus.
- This circuit modifies the ansa lenticularis pathway.
- In Parkinson's disease, the substantia nigra degenerates, and modification of the ansa lenticularis is lost.

Basal Ganglia Lesions

Symptoms

- Difficulty *initiating, continuing,* or *stopping movement*
- Problems with muscle tone (particularly rigidity)
- Increased involuntary, undesired movements (tremor, chorea)

Hemiballismus

- Hemiballismus is a lesion of the subthalamus and caudate causing a disinhibition of neuronal activity between the thalamus and the cortex.
- It results in *violent thrashing* of the *contralateral extremity*.
- Ballismus is the name of the disorder if the lesion is bilateral.

Athetosis

- Athetosis involves a *slow flailing* of the upper and lower extremities and *continuous movement*.
- Involves slow, aimless, purposeless movement—especially of the *distal musculature*
- Movements often have a *twisting* or turning quality to them.
- Such movements are caused by continuous and prolonged contraction of agonist and antagonist muscle groups.
- Caused by a lesion to the putamen and caudate
- Some cortical involvement
- Athetosis is a slower movement than chorea.

Chorea

- Choreiform means to dance. Chorea is *sudden, involuntary, jerky movements* that appear dancelike in quality. Movement may appear as more coordinated and graceful than in athetoid conditions.
- This usually involves axial and proximal limb areas—shoulder shrugs, moving hips, crossing and uncrossing the legs.
- Movements of the face may involve grimacing, eyebrow raising, eye rolling, and tongue protrusion.
- It is caused by a lesion to the caudate and putamen.

ANSA LENTICULARIS PATHWAY

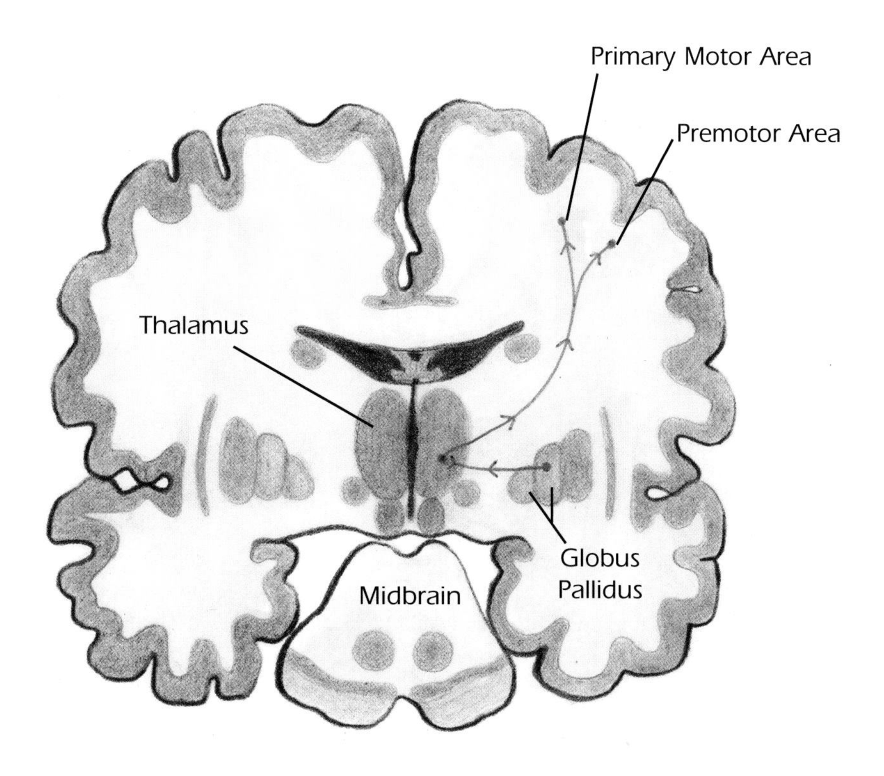

NIGROSTRIATAL PATHWAY

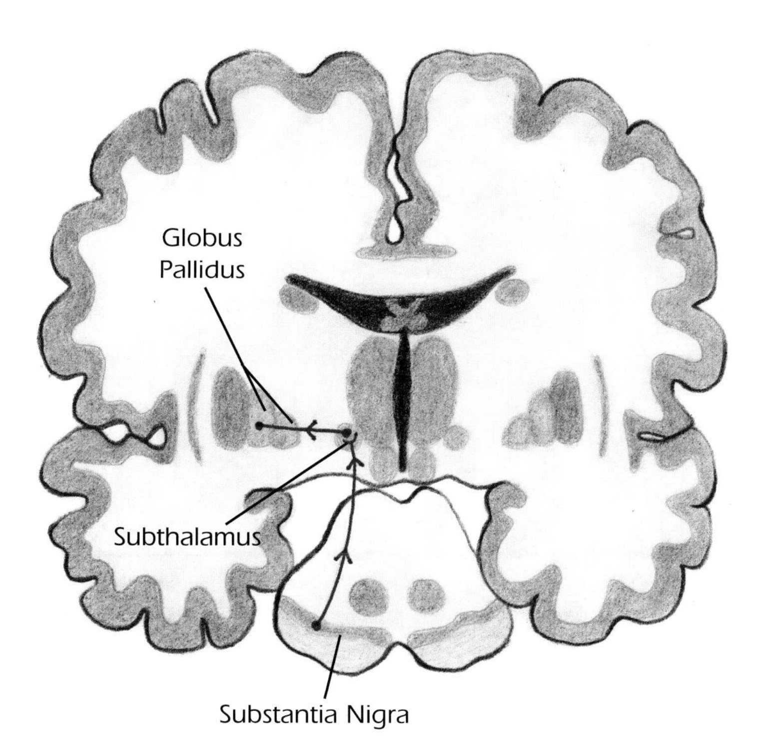

Huntington's Chorea

- Huntington's chorea is a *chronic degenerative disorder* that also involves a *progressive dementia*. The occurrence of dementia is likely related to the basal ganglia's role in cognition.
- It is an inherited disease that involves degeneration of the caudate nucleus and some areas of the cortex.
- Gamma-aminobutyric acid (GABA) levels are diminished in the striatum and globus pallidus.
- GABA is a neurotransmitter that runs in the nigrostriatal pathway (from the striatum to the globus pallidus).
- GABA in this pathway modulates the outflow of dopamine—which runs in the nigrostriatal pathway.
- Increased dopamine in the basal ganglia is believed to cause Huntington's disease.

Dystonia

- Dystonia involves muscle contractions producing twisting movements that are repetitive and often result in *abnormal postures*.
- The sustained muscle contractions of dystonia can last from seconds to hours.
- Such postures often result from simultaneous opposing movements that produce paralysis.
- Dystonic muscle contractions that occur or recur over a length of time can eventually cause joint degeneration and permanent fixed postures.
- Dystonia can occur in the limbs, face, neck, and trunk.
- Experimental research has shown some evidence for the effectiveness of deep brain stimulation in the treatment of dystonia. This procedure involves placing electrodes in the globus pallidus, which are stimulated by a battery surgically implanted in the abdomen. It appears that continuous stimulation of the globus pallidus can shut it down, thereby alleviating the major symptoms of dystonia.

Dyskinesia

- Dyskinesia involves *rhythmic, repetitive movements* that have an odd quality.
- It more frequently affects the face, mouth, jaw, and tongue (eg, grimacing, lip pursing, tongue protrusion, mouth opening and closing, jaw deviation) rather than the limbs.

Tardive Dyskinesia

- Tardive dyskinesia is a movement disorder related to treatment with dopamine receptor antagonists (neuroleptics and antiemetics).
- The term *tardive* refers to the fact that this movement disorder occurs after chronic use of these drugs.
- It is characterized by choreiform movements, dystonia, tics, and/or myoclonus.
- Example: tongue protrusions (orobuccolingual movements), chewing-type movements, facial grimacing, blepharospasm, lip smacking.
- Tardive dyskinesia is different from most disorders because the discontinuation of the causative agent (the neuroleptic) does not result in the amelioration of the movement disorder.

Pathology of Tardive Dyskinesia

- The chronic use of neuroleptics results in a chronic blockage of dopamine receptors in the basal ganglia.
- The sensitivity of the dopamine receptors becomes severely decreased.
- Also suspected is a decrease in the neurotransmitter GABA in the basal ganglia.

Parkinson's Disease

- This results from damage to the *nigrostriatal pathway* between the *substantia nigra* and the *basal ganglia. Dopamine* is the neurotransmitter that uses this pathway.
- Parkinson's is caused by cell death of the dopaminergic neurons of the substantia nigra—in other words, the substantia nigra degenerates.
- This leads to decreased levels of dopamine available for use by the basal ganglia.
- Depletion of dopamine then causes decreased modification of the *ansa lenticularis* pathway (the pathway that modifies the basal ganglia activity).
- Movement becomes disjointed and uncontrollable.

Glossary of Terms of Idiopathic (no known origin) Torsion Dystonia

Involving the Eye

- Blepharospasm: Eyes are involuntarily kept closed
- Oculogyric Crises: Attacks of forced deviation of gaze, often associated with a surge of Parkinsonism, catatonia, tics, and obsessiveness
- Opathalomoplegia: Paralysis of gaze

Involving the Throat, Jaw, Lips, or Tongue

- Oromandibular Dystonia: Involuntary opening and closing of the jaw, retraction or puckering of the lips
- Lingual Dystonia: Repetitive protrusion of the tongue or upward deflection of the tongue toward the hard palate
- Laryngeal Dystonia: Speech is tight, constricted, and forced. Smooth flow of speech is lost
- Pharyngeal Dystonia: Associated with dysphagia, dysphonia (hoarseness), and dysarthria (slurred words)

Involving the Neck

- Torticollis: Dystonic contractions of the neck muscles
- Retrocollis: Neck forced backward into hyperextension
- Anterocollis: Neck forced forward into hyperflexion
- Laterocollis: Neck forced laterally

Involving the Trunk

- Truncal Dystonia: Manifests as a lordosis, scoliosis, kyphosis, tortipelvis, opisthotonos (forced flexion of the head on the chest)

Involving the Legs

- Crural Dystonia: Dystonic movement of the legs

Dystonia of Single or Multiple Body Parts

- Focal Dystonia: Dystonia in a single body part
- Multifocal Dystonia: Dystonia of more than one body part
- Hemidystonia: Involvement of limbs on one side of the body (due to a space occupying lesion or infarction)
- Generalized Dystonia: Dystonia in a leg, the trunk, and one other body part; or both legs and the trunk

As with most hyperkinetic movement disorders, the abnormal movements usually disappear during sleep and worsen with anxiety, fatigue, temperature changes, and pain.

- As movement slows, mental processes also become delayed and halted. Problem solving drags, and language and communication become difficult.
- Half of all patients with Parkinson's also experience depression and anxiety disorders. This likely occurs because dopamine also regulates mood and affective processes.
- One-third of patients with Parkinson's develop dementia—again, this likely occurs because of the basal ganglia's collaborative role in cognition.

Parkinson's Symptoms

- Hypertonicity
- Cogwheel rigidity
- Bradykinesia (poverty of movement)
- Masklike face (facial muscles are affected by bradykinesia)
- Tremors at rest (pill-rolling)

- Cunctation-festinating gait—cunctation means to resist movement. Festination means to hurry. The individual experiences difficulty initiating and stopping movement.
- Loss of arm swing while walking
- Impaired postural reflexes (difficulty righting oneself and maintaining balance)
- Micrographia (handwriting becomes very small)

Pharmacologic Treatment for Parkinson's

L Dopa

- L dopa has been the most common treatment for Parkinson's.
- L dopa, or levadopa, is a precursor compound which the brain converts into dopamine.
- Unlike dopamine, however, L dopa is able to cross the blood brain barrier—the membrane that surrounds the brain and protects it from direct exposure to harmful substances.
- When L dopa is used and crosses the blood-brain barrier, it is synthesized into dopamine.
- With the increase in dopamine, the nigrostriatal pathway can run properly and modify the ansa lenticularis pathway.
- While the initial effects of L dopa are accompanied by a return of fluid movement, patients generally become less responsive to such drugs over time.

Dopamine Agonists

- Dopamine agonists—another class of drugs—imitate the function of dopamine. These include bromocriptine, cabergoline, pramipexole, and ropinirole.
- While initially such drugs are not as effective as L dopa, over time, their dosage in the body can be more accurately regulated.

Kinesia Paradoxa in Parkinson's

- Kinesia paradoxa is the sudden total conversion of Parkinsonism to normality or to hyperkinesia due to pharmacologic treatment.

Surgical Procedures for Parkinson's

- In the 1960s and 1970s, the only surgical option was *pallidotomy*—ablation of the globus pallidus.
- This was often a last resort treatment option and posed severe negative side effects.
- Today, *deep brain stimulation (DBS)* has become a more common procedure.
- DBS involves the implantation of electrodes into the basal ganglia or thalamic nuclei. A battery and controller implanted beneath the skin near the clavicle or abdomen send timed currents to the basal ganglia to improve motor performance.
- Patients often experience a significant reduction in Parkinsonian symptoms.
- DBS allows patients to decrease their drug intake and can provide improvement in motor function that can last for years.
- It does not, however, improve cognitive function or stop the progression of the disease.
- Undesirable side effects can include deafness and speech disorders.

Stem Cell Research

- Stem cells are structurally fundamental cells that are able to mature into any type of cell.
- They can be found in embryonic as well as some adult tissue—for example, the subventricular midbrain and the hippocampus.
- Researchers are attempting to transform stem cells into dopamine-producing neurons via natural growth factors.
- Such cells would then be transplanted into patients in an effort to reverse Parkinsonian symptoms.

Genetic Research

- Several researchers are involved in gene-based therapy in which modified viruses are used to carry genes into the midbrain.
- The genes would then activate specific enzymes that could release or transport dopamine.
- While initial animal experiments have produced promising results, greater research is needed to better understand the benefits and risks of gene-based treatment.

Tremor

- Tremor is involuntary oscillating movement resulting from alternating or synchronistic contractions of opposing muscles.
- There are different types of tremor.

Essential Tremor

- Most common form of tremor
- Characterized by tremor upon movement (not a resting tremor as in Parkinson's)
- Fingers and hands are affected first.
- Tremor moves proximally to the head and neck
- Slow progressive disorder

Tics

- Tics are repetitive, brief, rapid, involuntary, and purposeless movements involving single or groups of muscles.
- Tics can also be *fragments of movements or thoughts* that are split off from more integrated behavior.
- People with tics describe an inner tension that builds. This inner tension can be relieved by the ticcing behavior.
- Tics have a *conscious component*. The conscious experience of feeling compelled to make sound may imply that tics are a form of compulsion.
- Most people can *suppress their tics* for a brief period of time; however, suppression is often associated with a buildup of inner tension that causes the subsequent expression of tics to be more forceful than they would otherwise be.
- A tic can involve a *brief isolated movement* like eye blinks, head jerks, or shoulder shrugs. Tics involving movement are referred to as motor tics.
- A tic can also be a *variety of sounds*: throat clearing, grunting. Such tics are referred to as *phonic tics*.
- Tics can additionally be associated with *intrusive thoughts* that the individual feels compelled to express.
- Tics can be meaningful utterances:
    - *Echolalia* is an involuntary repetition of words just spoken by another person.
    - *Palilalia* is an involuntary repetition of words or sentences.
    - *Coprolalia* is an involuntary utterance of curse words.
- Tics frequently have a *waxing* and *waning course*. A specific tic can be present for weeks or even years. Other tics emerge and disappear with no predictable course.
- Tics will commonly increase during periods of stress and decrease during tasks requiring heightened concentration.
- While the long-term course of a tic disorder can be variable, most studies suggest that tics improve in late adolescence and early adulthood.
- Clinicians characterize tics by their anatomical location, number, frequency, intensity, and complexity.

Etiology of Tics

- The *dopamine system* is a form of modifying and modulating movement.
- Increased sensitivity to dopamine in the basal ganglia causes severe tics.
- People with a normal response to dopamine have mild tics—this explains why normal people sometimes have tics or mild obsessional thoughts, especially with stress.
- With increased sensitivity to dopamine, the caudate (which normally acts like a brake on certain extraneous movements) cannot suppress movements like tics.
- *Pediatric autoimmune neuropsychiatric disorders associated with streptococcal infections* (PANDAS) have been suggested as one possible cause for the childhood onset of tic disorders—in at least some percentage of children diagnosed with tics. In this subset of tic disorder, onset is abrupt, follows recovery from a streptococcal infection, occurs before puberty, and occurs in children with obsessive compulsive disorder (OCD) and motor hyperactivity.

Treatment

- Non-neuroleptic drugs (eg, clonidine, guanfacine, baclofen, and clonazepam) are indicated for the suppression of mild tics.
- Neuroleptic drugs, dopamine receptor antagonists, (eg, risperidone, olanzapine, ziprasidone, and quetiapine) are indicated for severe tic disorders.

Tourette's Syndrome

- Tourette's syndrome is a type of movement disorder involving motor and vocal tics that begin during childhood; persist for more than 1 year; and fluctuate with regard to type, frequency, and anatomical distribution over time.
- Symptoms of the disorder commonly first appear in childhood and adolescence. Approximately 10% of children have some form of tic disorder; 1% is diagnosed with Tourette's syndrome. Most tics fade by age 18; however, even when tics persist into adulthood, they tend to become less severe.
- The disorder is three times more common in boys than in girls.
- It also has a strong genetic component.
- Seventy percent of people with Tourette's syndrome also have obsessive–compulsive disorder (OCD).
- OCD may represent an alternative expression of Tourette's syndrome. Tourette's has been suggested as part of the obsessive–compulsive spectrum—a range of compulsions with simple tics at one end and complex rituals and obsessional thoughts at the other.
- The cause of Tourette's and OCD likely involves a dopamine and basal ganglia disorder. The orbitofrontal cortex—a center of judgment and decision making—has also been implicated, particularly in obsessions of thought.
- Sixty percent of people with Tourette's have attention deficit hyperactivity disorder (ADHD) involving short attention span, physical and mental restlessness, poor concentration, and diminished impulse control.
- A high percentage of people with Tourette's have concomitant learning disabilities, aggressiveness, anxiety, panic disorder, depression, mania, conduct disorder, oppositional defiant disorder, phobias, dyslexia, and stuttering. These disorders have been found to be 5 to 20 times more common in people having Tourette's than in the general population.
- Tourette's also involves a defect in motor pattern generators. Motor pattern generators are found in the brainstem and cortex and produce a variety of hard-wired movements (eg, reaching, grasping, walking, standing upright).
- These movement patterns are involved in all body postures and are present at birth; they are ready to mature as an infant begins to move.
- Increased sensitivity to dopamine causes faulty inhibition of motor pattern generators.
- The same drugs used to treat tic disorders (non-neuroleptic and neuroleptic drugs—as noted above) are used to treat Tourette's syndrome.

THE CEREBELLUM'S ROLE IN MOTOR CONTROL

Cerebellum

- The cerebellum has a role in the *coordination of movement*, the *maintenance of posture*, and *equilibrium.*
- While its major role is *proprioception*, the cerebellum has more recently been implicated in other functions—*attention shifting, practice-related learning, spatial organization*, and *memory*. The cerebellum also appears to work collaboratively with *cortically-based cognitive functions* to predict and prepare functional responses to environmental demands.
- With regard to motor performance, the cerebellum could be called an "error-correcting device" for the motor system. It receives proprioceptive information from the body and sends back information to modify muscle and joint activity for the achievement of precision motor control.

Three Cerebellar Lobes

Archicerebellum (Flocculonodular Lobe)

- Receives input from the vestibular nuclei through the inferior cerebellar peduncle

Paleocerebellum (Anterior Lobe)

- The paleocerebellum receives the posterior and anterior spinocerebellar tracts.
- It influences muscle tone by sending efferent fibers to the vestibular nuclei and the reticular formation through the superior cerebellar peduncle.

Neocerebellum (Posterior Lobe)

- The neocerebellum receives information from the cerebral hemispheres.
- Information from the cortex descends to the pontine nuclei and then travels to the cerebellum through the middle cerebellar peduncle.

CEREBELLAR LESIONS

Neocerebellar Lesions

- Neocerebellar lesions present with the following:
  - *Ipsilateral ataxia*—the posterior lobe receives input from the cortex. The dorsal columns mediate conscious proprioception. Loss of this cortical input results in ataxia.
  - *Ipsilateral hypotonia and hyporeflexia*
  - *Dysmetria*—inability to judge distance. Past-pointing, over-shooting, or underestimating one's grasp of objects occurs. People with dysmetria use visual cues for the readjustment of imprecise movement.
    - *Hypermetria* involves an overestimation of the distance needed to reach an object (overshooting).
    - *Hypometria* involves an underestimation (undershooting).
  - *Adiadochokinesia*—inability to perform rapid alternating movements. *Dysdiadochokinesia* involves an impaired ability. Rapid alternating movements involve the ability to switch back and forth between two joint positions—such as forearm supination and pronation. Drifting and lag commonly appear as the patient attempts to increase the speed of movement.
  - *Movement decomposition*—movement is performed in a sequence of isolated parts rather than as a smooth, singular motion. Also referred to as *dyssynergia*
  - *Asthenia*—Muscle weakness
  - *Intention tremors*—these occur during movement, as opposed to resting tremors—which occur in Parkinson's disease.
  - *Rebound phenomenon*—inability to regulate reciprocal movements. Example: The therapist gives resistance to the patient's flexed arm. When the therapist releases his or her resistance, the patient has no control or regulation over his or her speed of movement, and the patient's hand hits his or her chest. This occurs because immediate proprioceptive feedback is lost—the individual cannot regulate the speed of his or her arm movement quickly enough to prevent it from hitting his or her body.
  - *Ataxic gait*—the posterior lobe receives input from the cortex. The dorsal columns mediate conscious proprioception. Loss of this cortical input results in an ataxic gait.
  - *Staccato voice*—broken speech. The modulation of speech is a proprioceptive function.

Paleocerebellar Lesions

- Lesions in the paleocerebellar, or anterior, lobe produce severe *disturbances in extensor tone*.
- This is because this lobe receives the spinocerebellar tracts.
- When the spinocerebellar tracts are lost, an increase in extensor tone results.

Archicerebellar Lesions

- Lesions of the archicerebellar, or flocculonodular, lobe result in *uncoordinated trunk movements—ataxia*.
- The flocculonodular lobe receives input from the vestibular nuclei, the cuneocerebellar tract, and the rostral cerebellar tract.
- Loss of this vestibular input results in *balance deficits*—particularly in the trunk and upper extremities.
- *Nystagmus* also occurs because the cerebellum has connections to the oculomotor system via the medial longitudinal fasciculus.

SECTION 23

Sensory Functions and Dysfunctions of the Central Nervous System

SOMATOSENSORY CORTEX

Postcentral Gyrus (also called Primary Somatosensory Area)

- The primary somatosensory area (SS1) is responsible for the *detection* of *incoming sensory information* from the periphery.
- It is not responsible for the *interpretation* of sensory data.
- All sensory data go through the *thalamus* before reaching the *postcentral gyrus*.
- SS1 is also the site of the *sensory homunculus*.

Sensory Homunculus

- The sensory homunculus located in the *postcentral gyrus*.
- The sensory homunculus is the *cortical representation* of every body part's sensation.
- It is the somatotopic organization of body sensation from the *contralateral* side of the body.
- It is akin to the motor homunculus in the precentral gyrus.
- *Face*, *hands*, and *mouth* have a lot of representation for exploration of the external world.
- Like the motor homunculus, the sensory homunculus does *not* have a permanent organization.
- There is a plasticity to the homunculi should injury or disease occur.
- In people who read braille, the reading finger has a very large representation in the primary somatosensory cortex.
- In amputees, other body parts appropriate—or take over—the cortical representation or region that had been used for the amputated part.

Lesions to the Primary Somatosensory Area

- These result in a *loss of sensation* of the *contralateral body part*.
- The loss will depend on which part of the sensory homunculus was damaged.

Secondary Somatosensory Area

- The secondary somatosensory area (SS2) is responsible for the *interpretation* of sensory information.
- This is where meaning is attached to incoming sensory data.
- This area is located just posterior to the postcentral gyrus.

Lesions to the Secondary Somatosensory Area

Tactile Agnosia

- The umbrella term for the *inability to attach meaning to sensory data*
- Tactile agnosias include astereognosis, two-point discrimination disorder, agraphesthesia, and extinction of simultaneous stimulation (all described in the following sections).

Two Classifications of Sensation

Primary Sensation

- Primary sensation includes *pain and temperature*, *light touch*, *pressure*, *vibration*, and *proprioception*.

Oliver Sacks Story

- Sacks describes a 60-year-old woman who was blind at birth. She was overprotected and never encouraged to explore her environment with her hands.
- As an adult, she couldn't identify objects or sensations with her hands despite having intact sensory anatomy.
- Her sensory cortices were never stimulated.
- At 60, she had to recreate the region in the cortical map for her hands by learning to use her hands to identify objects/sensations.
- This is an example of how the cortex is use-dependent.

Adapted from: Sacks, O. *The Man Who Mistook His Wife for a Hat.* New York: Harper Perennial; 1987.

Impermanence of the Cortically Based Sensory and Motor Functions

- The cortical representation for all sensory and motor functions is impermanent and use-dependent.
- This impermanence and use-dependence is true not only for the homunculi but also for all areas of the cortex.
- For example, in people who become cortically deaf, regions of the auditory cortex may be reallocated for visual use.
- In people who become cortically blind, regions of the visual cortex may be reallocated for tactile and auditory use.

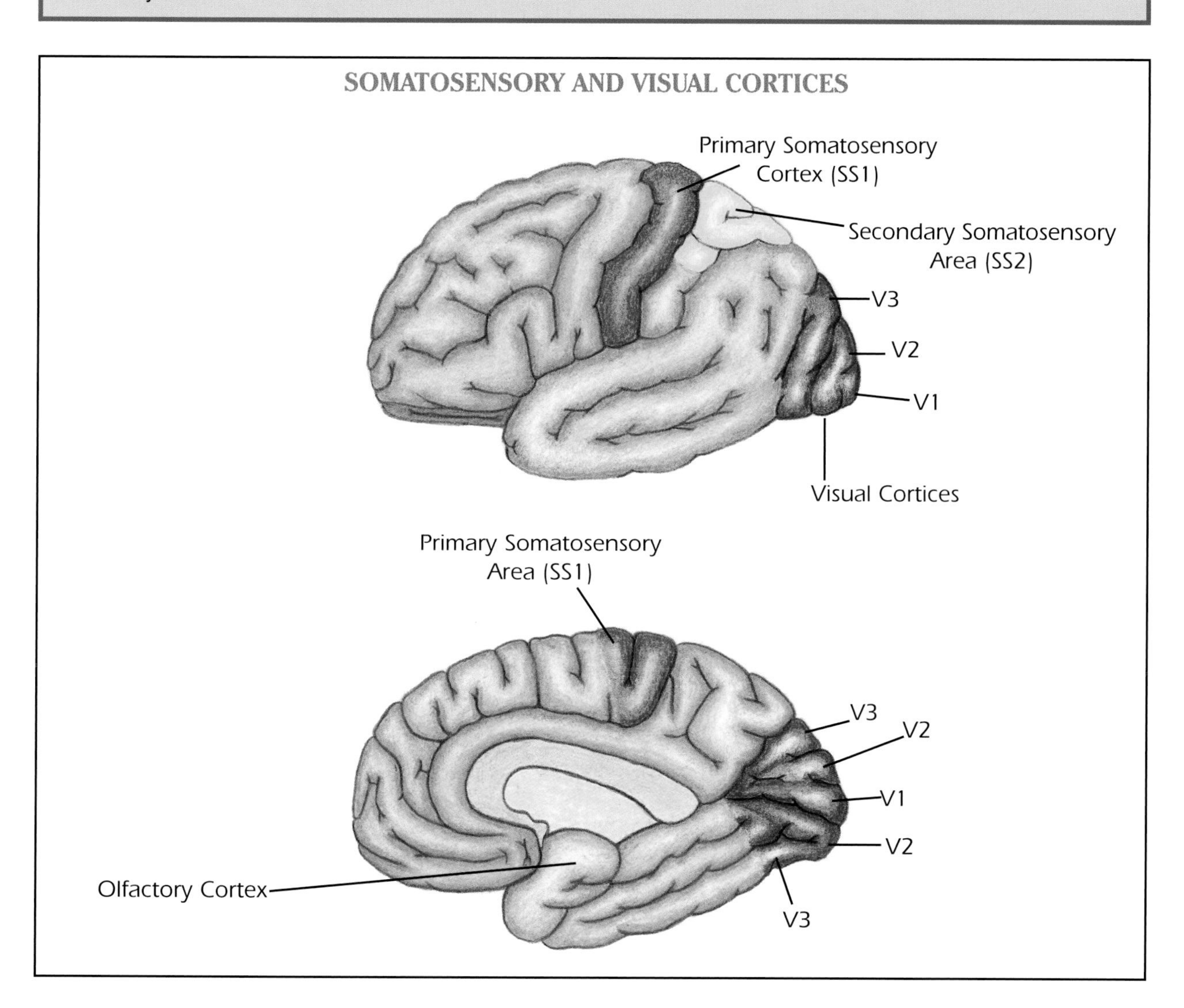

- Primary sensation is mediated by the *skin receptors*, *spinal nerves*, *spinal tracts* (dorsal columns and spinothalamic tracts), and *SS1*.

Cortical Sensation

- Cortical sensations are mediated by *SS2* and the *posterior multimodal association cortex*.
- Cortical sensations include the following *tactile agnosias*.

Astereognosis

- The inability to identify objects by touch alone (sensory anatomy is intact; cortical interpretation is impaired)

Two-Point Discrimination

- Loss of the ability to determine whether one has been touched by one or two points (sensory anatomy is intact; the cortical interpretation of sensation is impaired)
- Two-point discrimination is evaluated by a standard instrument called an *aesthesiometer*.

Agraphesthesia

- Loss of the ability to interpret letters written on the contralateral hand of the affected side—despite intact sensory anatomy. Cortical interpretation of sensation is impaired.

Extinction of Simultaneous Stimulation

- Also called *double simultaneous extinction*
- The patient is touched simultaneously on two different body regions—one on the involved side; one on the uninvolved side.
- Extinction occurs when the patient cannot feel the tactile sensation on the involved side (even though he or she could if the tactile sensation was applied *only* to the involved side).
- Extinction of simultaneous stimulation occurs because the neurons that carry tactile sensation from the involved side are cortically overridden by the neurons carrying sensation from the uninvolved side.

Abarognosis

- Loss of the ability to accurately estimate the weight of objects in relation to each other (sensory anatomy is intact; the cortical interpretation of sensation is impaired)

Atopognosia

- Loss of the ability to localize the exact origin of a sensation (sensory anatomy remains intact; cortical interpretation is impaired)

VISUAL CORTEX

Primary Visual Cortex

- Receives visual information from the *optic tracts*, *lateral geniculate nuclei* of the *thalamus*, and the *superior colliculi* of the *midbrain*
- Responsible for the *detection* of visual input—*not* interpretation
- Located at the most posterior region of the *occipital lobe*

Lesions to the Primary Visual Cortex

- Loss of sight (despite intact visual anatomy)

Lesions in the Primary Visual Cortex Produce Cortical Blindness

- When the primary visual cortex (V1) is lesioned, visual information travels through the *retina*, *optic nerves*, *optic tracts*, and to the *lateral geniculate nucleus* of the *thalamus*. Here the visual information may be rerouted to the *visual association areas* rather than first proceeding to V1.

Sensory Signs and Symptoms	*Suspect*
Astereognosis, agraphesthesia	Contralateral parietal lobe impairment
Decreased sensation to pin prick (hemihyperesthesia)	Spinal cord tract damage
Hyperesthesias only in the distal limbs	Peripheral neuropathy

- Visual information may be processed unconsciously by the lateral geniculate nuclei. These nuclei form a *primitive visual system* that may have evolved to allow organisms the ability to respond immediately to visual data without the time needed for higher level cortical processing.
- The lateral geniculate nuclei are responsible for much of the visual processing of *primitive organisms.*
- In humans, this type of sight is known as *blind sight* or *unconscious sight.*
- When patients with V1 damage are asked to guess about observed objects in motion, or to distinguish colors, their ability to do so is better than chance.
- This suggests that visual information is in fact processed but on a subcortical (or unconscious) level.
- Patients, however, are not consciously aware of having seen anything and are often surprised that their guesses are so accurate—their sight was not processed at a conscious level.

Visual Association Areas (V2 and up)

- The visual cortex has *more than 30 specialized areas.*
- V2 (and up) is responsible for the *interpretation* of visual input.
- This is where *meaning is attached* to incoming visual data.
- V1 has a mature appearance at birth. The visual association areas do not. They appear to be dependent upon the acquisition of experience.

Lesions to the Visual Association Areas

- Lesions to the visual associations areas result in *visual agnosias*—an umbrella term that denotes the inability to attach meaning to visual data.

V2

- V2 sends visual information from V1 to the appropriate visual association area for interpretation.
- V2 processes information for further refinement by the specialized visual association areas.

V3

- V3 is responsible for interpreting *form discrimination*—the ability to recognize identifiable shapes.

Lesion to V3

- No one has ever reported a complete and exclusive loss of form vision.
- Area V3 forms a ring around V1 and V2. A lesion large enough to destroy all of V3 would almost certainly destroy V1, causing total blindness.
- V4 would also have to be knocked out since V4 plays a role in form discrimination (with regard to line orientation).

V4

- V4 is responsible for the interpretation of *color vision* and *line orientation.*
- When people view an abstract color painting with no recognizable shapes, V4 has the highest cerebral blood flow on positron emission tomography (PET) scans.

Lesion to V4

- A lesion to V4 produces *achromatopsia* in which people *only see shades of gray.*
- This differs from color blindness due to damaged cone receptors in the retina.
- In achromatopsia, people *cannot recall* or bring up in memory *what colors look like.*
- If their retinas and V1 regions remain intact, their knowledge of form, depth, and motion are preserved.

V5

- V5 is responsible for interpreting *visual motion* (identifying objects that are in motion).
- When people view a pattern of moving black and white squares on a computer screen, V5 has the highest cerebral blood flow on PET scans.

Lesion to V5

- A lesion to V5 produces *akinetopsia* in which people *neither see nor understand the world in motion.*
- While at rest, objects may be perfectly visible and understandable.
- Objects in motion appear to vanish.
- The other attributes of vision (color and form) remain intact.

Oliver Sacks Story

- Sacks described a man who was legally blind since early childhood. When the man's sight was surgically restored in his 40s, he could not integrate perception of color, form, and motion into a coherent visualization that had meaning—even though his visual anatomical structures were now intact.
- This suggests that the visual association areas only develop with direct experience and use.

Adapted from: Sack, O. *An Anthropologist on Mars.* New York: Vintage; 1995.

Depth Perception Develops Only from Experience

- Depth perception is the ability to interpret whether objects are closer or farther in relation to another object.
- Research has found that depth perception—like other cortical skills—develops only from experience.
- In cultures living in deep rain forests—where expansive views are prevented by the immense density of flora—people do not develop a sophisticated level of depth perception.

Sequence of Color Vision

1. V1, the primary visual area, detects some type of visual information and sends it to V2 for further processing.
2. V2 identifies that the visual information contains color and sends it to V4 for further processing.
3. V4 interprets the color visual information. If V4 is damaged, people see the world in shades of gray.

Visual Regions Connect Together and Project to Surrounding Structures

- All visual areas connect *directly* and *reciprocally* with each other.
- The visual association areas project to the *posterior multimodal association area*—a cortical region where somatosensory, visual, and auditory information are integrated and interpreted in relation to each other. Here, visual information is integrated with other sensory data (taste, sound, touch, smell, etc)
- This information then travels to the *hippocampus* for the *storage of visual memories*. In the hippocampus, past visual memories are compared to new visual data for decision-making processes.
- Visual data also travel to other areas of the *limbic system* for *emotional association*. Here, visual information is compared to similar visual data that already have emotional associations. When a new visual stimulus is similar to stored visual memories having unhappy emotional associations, the novel visual stimulus can elicit the same sad feelings.

Auditory Cortex

Primary Auditory Cortex

- The primary auditory cortex (A1) is located within the *insula* in the temporal lobe.
- It is responsible for *detecting* sounds from the environment.
- A1 sends auditory information to the *auditory association areas* for *interpretation*.

Lesion to A1

- *Unilateral deafness* occurs if only one hemisphere is lesioned.
- *Bilateral deafness* occurs if both hemispheres are involved.
- Patients with *cortical deafness* resulting from a lesion to A1 may still have the ability to process auditory data on an unconscious level.
- This ability may be derived from the auditory pathway of the *inferior colliculi* (of the midbrain) and the *medial geniculate nucleus* (of the thalamus).
- Researchers have found that some patients with cortical deafness still retain the ability to startle in response to a loud noise. Such patients, however, are not consciously able to hear the noise.

Auditory Association Areas

- There are several *auditory association areas located throughout the cortex.*
- They are responsible for the *interpretation* of auditory data.
- These areas have not as yet been mapped as precisely as the visual association areas.
- Different auditory association areas interpret specific auditory information (eg, animal sounds, human language, the sounds of machinery, etc).

Lesion to the Auditory Association Areas

- A lesion to the auditory association areas results in *auditory agnosia*—an umbrella term that denotes the inability to attach meaning to specific sounds.

Broca's Area

- Broca's area is located *only* in the left hemisphere.
- It mediates the *motoric aspects of speech* and is responsible for the *verbal expression* of language.
- Broca's area is located *just above the lateral fissure* and sits within the *inferior frontal gyrus* of the *premotor area.*

Lesion to Broca's Area

- Lesions to Broca's area produce *expressive aphasia* or *nonfluent speech.*
- Patients can understand what is spoken to them, but they cannot form meaningful sentences.

Wernicke's Area

- Also located *only* in the left hemisphere
- Sits within the *superior temporal gyrus* in the *temporal lobe*
- Responsible for the *comprehension* of the *spoken word*

Lesion to Wernicke's Area

- Patients cannot understand what is spoken to them.
- However, they can produce intact sentences (although the meaning of their words does not relate to anything that others have said to them).

AUDITORY CORTICES

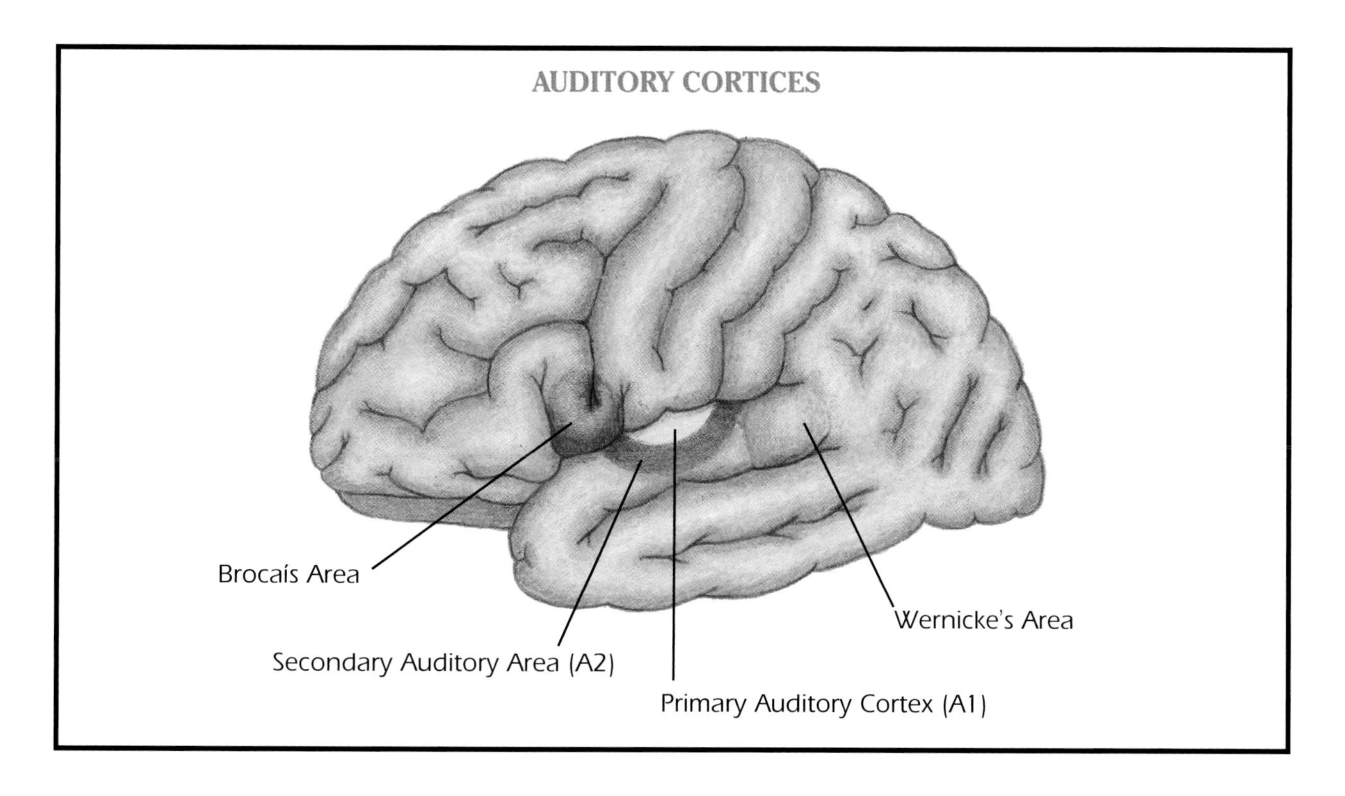

SECTION 24

Thalamus and Brainstem Sensory and Motor Roles: Function and Dysfunction

THALAMUS

- The thalamus is considered to be part of the diencephalon.
- This structure is a major *relay* and *processing center* for all types of *sensory* and *motor* information.
- There are two thalamic lobes—one in each hemisphere.
- The thalamus contains 26 pair of nuclei.
- Almost every major structure for sensory and motor data has connections with the thalamus. These include the *cortex* (all lobes), *brainstem*, *reticular formation*, *hypothalamus*, *limbic system structures* (eg, amygdala and hippocampus), *basal ganglia*, and *cerebellum*.
- Four of the most studied thalamic nuclei follow:
    - *Ventrolateral nucleus*—projects to the primary motor area (M1)
    - *Lateral geniculate nucleus*—projects to the primary visual area (V1)
    - *Medial geniculate nucleus*—projects to the primary auditory area (A1)
    - *Ventral posterolateral nucleus*—projects to the primary somatosensory area (SS1)

Thalamic Pathways

Sensory Afferent Pathways

- Sensory receptors in the *peripheral nervous system (PNS)* project sensory messages to the spinal nerves. The *spinal nerves* carry sensory information to the *spinal cord tracts*. Sensory information is then projected to the *brainstem*, where it is processed by the *reticular formation*. Sensory information then travels to the *thalamus* and finally to the *cortex* (SS1).
- Sensory information may also be carried to the *spinal cord (SC)* from the *spinal nerves* in the PNS. Sensory information then travels from the SC to the *brainstem* where it is projected through the *middle cerebellar peduncle* to the *cerebellum*. The cerebellum then sends the sensory information through the *superior cerebellar peduncle* to the *thalamus*. At the thalamic level, sensory information is either rerouted to the *cortex* or back down through the *brainstem*.

Motor Efferent Pathways

- Motor messages are sent from the *cortex* (M1) to the *thalamus*. At the thalamic level, motor messages are sent down through the *brainstem*, *SC*, and to the *muscles* in the PNS.
- Motor messages may also be projected from the *thalamus* to the *cerebellum*, *SC*, and to the *motor neurons* in the *ventral horn*.

Ansa Lenticularis Pathway

- The ansa lenticularis pathway carries motor messages from the *basal ganglia* to the *ventrolateral nucleus* of the thalamus. The information is then sent to *M1*.
- This pathway allows the *basal ganglia*, *thalamus*, and *cortex* to communicate with each other.

Superior Colliculi Pathway

- The superior colliculi of the midbrain receives sensory messages from the optic pathways and the *thalamic lateral geniculate nucleus*.

- The superior colliculi then sends this information to the thalamus via the *medial longitudinal fasciculus*.
- This pathway carries information that controls the *position* of the *eyes* and *head* in response to visual information.
- The pathway that carries visual information between the superior colliculi, and the lateral geniculate nucleus forms a *primitive visual system* that is likely responsible for *cortical sight* or *blind sight*. This is the phenomenon in which patients with cortical blindness still retain the ability to process some visual information on an unconscious or subcortical level.

Inferior Colliculi Pathway

- The inferior colliculi of the midbrain receives sensory information from the thalamic *medial geniculate nucleus* and the *auditory cortex*.
- The inferior colliculi then projects this sensory information back to the *thalamus* and *auditory cortex* for the further processing of auditory information.
- This pathway is a *primitive auditory pathway* that allows sounds to be processed on an unconscious or subcortical level. For example, patients with *cortical deafness* may still startle in response to the sound of a slammed door without consciously hearing the door slam shut.

Thalamic Mediodorsal Nucleus Pathway

- The thalamic mediodorsal nucleus receives and projects sensory information to and from the *amygdala*, *substantia nigra*, and the *temporal cortex*.
- When the mediodorsal nucleus is lesioned, *memory loss* occurs.

Lesions to the Thalamus

Thalamic Syndrome

- Thalamic syndrome results from *vascular insufficiency* (eg, cerebrovascular accident) and involves an alteration of sensory perception.
- Patients may become either *hyper-* or *hyposensitive to sensation* (particularly pain and noxious stimuli) on the contralateral side of the lesion.
- Several weeks after onset, patients develop burning, agonizing pain in the affected body parts.

Thalamic Tumors

- Sometimes patients present with specific symptoms depending on which thalamic nuclei are involved:
    - A lesion to the *ventrolateral nucleus* destroys communication with M1. This results in paralysis of the involved body part

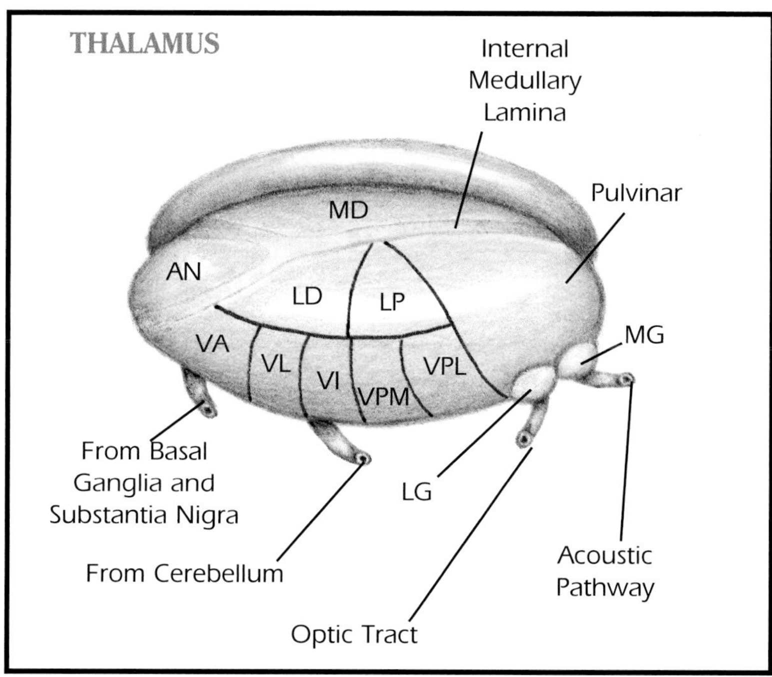

- AN—Anterior Nuclei
- LD—Lateral Dorsal Nucleus
- LG—Lateral Geniculate Nucleus
- LP—Lateral Posterior Nucleus
- MD—Mediodorsal Nucleus
- MG—Medial Geniculate Nucleus
- VA—Ventral Anterior Nucleus
- VI—Ventral Intermedial Nucleus
- VL—Ventrolateral Nucleus
- VPL—Ventral Posterolateral Nucleus
- VPM—Ventral Posteromedial Nucleus

- A lesion to the *lateral geniculate nucleus* destroys communication with V1. This can result in cortical blindness.
- A lesion to the *medial geniculate nucleus* destroys communication with A1. Cortical deafness or hyper/hyposensitivity to sound can occur.
- A lesion to the *posterolateral nucleus* destroys communication with SS1. Paresthesias, hypoesthesia, and/or causalgia can occur.
- Damage to the *ventral posteromedial nucleus* and the *ventral posterolateral nucleus* results in a loss of all forms of sensation—including light touch, tactile localization and discrimination, and proprioception from the contralateral side of the body.

- In other cases a thalamic lesion results in dysfunction of neighboring structures, producing clinical symptoms that overshadow those produced by the thalamic disease. For example, a thalamic lesion may also involve the midbrain and, thus, result in coma.

Brainstem

- The brainstem is a phylogenetically old part of the brain that controls *vegetative functions* essential for survival—respiration and primitive stereotyped reflexes.
- The *primitive stereotyped reflexes* include the following:
  - Cough and gag reflex
  - Pupillary response
  - Spontaneous swallowing reflex
- The brainstem also controls the reticular formation.

Reticular Formation

- The reticular formation is a poorly localized area of the brainstem.
- It resembles a net that is comprised of nerve cells and fibers.
- The net extends up through the *medulla*, *pons*, and *midbrain*. Fibers connect it to the SC, hypothalamus, thalamus, and cortex.
- The *reticular activating system* is thought to be located diffusely in the rostral midbrain.
- The *reticular inhibiting system* is believed to be located diffusely from the caudal midbrain to the caudal medulla.
- Simplistically, the reticular formation acts as a screen for incoming sensory data to the cortex and outgoing motor information from the cortical and subcortical areas.

Major Functions of the Reticular Formation

- Because the reticular formation has vast connections to all regions of the nervous system, it is believed to have many different types of function.
- Following are several of the most important and understood functions of the reticular formation.

Regulation of Consciousness

- The *reticular activating system* is responsible for states of *wakefulness*.
- *Lesions* to the *reticular activating system* result in a *stuporous state*. When the reticular activating system is lost, the reticular inhibitory system becomes dominant and causes heightened somnolence.
- Conversely, the *reticular inhibiting system* is responsible for states of *sleep* and *unconsciousness*.
- Lesions to the *reticular inhibitory system* result in *constant wakefulness*. When the reticular inhibiting system is lost, the reticular activating system becomes dominant and causes heightened arousal.

Control of Muscle Tone

- The reticulospinal tracts can influence the activity of the alpha and gamma motor neurons.
- This allows the reticular formation to influence muscle tone and reflex activity.
- Along with the vestibulospinal and reticulospinal tract, the reticular formation influences tone in *antigravity muscles*.

Control of Pain

- The reticular formation may have a key role in the gating mechanism for the control of pain.
- The reticulospinal tracts are descending sensory tracts that receive pain information from the periphery through afferent spinal nerves that synapse in the reticular formation of the brainstem.
- The reticulospinal tracts have their origin in the *medullary reticular formation.*
- There they travel to the *raphe nuclei of the brainstem.* The raphe nuclei are a group of nuclei located in the medulla. They are situated along the midline.
- When the raphe nuclei become excited, they *release endorphins* through a descending pathway to the place of pain origin to decrease the pain sensation.

Regulation of Circadian Rhythms

- The reticular formation and the *hypothalamus* work conjointly in the regulation of circadian rhythms or sleep and wake cycles.

Reticular Activating System

- The reticular activating system is a subsystem of the reticular formation.
- The activating system *alerts the cortex to attend to important incoming sensory information.*
- This process is important for *filtering unnecessary data* so that people can focus on other activities more intently.
- When important sensory data is detected by the reticular activating system, it alerts the cortex—which then arouses the body and prepares it for activity.
- The cortex uses the *thalamus, hypothalamus, limbic system, vestibular system,* and the *autonomic nervous system (ANS)* to arouse the body.

Reticular Inhibiting System

- The reticular inhibiting system receives messages from the *cortical* and *subcortical centers* to *calm the body* in response to specific sensory data.
- The reticular inhibiting system works conjointly with the *vestibular system* to *calm the body* via such sensory stimulation as *slow rocking, deep pressure,* and *vibration.*

Sensory Processing Disorders of the Reticular Formation

- In children with attention deficit hyperactivity disorder (ADHD), learning disability, autism, and tactile defensiveness, the reticular activating system may be functioning in overdrive.
- These children cannot easily screen or filter extraneous sensory data. Thus, they react to almost all sensory stimulation in the environment in a random and disorganized fashion.
- Conversely, some children's reticular inhibiting system may be functioning in overdrive. These children have difficulty registering the sensory information attempting to enter the central nervous system (CNS).
- Children with sensory processing disorders—or regulatory disorders—may seek specific types of stimulation in order to obtain the type of sensory data that their nervous systems require in order to develop and function optimally.
- Sensory integration treatment provides the child with sensory stimulation, which over a period of time, may enhance the brain's ability to process sensory data more functionally.
- For children with hyper-reticular activating systems, slow rocking, deep pressure, and vibration may be the kind of vestibular sensation needed to calm the body.
- For children with hyper-reticular inhibiting systems, sensory stimulation that arouses the brain is likely needed.

Pharmacologic Treatment for Sensory Processing Disorders

- Ritalin and other brands of methylphenidate are mild CNS stimulants used in the treatment of ADHD in children.
- Ritalin attempts to balance the divisions of the reticular formation. It can calm an overactive reticular activating system. Or stimulate the reticular activating system if it is sluggish.

Brainstem Damage and Coma

- Severe brainstem damage can result in coma.
- There are two basic types of comatose states involving the brainstem: *persistent vegetative state* and *brain death.*

Persistent Vegetative State

- In a persistent vegetative state the *brainstem*—including the reticular activating system—*remains intact.*
- The brunt of *neurologic destruction* is located in the *cerebral hemispheres.*
- A persistent vegetative state often results from *cardiac* or *respiratory arrest* causing a *lack of blood flow* (ischemia) and the *loss of oxygen* (hypoxia) *to the brain* for several minutes.
- The brainstem is fairly resistant to ischemia and hypoxia. Four to six minutes of complete blood and oxygen loss to the brain can result in extensive cortical destruction while sparing the brainstem.
- After a period of days to a month, the patient will awaken into a condition of *eyes-opened unconsciousness.* This is called a *vegetative state.*
- Because the brainstem is spared, the *cough, gag,* and *swallowing reflexes remain intact.*
- This decreases the likelihood of respiratory infections and substantially lengthens life expectancy (particularly with life support technological systems).

Brain Death

- Brain death occurs when *all brainstem functions are lost.*
- The *heartbeat can continue* because *it is semi-autonomous* from ANS regulation.
- This type of coma is a state of *sleeplike (eyes-closed) unarousability due* to *extensive damage* to the *reticular activating system.*
- The *cough, gag,* and *swallowing reflexes are lost,* leading to fatal respiratory infections in 6 months to a year.

Cranial Nerve Nuclei Damage in the Brainstem

- The cranial nerve nuclei are located in the brainstem.
- They innervate ipsilateral body structures.
- Brainstem disorders depend upon which cranial nerve nuclei have been lost.
- See Section 8 for disorders of the cranial nerve nuclei.

Spinal Cord Tract Damage in the Brainstem

- Descending motor tracts and ascending sensory tracts travel through the brainstem and mediate motor and sensory functions of structures on the contralateral side of the body.
- When spinal cord tracts are lost due to brainstem damage, hemiplegia and hemiparesthesia on the contralateral side of the body result.
- See Sections 18 and 19 for spinal cord injuries and diseases.

Locked-In Syndrome

- Locked-in syndrome appears much like coma; but instead of losing consciousness, the patient remains aware but has no means of communication other than eye movement.
- Such patients have lost the ability to speak and control muscles other than eyeball movement.
- Patients are able to move their eyes vertically and blink voluntarily. These movements can be used to respond to questions.
- Locked-in syndrome can result from infarction or hemorrhage in the ventral pons.
- The syndrome can be caused from traumatic brain injury, vascular diseases, demyelinating diseases, or drug overdose.
- The majority of patients with locked-in syndrome do not regain function; however, experimental research is attempting to help patients regain the ability to communicate.
- Researchers have successfully implanted electrodes into the motor cortex of several patients. The electrodes send messages to a computer that translates signals into cursor movements (on a computer screen). Patients can manipulate the cursor to spell or activate icons representing specific activities. While preliminary results have been successful, further research is needed before such research can become standard protocol.

SECTION 25

Right vs Left Brain Functions and Disorders

Hemisphere Dominance

- Generally, one side of the brain has dominant control.
- Specific individuals may be left or right brain dominant although the left hemisphere is dominant in most people.
- Hemisphere dominance may be reversed in some people who are left-handed, or only certain brain functions may be dominant in the right hemisphere while other functions are within the domain of the left hemisphere.

Left Brain Functions

- The left hemisphere controls *motor function* on the *right side* of the body.
- This hemisphere also receives *sensory information* from the *right side* of the body.
- The left hemisphere has a role in language—specifically the *interpretation and expression of aural and written words*.
- The hemisphere is specialized for the interpretation of the *concrete meanings* of words (as opposed to the abstract or symbolic meaning).
- For example, the left hemisphere interprets the *literal meaning* of a story—but *not* the hidden or symbolic meaning.
- The left hemisphere controls concrete functions that can be easily observed and measured:
  - Interpreting the concrete meaning of written or spoken words
  - Math calculations
  - Writing the letters of the alphabet
  - Reading a sentence
  - Categorizing shapes
  - Sequencing steps in a task

Right Brain Functions

- The right hemisphere controls *motor function* on the *left side* of the body.
- This hemisphere also receives *sensory information* from the *left side* of the body.
- It has a role in the *interpretation of perception*—how humans perceive their environment.
- The right hemisphere also has a role in the interpretation of information that is *abstract* and *creative* (as opposed to concrete and logical).
- The hemisphere controls abstract functions that cannot be easily observed—functions that relate to *the perception of oneself in relation to the environment.*
- It also has a role in *language*, but it is the interpretation of the *abstract* or *symbolic meaning* of a story or joke.
- This hemisphere is similarly responsible for interpreting someone's *verbal tone* and *gestures*—in other words, understanding the meaning behind the words.
- Perception includes *visual* and *spatial perception*, *language perception*, *motor planning perception*, *body schema perception*, *tactile perception*, and *auditory perception*.

Left Hemisphere Disorders

- Wernicke's and Broca's aphasia
- Contralateral motor and sensory problems
- Acalculia (the inability to calculate math problems)
- Agraphia (the inability to write words that had been familiar preinjury/disease)
- Alexia (the inability to read written words that had been familiar preinjury/disease)

Right Hemisphere Disorders

- Right hemisphere disorders involve impairment in the recognition of physical reality.
- Such disorders *distort physical reality*—they distort the environment and/or one's own body perception:
  - Visual-spatial disorders
  - Body schema perception disorders
  - Apraxias: motor planning perceptual problems
  - Perceptual language disorders
  - Tactile perceptual disorders
  - Auditory perceptual disorders
  - Contralateral motor and sensory problems

The Correlation Between Anatomical Damage and Symptomatology

- It is difficult to correlate precise anatomical damage with specific symptomatology.
- This is due to individual differences in human brains.
- Each human brain varies with regard to sulci and gyri patterns.
- It is also difficult to correlate damage and symptomatology because of neuroplasticity.

Savant Syndrome

- Savant syndrome generally occurs in people with IQs between 40 and 70 who have marked impairment in most daily living skills. Despite such impairment, however, people with savant syndrome possess highly sophisticated skills in areas such as music, art, mathematics, and memory.
- For example, most people with musical savant syndrome have perfect pitch and can play musical instruments at an amazing skill level without any musical training.
- People with artistic savant syndrome are able to paint and sculpt perfect replicas of objects and people whose images are stored in their memory.
- People possessing mathematical savant syndrome can calculate with incredible speed and accuracy.
- Other people with savant syndrome can memorize many languages (but without understanding) or possess outstanding knowledge in areas such as history, statistics, and navigational abilities. Such people seem to have superior memory capabilities.
- One explanation for savant syndrome involves an overspecialization of the right hemisphere. Such overspecialization appears to occur concomitantly with damage to the left hemisphere.
- In the book *Cerebral Lateralization*, the authors Geschwind and Galaburda suggest that in utero, the left hemisphere completes its development later than the right hemisphere. Because of this lag in development the left hemisphere is more vulnerable to prenatal influences for a longer period of time.
- Geschwind and Galaburda suggest that in male fetuses, circulating testosterone can slow brain growth and impair the development of neuronal connections in the left hemisphere. The right brain then becomes larger and more dominant in males in order to compensate for impairment in the left hemisphere.
- This theory seems to support the finding that savant syndrome occurs disproportionately in more males than females at a 6:1 ratio. This greater male to female ratio is also seen in other forms of CNS disorders such as dyslexia, delayed speech, delayed hand dominance, stuttering, attention deficit hyperactivity disorder (ADHD), learning disability, and autism.
- The skills that are associated with savant syndrome tend to be limited to right brain functions (ie, creativity, artistic skills, visuospatial abilities, and abstract functions), further supporting the above theory.
- Researchers are beginning to use neurodiagnostic imaging scans to compare the left and right hemispheres in people with savant syndrome. Early studies have supported the idea that the left hemisphere has been impaired, thus giving rise to overdevelopment of the right hemisphere.
- Other researchers have found the emergence of savant-like skills in some patients with adult onset dementia. Such patients developed the ability to paint, draw, or play music with incredible precision. As in savant syndrome, single photon emission computed tomography (SPECT) scans showed that these patients had sustained damage to their left hemisphere as a result of the disease process.
- Some researchers suggest that savantlike skills may lie dormant in the larger population. These researchers contend that the same neural circuitry that underlies savant skills is present in the larger population, but that the ability to access such skills becomes lost in a society that encourages left brain functions.

Neuroplasticity

- Neuroplasticity occurs when other areas of the brain assume the functions once mediated by regions that have been damaged.
- Human brains appear to possess a vast amount of brain matter that does not become active until damage occurs to other areas. At this time, those previously unused regions may become active and take over the function of damaged areas.
- It is possible that the same brain function may be shared by several, separate brain regions that lie dormant until injury/disease occurs. This may be the brain's evolutionary attempt at compensation.
- Neuroplasticity is most viable in children because the central nervous system (CNS) is not fully mature, but rather is still developing.

Neuroplasticity Is Dependent Upon

- Severity of neurologic damage
- Age (younger brains tend to have greater neuroplasticity)
- Premorbid health status
- Preinjury use of the damaged brain area: brain areas that were frequently used preinjury—and then became damaged—have a greater capacity to regain function. The more a brain region is used, the more it develops neuronal connections that can help in the recovery process.

Constraint-Induced Therapy

- Constraint-induced therapy (CIT) is a rehabilitation technique in which a patient's unaffected limb is restricted in order to facilitate functional use of the affected limb.
- A patient's uninvolved extremity may be constrained for 30 minute practice sessions, 5 hours per day, over 5 days per week. During this time, patients participate in activity sessions that force them to use their affected limb.
- Repeated, task-specific practice with the affected extremity appears to enhance the growth and branching of dendrites and the remodeling of synaptic connections—or the mechanisms underlying neuroplasticity.
- Research regarding CIT has demonstrated its effectiveness in patients having sustained neurologic damage (eg, traumatic brain injury and stroke).
- Patient compliance can sometimes be difficult to obtain because patients with neurologic damage may fatigue easily or display anger in response to frustration caused by the restraint device.

SECTION 26

Perceptual Functions and Dysfunctions of the Central Nervous System

Multimodal Areas of the Central Nervous System

- The multimodal areas of the CNS are also called convergence association areas.
- All sensory areas (as listed below) send information to the multimodal areas.
    - Somatosensory Association Area
    - Auditory Association Area
    - Visual Association Area
    - Motor Association Area (prefrontal cortex)
    - Olfactory Cortex (pyriform cortex in the temporal lobe)
    - Gustatory Cortex (frontal insula)
- This information converges and becomes integrated together.
- There are two multimodal association areas in the brain:
    - Posterior Multimodal Association Area
    - Anterior Multimodal Association Area (the prefrontal cortex)

Posterior Multimodal Association Area

- The posterior multimodal association area is located in the posterior of the brain in a region where the parietal, occipital, and temporal lobes meet.
- This posterior multimodal association area integrates sensory information processed by the somatosensory, visual, and auditory association areas, and the olfactory and gustatory cortices.
- This is where scent, vision, touch, taste, and sound are all connected to form a sensory experience.
- For example, the posterior multimodal association area combines the sound of the ocean, the color of the waves and crests, the feel of sand on one's feet, the taste of salt water in the air, and the smell of the sea—all into one cohesive sensory experience.
- After integrating the five types of sensory data, the posterior multimodal association area then sends this data to the anterior multimodal association area (the prefrontal cortex).

Anterior Multimodal Association Area

- The anterior multimodal association area is located in the prefrontal cortex—one storage area for motor plans.
- The anterior multimodal association area takes the integrated sensory data from the posterior multimodal association area and uses it to make decisions about which motor plan to implement.
- Once a decision is made, the anterior multimodal association area sends this information to the premotor area to access the appropriate motor plan.
- Once the appropriate motor plan is accessed, the primary motor area (M1) implements the motor plan.
- For example, in response to the sensory data described above, one may decide to implement the motor plans required for swimming in the ocean.

Impairment of the Multimodal Association Areas

- Because the multimodal association area in the right hemisphere plays an important role in perception (how one perceives the environment and one's relationship to the environment), damage to the right multimodal association area often results in perceptual disorders.

Perceptual Impairment

- Perceptual impairment more often involves dysfunction of the right hemisphere (specifically the right posterior multimodal association area), rather than the left.
- Right hemisphere disorders of the posterior multimodal association area involve impairment in the recognition of physical reality.
- Physical reality becomes distorted.
- One's relationship to the environment becomes distorted. One's relationship to one's own body also becomes distorted.
- There are several classifications of perception:
  - Visual Perception (including spatial perception)
  - Language Perception (including expressive and receptive language perception)
  - Body Schema Perception
  - Motor Planning Perception (or praxis)
  - Tactile Perception
  - Auditory Perception

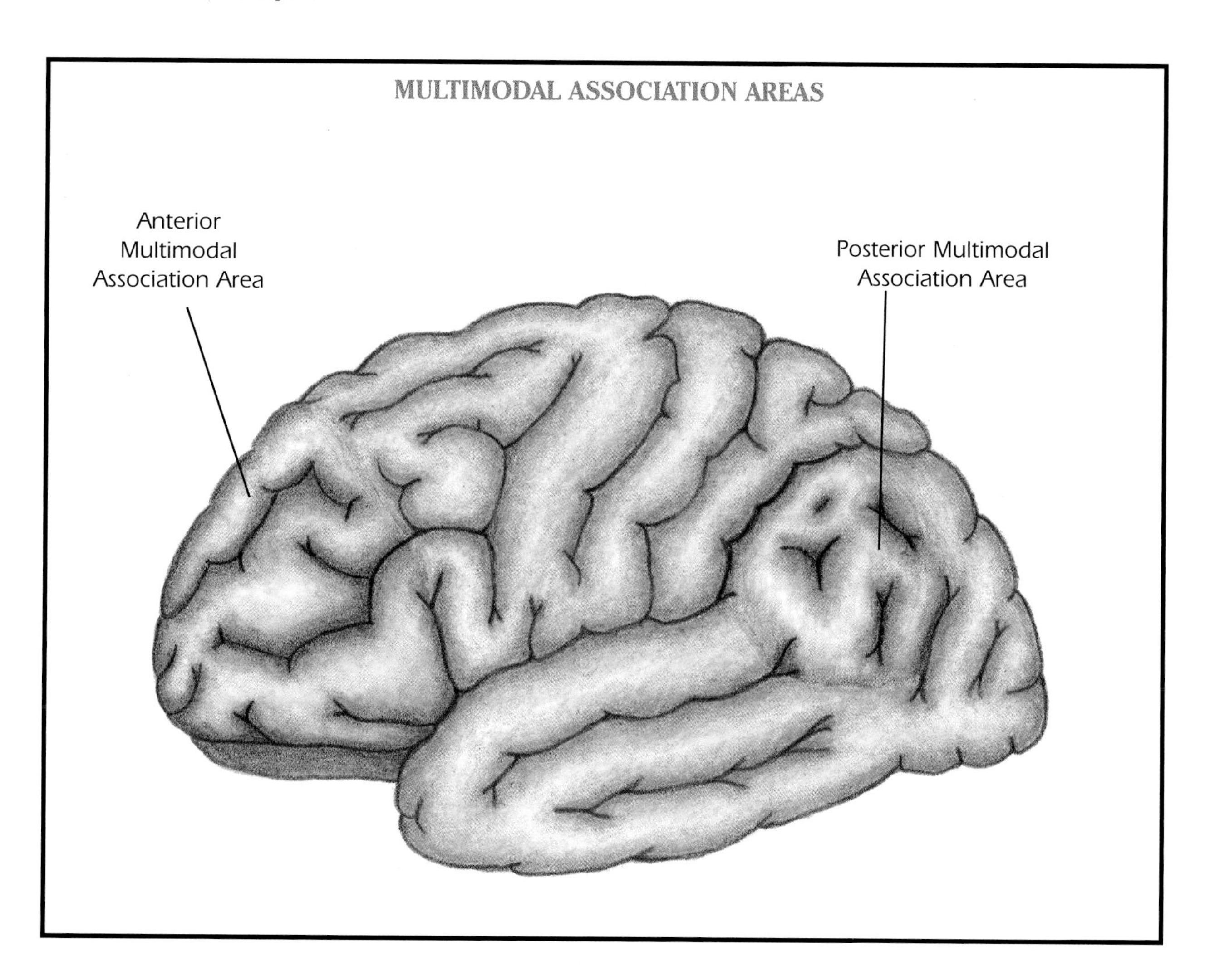

Visual Perceptual Disorders

Visual Agnosia

- Visual agnosia is an umbrella term for the inability to identify and recognize familiar objects and people (although the visual anatomy remains intact).
- Lesions that cause visual agnosias are usually located in the right hemisphere in the posterior multimodal association area.

Prosopagnosia

- Prosopagnosia is the inability to identify familiar faces because the individual cannot perceive the unique bone structure and facial muscle position that make each human face different from each other (a function of the right hemisphere).
- The left hemisphere specializes in identifying faces through facial details. A person with prosopagnosia may not be able to recognize friends and family, but he could recognize a particular face if it was characterized by a unique feature—such as a mole or scar. For example, a person may be able to identify his wife because she has a mole on her nose. This form of pattern recognition, however, causes the individual to think that everyone with a mole on her nose is his wife.

Oliver Sacks Story

Prosopagnosia

In *The Man Who Mistook His Wife For a Hat,* Oliver Sacks describes a 32-year-old patient who remained comatose for 3 weeks after a severe motor vehicle accident. As the patient regained consciousness and function he began to complain of an inability to recognize familiar faces—even family members and close friends. There were three faces, however, which he could identify. All were work colleagues with unique distinguishing characteristics. One man had an eye blinking tic; another had a large mole on his cheek; and the third was a tall thin man whose skeletal structure made him standout in crowds. The patient could recognize these men because of their single dominant feature. All other people could only be identified by their voices. In fact the patient even had difficulty recognizing his own face in a mirror.

Simultanagnosia

Sacks also describes a patient with simultanagnosia. When asked to view a *National Geographic* cover, the patient could only pick out isolated, small features—a bright color or shape. But he was unable to integrate all pieces of the cover as a visual whole representing a specific scene—the Sahara Dunes. Instead, the patient began compensating for his inability by inventing nonexistent images. He began describing a landscape of tiny guest houses with parasols lining the beach—all the while looking past the magazine cover into midair.

Adapted from: Sacks, O. *The Man Who Mistook His Wife for a Hat.* New York: Harper Perennial; 1987.

Simultanagnosia

- Simultanagnosia involves difficulty interpreting a visual stimulus as a whole.
- Patients often confabulate to compensate for what they cannot interpret visually.
- For example, when asked to look out a window and describe the view, a person may be able to visually interpret isolated objects in her visual field (such as a person walking by or flowers in a window box) but would be unable to integrate all visual data in order to understand that she was viewing a city street with moving cars, shops, and people.

Metamorphopsia

- Metamorphopsia involves a visual distortion of the physical properties of objects, so that objects appear bigger, smaller, or heavier than they really are.

- For example, a person with metamorphopsia may cut his food into tiny pieces because he perceives them as larger. He may open his mouth wide to receive each piece, perceiving that the food is larger on his fork.

Color Agnosia

- In color agnosia patients display an inability to attach appropriate colors to specific objects.
- In other words, they do not appear to know the color of common objects.
- For example, a patient might think that bananas are blue.

Color Anomia

- Patients have lost the names for colors.
- But they would continue to recognize that a blue banana was strange.

Cerebral Achromatopsia

- Cerebral achromatopsia is a condition that occurs when V4 is lost and the world appears in shades of gray.
- The memory of color is erased.
- Color agnosia and color anomia differ from cerebral achromatopsia. In color agnosia, patients remember colors but can no longer associate color and objects correctly. Similarly, in color anomia, patients remember colors but can no longer remember the names for specific colors.

Visual-Spatial Perceptual Dysfunction

Right-Left Discrimination Dysfunction

- Involves difficulty understanding and using the concepts of right and left.
- Quick Screen:
  - Ask the patient to point to left and right body parts.
  - Give the patient left/right directions around the halls of the treatment facility. Assess how well patients are able to use left/right commands as they negotiate changes in direction.

Figure-Ground Discrimination Dysfunction

- Involves difficulty distinguishing the foreground from the background.
- Quick Screen:
  - Ask the patient to pick out forks from a drawer of disorganized utensils.
  - Using a plate of food, observe whether a patient can distinguish the potatoes from the background of the white dish.

Form-Constancy Dysfunction

- Involves difficulty attending to subtle variations in form, or changes in form such as a size variation of the same object. The ability to correctly observe that an object has constant physical attributes—when viewed from different distances, vantage points, or under different light sources—is impaired.
- Quick Screen:
  - Can the patient still identify an object when it is turned on its side or placed upside down?
  - Can a patient identify a group of shapes as triangles when they are placed at different distances from the viewer?
  - Can the patient identify the same object when it is placed under different light sources?

Position in Space Dysfunction

- Involves difficulty with concepts relating to positions such as up/down, in/out, behind/in front of, before/after.
- Quick Screen:
  - Give the patient directions using the above terms.
  - Place the pencil on top of the box and place the box inside the drawer.
  - Take the pot from underneath the sink and place it on the counter. Put the rice inside the pot.

Topographical Disorientation

- Involves difficulty comprehending the relationship of one location to another.
- Quick Screen:
  - Can the patient find his way around the hospital using written directions or a pictorial map?

Depth Perception Dysfunction (Stereopsis)

- Involves difficulty determining whether one object is closer to the patient than another object.
- Quick Screen:
  - Can the patient determine whether objects in the natural environment are near or far in relation to each other and in relation to the patient?

Body Schema Perceptual Dysfunction

- Body schema is the awareness of spatial characteristics of one's own body—an awareness formed by current and previous sensory input.
- It is the neural perception of one's body in space.
- This neural perception is derived from a synthesis of tactile, proprioceptive, and pressure sensory associations about the body and its parts.
- This perceptual disorder is also referred to as *somatagnosia*.
- Body schema perceptual dysfunction is more likely to occur as a result of right hemisphere lesions in the posterior multimodal association area.
- However, as with all of the perceptual disorders, body schema perceptual dysfunction can also result from left hemisphere lesions.

Body Schema Differs from Body Image

- Body image is the emotional and cognitive assessment one holds about one's body.
- For example, patients with an amputated hand may have a body image of impairment—they may emotionally feel that their body is now damaged. Their body schema, however, may involve phantom pain in which they can feel their amputated hand. The pain of the phantom sensation may be experienced as cramping and burning.
- Patients with anosognosia (a severe neglect syndrome) often have positive thoughts about their body image, believing that all of their limbs are intact and functional—despite having a hemiplegic left side. Their body schema tells them that they have no left side. Because they are not processing the neural input from their left limbs, they have no awareness that the left side of their body exists and is hemiplegic.

Finger Agnosia

- Finger agnosia involves an impaired perception concerning the relationship of the fingers to each other.
- It also involves an impaired identification and localization of one's own fingers.
- Quick Screen:
  - Ask the patient to tap her index finger or to touch her ring finger with her thumb.

Unilateral Neglect

- Involves the inability to integrate and use perceptions from the left side of the body or the left side of the environment.
- The awareness of the left side of the environment is lost temporarily.
- Patients more often have a left neglect rather than a right neglect; although right neglect syndromes occur as well.
- Right neglect syndromes tend to resolve more quickly than left neglect syndromes.
- A left neglect often results from a lesion to the right hemisphere's posterior multimodal association area.
- Patients with unilateral neglect can be trained to heighten their awareness of the left (or right) side of their bodies and environments.

- Quick Screen:
  - Ask the patient to (a) draw a clock and a human figure, (b) read a paragraph, or (c) perform a "cross-out the Hs" worksheet.
  - Observe the patient while eating a meal. Check whether the patient attends to both sides of the plate or ignores one half.

Anosognosia

- Anosognosia is *extensive neglect* and *failure to recognize one's own body paralysis on one side.*
- While patients with unilateral neglect can usually be taught to enhance their awareness of the left side of the body and environment, patients with anosognosia cannot be taught in the same way.
- Anosognosia is accompanied by a strange affective dissociation. Patients show extraordinary indifference to their affected limb.
- They have no concept that they have a paralyzed limb.
- Patients may ask the hospital staff to "take the arm away with the lunch tray."
- Anosognosia is usually a transient state of the acute CVA patient. Anosognosia will usually resolve as the patient recovers.
- Right hemisphere lesions cause more severe neglect syndromes than do left hemisphere lesions.
- Left neglect syndromes resolve less readily than right neglect.
- Patients who are several months post neurologic damage are more likely to display left neglect. Right neglect syndromes more commonly clear by several months post neurologic insult.
- Sometimes patients with severe left neglect must live with their neglect syndrome for the remainder of their lives—particularly if the patient is elderly or the neurologic damage is severe.

Double Simultaneous Extinction (Extinction of Simultaneous Stimulation)

- Double simultaneous extinction is categorized as a cortical somatosensory association disorder.
- However, it is also considered to be a form of attentional neglect.
- The therapist touches the patient on two body regions: (a) first, on the involved limb, and (b) second, on the uninvolved limb. The patient's eyes are closed.
- The patient is able to feel the tactile stimulus on both the involved and uninvolved limbs.
- Then, the therapist touches the patient on the same body regions but simultaneously.
- Extinction occurs when the patient cannot feel the tactile sensation on the involved side (even though he could when the tactile sensation was only applied to the involved side).
- As the neurons mediating tactile sensation recover on the involved side, they can be overridden by tactile stimulation on the uninvolved side.
- This is called a *limited attention recovery phase.*
- As the neurologic damage resolves, the phenomenon of double simultaneous extinction disappears.

Language Perception Dysfunction

The Aphasias

- Aphasia means impairment in the expression and/or the comprehension of language.
- There are two classifications of aphasia: Receptive and Expressive.

Receptive Aphasia

- Receptive aphasia is impairment in the comprehension of language.
  - **Wernicke's Aphasia** involves difficulty comprehending the literal interpretation of language. Wernicke's aphasia always results from a left hemisphere lesion in the brain region referred to as Wernicke's area.
  - **Alexia** is the inability to comprehend the written word; also the inability to read. Alexia can occur as a result of lesions to either hemisphere; more often the left hemisphere is lesioned.

- **Dyslexia** is the impaired ability to read. It is a language problem in which the ability to break down words into their most basic units—phonemes—is impaired.
- **Asymbolia** is difficulty comprehending gestures and symbols. Usually the left hemisphere is lesioned.
- **Aprosodia** is impaired comprehension of tonal inflections used in conversation. Patients have difficulty perceiving the emotional tone of someone's conversation. Aprosodia usually results from a right hemisphere lesion.

Receptive Aphasia in the Right vs Left Hemisphere

- Patients with receptive aphasia from a *left hemisphere lesion* can still perceive and accurately interpret the emotional tones of a conversation, even though they cannot understand the concrete meaning of the words.
- Patients with receptive aphasia from a *right hemisphere lesion* can understand the concrete meaning of the words but not the emotional tone of conversations.

Expressive Aphasia

- Expressive aphasia involves difficulty expressing clear, meaningful language.
  - **Broca's Aphasia** is an expressive language disorder in which patients can understand what is spoken to them, but they cannot express their ideas in an understandable way. Often they speak gibberish or sentences that do not make sense. Patients with Broca's aphasia commonly display word finding difficulties in which they either (a) cannot complete sentences because they cannot retrieve the correct words, or (b) they mistakenly use the wrong word. Broca's aphasia always results from a left hemisphere lesion in the brain region referred to as Broca's area.
  - **Anomia** is the inability to remember and express the names of people and objects. The individual may know the person but cannot remember his name. Anomia differs from prosopagnosia; in prosopagnosia individuals do not recognize familiar faces. Lesions resulting in anomia can occur in either hemisphere.
  - **Agrommation** is the inability to arrange words sequentially so that they form intelligible sentences. It occurs as a result of left hemisphere lesions.
  - **Agraphia** is the inability to write intelligible words and sentences. It occurs as a result of left hemisphere lesions.
  - **Acalculia** is the inability to calculate mathematical problems. Dyscalculia is difficulty calculating math. Lesions that result in acalculia occur in the left hemisphere.
  - **Alexithymia** is the inability to express one's emotions through words. Dyslexithymia involves difficulty attaching words to feelings. These can occur as a result of either left or right hemisphere damage.

Differences in Male and Female Language Processing

- Research has shown that *females* tend to use *both hemispheres* in the processing of language. They use the left hemisphere to interpret the concrete meaning of words and sentences, and the right hemisphere to interpret the emotions attached to those words and sentences.
- Females are also more able to attach words to their emotions. This requires the ability to use the left hemisphere to attach words to emotions generated in the right hemisphere.
- *Males* predominantly use the *left hemisphere* to process language. Because they do not as readily integrate both hemispheres in language processing, they may have more difficulty attaching words to their emotions—referred to as dyslexithymia.

Perceptual Motor Dysfunction

- Perceptual motor dysfunction involves the *apraxias* or *motor planning impairments*.
- Apraxia can result from either right or left hemisphere lesions—but usually result from right hemisphere lesions of the anterior multimodal cortex, the premotor area, and/or the primary motor cortex.
- Patients with apraxia have a distorted perception of the motor strategies required to negotiate their environment.
- There are several classifications of apraxia:
  - Ideational Apraxia
  - Ideomotor Apraxia I and II
  - Dressing Apraxia
  - Two- and Three-Dimensional Constructional Apraxia

Ideational Apraxia

- Ideational apraxia involves an inability to cognitively understand the motor demands of the task.
- For example, a patient may not understand that a shirt is an article of clothing to be worn on the torso and upper extremities.

Ideomotor Apraxia I

- Ideomotor apraxia involves the *loss of the kinesthetic memory of motor patterns*—essentially the motor plan for a specific task is lost.
- In ideomotor apraxia I the patient cannot access the appropriate motor plan. She can cognitively understand the motor demands of the task but cannot translate that understanding into appropriate motor movements. For example, a patient can understand that a shirt is an article of clothing to be worn on the torso and upper extremities but she cannot access these motor plans.

Ideomotor Apraxia II

- In ideomotor apraxia II the patient *cannot implement the appropriate motor plan*. She understands the motor demands of the task but when she attempts to implement the appropriate motor plan, an inappropriate one is activated. For example, when a patient is shown a toothbrush, she understands that this is a self care item used for cleaning teeth. When she attempts to activate the motor plan for teeth brushing, however, another motor plan becomes activated instead and she attempts to brush her hair with the toothbrush.

Dressing Apraxia

- Dressing apraxia is a form of ideomotor apraxia involving an inability to dress oneself due to impairment in either (a) body schema perception or (b) perceptual motor functions.
- Example: Patients may attempt to put their arms through pant legs or dress only one half of their body.

Two- and Three-Dimensional Constructional Apraxia

- This type of apraxia involves an inability to copy two- and three-dimensional designs or models.
- A patient who is an architect may be unable to draw two-dimensional blueprints.
- A patient who is a building contractor may be unable to put together a wooden birdhouse kit.
- Patients with constructional apraxia due to a right hemisphere lesion draw objects or put pieces of a kit together in a spatially disorganized way.
- Patients with constructional apraxia due to a left hemisphere lesion draw objects that lack detail. Three-dimensional objects are correctly spatially organized, but pieces are often left out.

TACTILE PERCEPTUAL DYSFUNCTION

Tactile Agnosia

- Tactile agnosia is the umbrella term for the *inability to attach meaning to somatosensory data.*
- Tactile agnosia commonly results from lesions to the secondary somatosensory area (SS2).
- The anatomy for touch and pain/temperature receptors remains intact.
- The ability to attach meaning to somatosensory data is referred to as cortical sensation (as opposed to primary sensation).

Cortical Sensation

- Cortical sensation includes the following forms of tactile agnosia:
  - **Astereognosis** is the inability to identify objects by touch alone. Astereognosis can be further broken down into ahylognosia and amorphognosia.
  - **Ahylognosia** is the inability to discriminate between different types of materials by touch alone.
  - **Amorphognosia** is the inability to discriminate between different forms/shapes by touch alone.
  - **Two-Point Discrimination** is the loss of the ability to determine whether one has been touched by one or two points. An aesthesiometer is the instrument used to assess two-point discrimination.
  - **Agraphesthesia** is the loss of the ability to interpret letters written on the contralateral hand.
  - **Double Simultaneous Extinction** is the inability to determine that one has been touched on both the involved side and the uninvolved side—the neural sensation of the uninvolved side overrides the ability to perceive touch on the involved side.
  - **Abarognosis** is the inability to accurately estimate the weight of objects—particularly in comparison to each other.
  - **Atopognosia** is the inability to accurately perceive the exact location of a sensation.

AUDITORY PERCEPTUAL DYSFUNCTION

Auditory Agnosia

- Auditory agnosia is the umbrella term for the *inability to attach meaning to sound.*
- This condition commonly occurs concomitantly with other communication disorders.
- The external anatomy responsible for hearing remains intact. Instead, lesions involve the posterior multimodal association area—particularly the association areas of the temporal lobe.
- Lesions to the left auditory association areas result in an inability to attach meaning to language. Language can be heard, but not interpreted.
- Lesions to the right auditory association areas produce an inability to attach meaning to non-language sounds. For example, a patient may be unable to interpret the sound of a train whistle.
- Sometimes patients cannot distinguish between sounds. For example, a patient with auditory agnosia may be unable to distinguish the sound of a train from the sound of thunder and wind.

Other Perceptual Phenomena

Synesthesia

- Synesthesia is a perceptual phenomenon involving the ability to combine senses in response to specific stimuli.
- For example, some people have the ability to see colors when they hear music. These people report that specific musical notes are associated with specific colors. Such colors are always elicited whenever certain musical notes are heard.
- Others have the ability to see colors when reading the alphabet. These people report that specific letters are always associated with specific colors.

Migraine-Induced Auras

- Fifteen percent of people who experience migraines also experience auras that precede migraine headaches.
- Auras are cortically generated perceptions or hallucinations; they occur as a precursor to migraine headaches.
- Auras can involve all types of sensory phenomena (eg, odors, colors, sounds, tastes), but most often occur in the form of visual symptoms—such as flashing lights, blotting out of vision, and sparkling colored moving lines.
- Auras involve a sequence of neurologic events leading from cortical excitation of nerve cells to activation of pain-sensitive structures. While the actual aura is a cortically-generated hallucination, the pain of the headache likely involves dysfunction in the nerves to major blood vessels in the head.
- Auras tend to develop gradually over 5 to 20 minutes and last for approximately 60 minutes. They often serve as a warning to patients that a migraine headache will occur shortly. The headache itself can last up to 72 hours.
- Many people with creative talents who experience migraine-induced auras report that the aura can be inspirational to their work. The author of Alice in Wonderland, Lewis Carroll, was reported to have experienced migraine-induced auras. Such auras purportedly served as inspiration for many of the unique phenomena written about in the book.

SECTION 27

Blood Supply of the Brain: Cerebrovascular Disorders

Major Arteries

Internal Carotids (2)

Route

- The internal carotids rise from the common carotid artery and enter the brain at the level of the optic chiasm.

Supply

- They are the major arteries that supply the brain.

Vertebral Arteries (2)

Route

- The vertebral arteries run along the lateral aspect of the medulla.
- They connect to form the basilar artery at the base of the pons-medulla junction.
- They give rise to the anterior spinal artery.

Supply

- These arteries supply the lateral medulla areas.

Anterior Spinal Artery (1)

Route

- The anterior spinal arteries begin as two small branches that become one main artery.
- The two anterior spinal branches rise off of the vertebral arteries and become one main artery that travels along the anterior surface of the medulla and spinal cord.

Supply

- The spinal artery supplies the anterior portion of the medulla and spinal cord.

Three Arteries that Supply the Cerebellum

Posterior Inferior Cerebellar Arteries (2)

Route

- The posterior inferior cerebellar arteries rise from the vertebral arteries at the medulla level.

Supply

- They supply part of the dorsolateral medulla (including the cerebellar peduncles), the inferior surface of the cerebellum, and the deep cerebellar nuclei.

Anterior Inferior Cerebellar Arteries (2)

Route

- The anterior inferior cerebellar arteries rise from the vertebral arteries at the pons-medulla junction.

Supply

- They supply the inferior surface of the cerebellum and the deep cerebellar nuclei.

Superior Cerebellar Arteries (2)

Route

- The superior cerebellar arteries rise from the basilar artery at the pons-midbrain junction.

Supply

- They supply the superior aspect of the cerebellum and parts of the deep cerebellar nuclei.

Basilar Artery (1)

- The basilar artery does not supply the cerebellum. But it does give rise to the superior cerebellar arteries.

Route

- Travels along the anterior aspect of the pons.
- Gives rise to the superior cerebellar arteries.

Supply

- Supplies the anterior and lateral aspects of the pons.

Blood Supply to the Brain

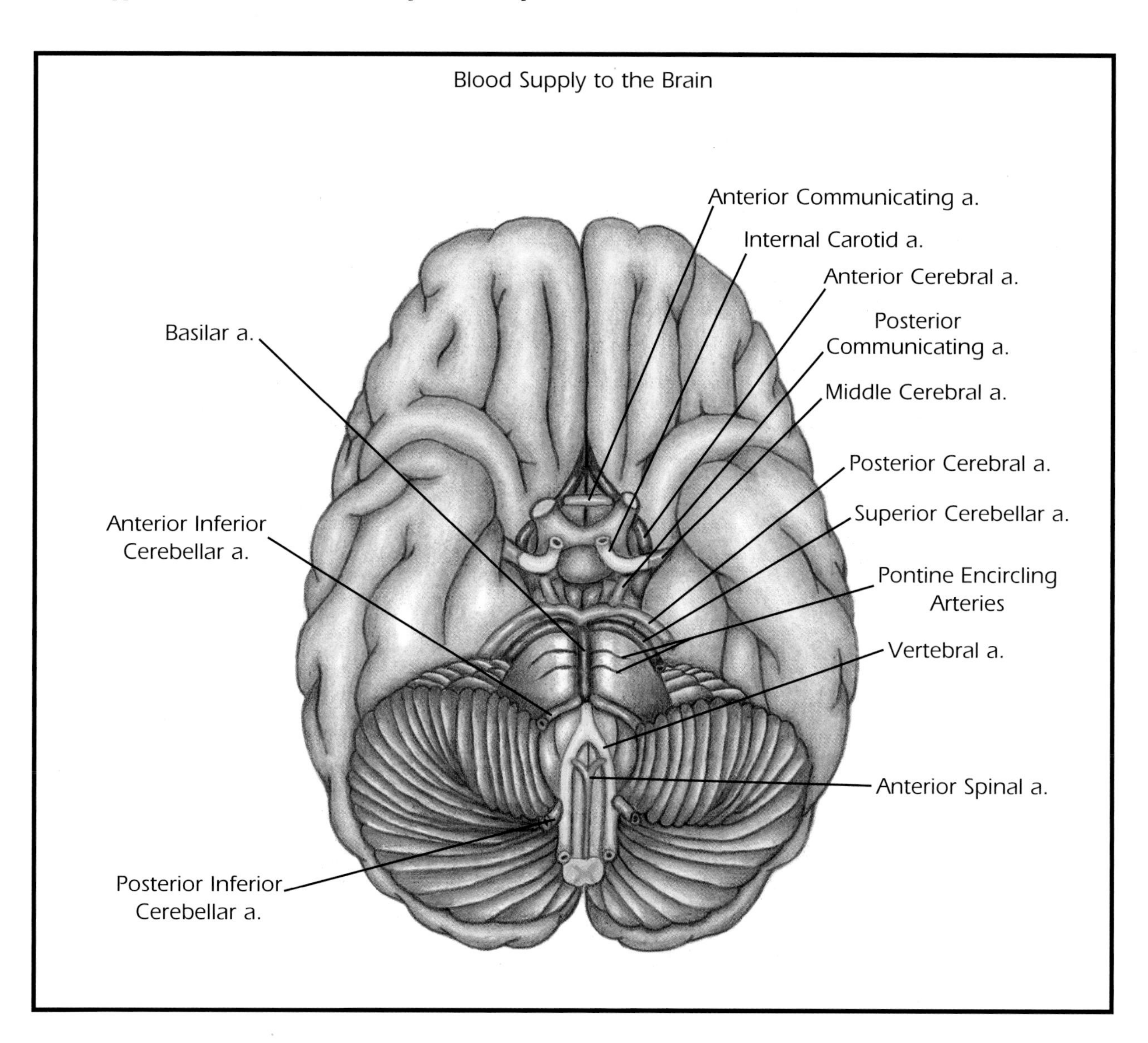

Three Main Cerebral Arteries

Posterior Cerebral Arteries (2)

Route

- The posterior cerebral arteries rise from the basilar artery.

Supply

- They supply the medial and inferior surfaces of the temporal and occipital lobes, the thalamus, and the hypothalamus.

Middle Cerebral Arteries (2)

Route

- The middle cerebral arteries rise from the internal carotids and travel through the lateral fissure to the brain's surface.

Supply

- These arteries supply the lateral surfaces of the frontal, temporal, and parietal lobes.
- They also supply the inferior surface of part of the frontal and temporal lobes.

Anterior Cerebral Arteries (2)

Route

- The anterior cerebral arteries rise from the internal carotids.

Supply

- These arteries supply the superior, lateral, and medial aspects of the frontal and parietal lobes.
- They also supply part of the basal ganglia and the corpus callosum.

Communicating Arteries and Multiple Encircling Arteries

- The communicating arteries provide blood supply pathways to the major cerebral arteries.
- The multiple pontine encircling arteries provide a blood supply pathway to the pons.

Posterior Communicating Arteries (2)

Route

- The posterior communicating arteries connect the internal carotids and the posterior cerebral arteries.

Supply

- They supply the diencephalon, internal capsule, and the optic chiasm.

Anterior Communicating Artery (1)

- The anterior communicating artery connects the two anterior cerebral arteries.

Pontine Encircling Arteries (Multiple)

Route

- These arteries rise from the basilar artery and wrap around the pons.

Supply

- They supply the lateral and posterior portions of the pons.

Circle of Willis

- The circle of Willis is a circuit of interconnecting arteries that function to prevent lack of blood flow to the brain due to occlusion.
- Components of the circle of Willis include the following:
  - Posterior cerebral arteries
  - Posterior communicating arteries
  - Internal carotids
  - Anterior cerebral arteries
  - Anterior communicating artery

CIRCLE OF WILLIS

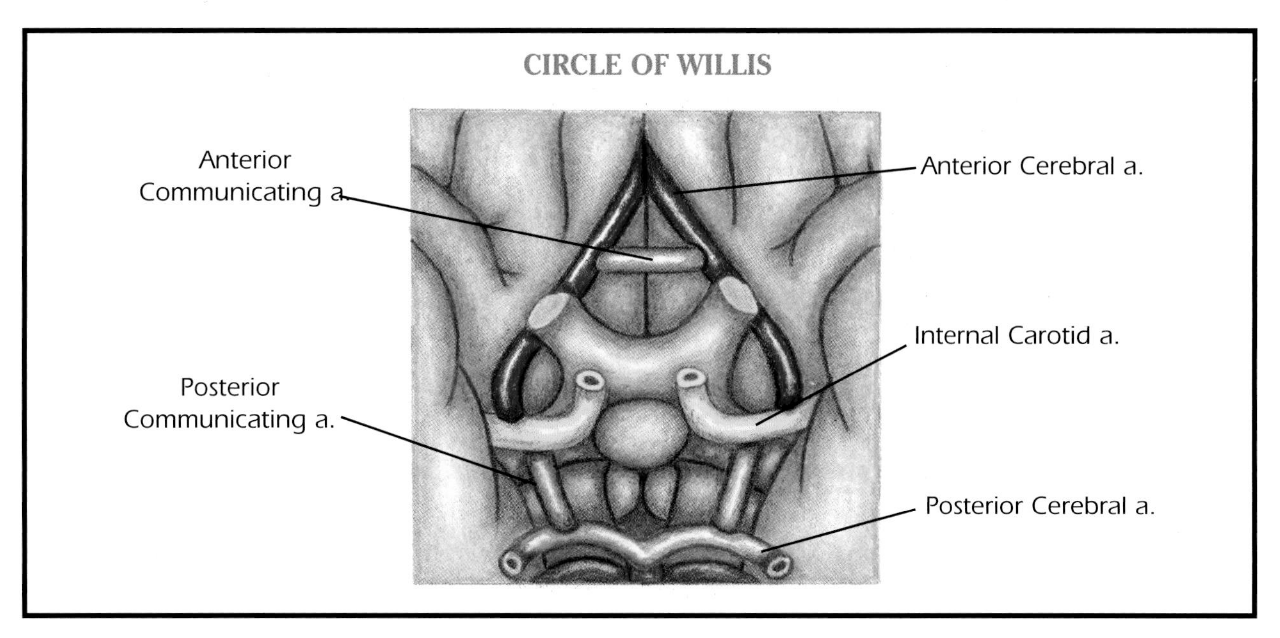

BLOOD SUPPLY OF SPECIFIC BRAIN REGIONS

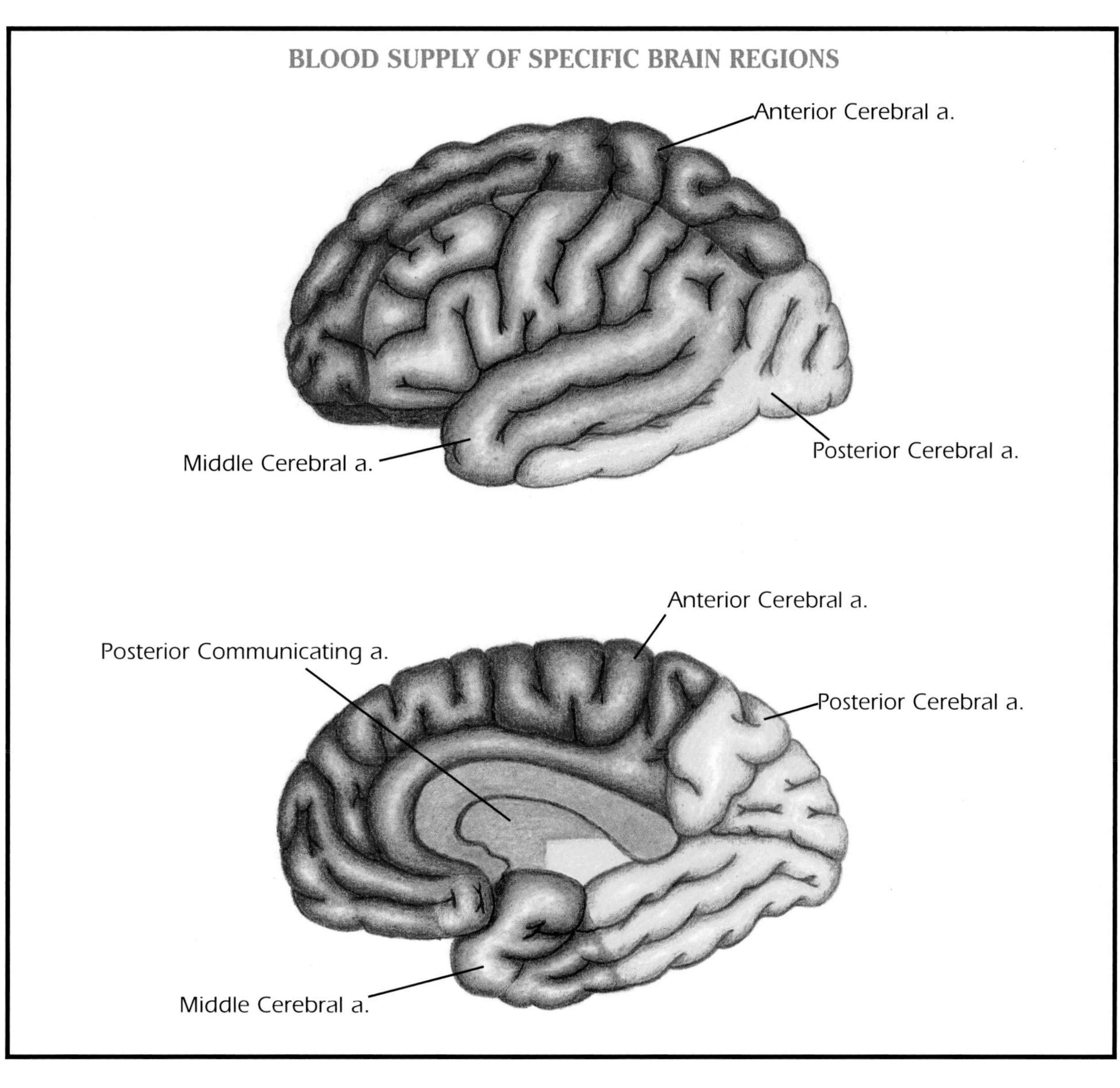

Common Areas of Arterial Occlusion in the Cortex

Middle Cerebral Arterial Occlusion

- The middle cerebral arteries are the most common site of occlusion resulting in cerebrovascular accident (CVA).
- The middle cerebral arteries supply the lateral surfaces of the frontal, temporal, and parietal lobes, and the inferior surface of portions of the frontal and temporal lobes.

Middle Cerebral Arterial Occlusion in the Left Hemisphere

- An occlusion in the middle cerebral artery in the left hemisphere may result in the following:
  - **Contralateral Hemiplegia** (on the right side of the body)—The primary motor area is lesioned.
  - **Contralateral Hemiparesthesia** (on the right side of the body)—The primary somatosensory area is lesioned.
  - **Aphasia**—Broca's or Wernicke's area may be lesioned. Other language areas may also be damaged, resulting in different types of aphasia.
  - **Cognitive Involvement**—Impairment in cognitive function results from a frontal lobe lesion.
  - **Affective Involvement**—Often when the left hemisphere is lesioned, the patient may display emotional lability and depression. This is sometimes referred to as a catastrophic response.

Middle Cerebral Arterial Occlusion in the Right Hemisphere

- An occlusion in the middle cerebral artery in the right hemisphere may result in the following:
  - **Contralateral Hemiplegia** (on the left side of the body)—The primary motor area is lesioned.
  - **Contralateral Hemiparesthesia** (on the left side of the body)—The primary somatosensory area is lesioned.
  - **Perceptual Deficits**—Left neglect syndromes are common with damage to the right hemisphere, particularly to the posterior multimodal association area.
  - **Apraxia**—The anterior multimodal association area, premotor area, and/or the primary motor cortex may be lesioned.
  - **Cognitive Involvement**—Impairment in cognitive function results from a frontal lobe lesion.
  - **Affective Involvement**—Often when the right hemisphere is lesioned, the patient may display euphoria or report a sense of well-being. If a neglect syndrome is present, the patient is often unaware of his or her deficits.

Posterior Cerebral Arterial Occlusion

- The posterior cerebral arteries supply the medial and inferior surfaces of the temporal and occipital lobes.
- These arteries also help to supply the thalamus and the hypothalamus; however, a lesion to a posterior cerebral artery will likely not affect the thalamic and hypothalamic functions.
- A lesion to one of the posterior cerebral arteries may result in the following:
  - **Memory Loss**—Due to temporal lobe involvement
  - **Visual Perceptual Deficits** result from damage of the occipital lobe and the posterior multimodal association area.
  - **Visual Field Cuts** result from occlusion to the optic chiasm. The optic chiasm is supplied by the posterior communicating arteries, which connect to the posterior cerebral arteries.

Anterior Cerebral Arterial Occlusion

- The anterior cerebral arteries supply the superior, lateral, and medial aspects of the frontal and parietal lobes.
- These arteries also help to supply portions of the basal ganglia and corpus callosum.

- A lesion to one of the anterior cerebral arteries may result in the following:
  - **Contralateral Hemiplegia**—The primary motor area is lesioned.
  - **Contralateral Hemiparesthesia**—The primary somatosensory area is lesioned.
  - **Cognitive Involvement**—Due to frontal lobe involvement.
  - **Apraxia**—The anterior multimodal association area, premotor area, and/or the primary motor area may be lesioned.
  - **Affective Involvement**—If the left hemisphere is lesioned, emotional lability and depression may occur. If the right hemisphere is lesioned, euphoria or emotional dissociation may occur.

CEREBELLAR ARTERIAL OCCLUSION

- The three major symptoms of cerebellar disorders include:
  - Incoordination
  - Ataxia
  - Intention tremors

Posterior Inferior Cerebellar Arterial Occlusion

- The posterior inferior cerebellar arteries supply the cerebellar peduncles.
- Cerebellar arterial occlusion often involves the brainstem structures that are supplied by the cerebellar arteries.
  - **Ipsilateral Hypertonicity and Hyperactive Reflexes**—Because the posterior and anterior spinocerebellar tracts travel through the superior cerebellar peduncle, damage to the superior cerebellar peduncles may result in ipsilateral hypertonicity and hyperactive reflexes.
  - **Vertigo, Nausea, Nystagmus**—Because the posterior inferior cerebellar arteries also supply blood to the medulla, an occlusion to this artery may also result in vertigo, nausea, and nystagmus as a result of vestibular nerve nuclei loss.

Anterior Inferior Cerebellar and Superior Cerebellar Arterial Occlusion

- Occlusion to either of these two arteries may result in the following:
  - **Ipsilateral Ataxia**
  - **Ipsilateral Hypotonicity and Hyporeflexia**
  - **Dysmetria**
  - **Adiadochokinesia (and dysdiadochokinesia)**
  - **Movement Decomposition**
  - **Asthenia**
  - **Rebound Phenomenon**
  - **Staccato Voice**
  - **Ataxic Gait**
  - **Intention Tremors**
  - **Incoordination**
- Occlusion to these arteries may also result in the following due to the arteries' connection to the blood supply of the medulla:
  - **Vestibular Signs (nystagmus, vertigo, nausea)**
  - **Facial Sensory Impairment**
  - **Dysphagia**
  - **Dysarthria**
  - **Bell's Palsy**

Occlusion of Arteries that Supply the Brainstem

Anterior Spinal Artery Occlusion

- The anterior spinal arteries supply the medulla—including the pyramids and the vestibular, hypoglossal, glossopharyngeal, and vagal nerve nuclei.
  - **Contralateral Hemiplegia** can result if the pyramids are lost.
  - **Deviation of the Tongue to the Affected Side** can result if the hypoglossal nerve nuclei are lost.
  - **Dysphagia and Loss of the Gag Reflex** can result if the glossopharyngeal and/or vagal nerve nuclei are lost.
  - **Nystagmus and Balance Disturbances** can result if the vestibular nuclei are lost.

Vertebral Arterial Occlusion

- The vertebral arteries supply the lateral aspect of the low medulla including the accessory nuclei.
- **Dysphagia** may occur if the accessory nerve nuclei are lost.

Basilar Arterial Occlusion

- The basilar artery supplies the pons including the corticospinal tracts, and the abducens, trigeminal, and facial nerve nuclei.
  - **Contralateral Hemiplegia** can occur if the corticospinal tracts are lost.
  - **Medial or Internal Strabismus** can occur if the abducens nerve nuclei are lost.
  - **Ipsilateral Sensory Loss of the Face** can occur if the trigeminal nerve nuclei are lost.
  - **Loss of the Masseter Reflex and the Corneal Reflex** can also occur if the trigeminal nerve nuclei are lost.
  - **Bells' Palsy and Hyperacusis** can occur if the facial nerve nuclei are lost.

Cerebrovascular Accident

- Cerebrovascular disorder (CVA), or stroke, is an umbrella term applied to conditions in which blood flow to cerebral vessels becomes disrupted, either from clotting or rupture.
- There are two primary types of CVA: ischemic and hemorrhagic.
  - Ischemic strokes are the most common type of CVA and result from thrombosis (static clot) or emboli (traveling clot).
  - Hemorrhagic strokes involve bleeding into brain tissue and can result from hypertension, aneurysms, or head injury. Hemorrhagic strokes are the most fatal type.
- Risk factors include the following:
  - Age
  - Sex (men have a 19% greater risk of stroke than women)
  - Race (African Americans have a 60% greater risk than the general population)
  - Hypertension
  - High cholesterol levels
  - Cigarette smoking
  - Diabetes mellitus
  - Prior stroke
  - Obesity
  - Heart disease

Thrombotic Strokes

- Thrombi are clots formed by plaque development in a vessel wall.
- These are the most common type of ischemic strokes and occur in atherosclerotic blood vessels.
- Common sites of plaque formation include larger vessels of the brain—including the origin of the internal carotid arteries, the vertebral arteries, and the junction of the basilar and vertebral arteries.
- Thrombotic strokes usually occur gradually over several days.
- They are frequently seen in older persons with arteriosclerotic heart disease.
- This type of stroke is not associated with exertion or activity and can occur when the person is at rest.

(Continued)

Cerebrovascular Accident

(Continued)

Lacunar Infarcts

- Lacunar infarcts are small clots located in the deep regions of the brainstem and subcortical structures.
- They are often found in single deeply penetrating arteries that supply the internal capsule, basal ganglia, and brainstem.
- They commonly result from occlusion of the smaller branches of the large cerebral arteries—most notably the middle and posterior cerebral arteries. Sometimes lacunar infarcts can also occur in the anterior cerebral, vertebral, and basilar arteries.
- Because they are small, lacunar infarcts usually do not cause severe impairment.

Embolic Strokes

- Emboli are clots that dislodge from their site of origin and travel to a cerebral blood vessel, where they become trapped and interrupt blood flow.
- Embolic strokes often affect the smaller cerebral vessels. The most frequent site is the middle cerebral artery.
- This type of stroke commonly has a sudden onset and is associated with the presence of cardiac disease (eg, rheumatic heart disease, ventricular aneurysm, and bacterial endocarditis). Cardiac embolism can also occur after a recent myocardial infarction.

Hemorrhagic Stroke

- Hemorrhagic strokes are frequently fatal.
- However, if patients can survive the initial hemorrhagic damage, prognosis is generally good.
- Hemorrhagic strokes involve bleeding into brain tissue after the rupture of a blood vessel wall.
- This type of stroke results in edema and compression of brain tissue that, if not medically treated immediately, can be fatal.
- Hemorrhagic strokes commonly occur suddenly and are associated with exertion and activity.
- **Aneurysmal Subarachnoid Hemorrhage**
  - An aneurysm is a bulge occurring in a blood vessel wall as a result of clot formation.
  - Most aneurysms are small saccular structures called berry aneurysms.
  - Berry aneurysms commonly occur in the circle of Willis or the junction of two vessels.
  - Aneurysms tend to enlarge with time and weaken vessel walls until rupture occurs.

Transient Ischemic Attack

- Transient ischemic attacks (TIAs) are sometimes referred to as mini strokes.
- They are characterized by focal ischemic cerebral incidents that last less than 24 hours—most TIAs usually last less than 1 – 2 hours.
- The causes of TIAs include atherosclerotic disease and emboli.
- TIAs may provide a warning of an impending larger stroke.
- Signs include the following:
  - Numbness and mild weakness on one side of the body
  - Transient visual disturbances (eg, blurred vision, fading vision)
  - Dizziness
  - Falls
  - Confusion and possible blackout

SECTION 28

Commonly Used Neurodiagnostic Tests

Lumbar Puncture (Spinal Tap)

- A lumbar puncture is an *invasive* procedure used to remove a sample of cerebrospinal fluid (CSF) from the subarachnoid space for diagnostic evaluation.
- As a result of the technological advancement of other diagnostic evaluations, lumbar punctures are no longer commonly used in the assessment of central nervous system (CNS) disorders.
- Lumbar punctures are used to:
  - *Collect CSF to detect pathological conditions*
  - *Measure CSF pressure in the diagnosis of hydrocephalus*
  - *Administer contrast dye to the spinal canal during imaging scans*
  - *Administer anesthesia to the spinal cord for surgical procedures*
  - *Relieve hydrocephalus*
- Disorders in which lumbar punctures are indicated include:
  - *CNS infection (eg, meningitis, encephalitis)*
  - *Space occupying lesions of the subarachnoid space*
  - *Multiple sclerosis*
  - *Acute inflammatory demyelinating polyneuropathy (Guillain-Barré Syndrome)*
  - *Neuroimmunologic disorders*
- The procedure involves the insertion of a hollow needle between the third and fourth vertebrae into the spinal canal; however, the needle does not contact spinal nerves.
- Most often 5 to 20 ml of CSF is collected and measured with a manometer attached to the needle.
- The CSF is then evaluated for cell counts, biochemical and immunologic studies, and microbiologic analysis.
- While the procedure is safe, some patients experience a spinal headache afterwards, caused by minimal CSF leakage.

Myelogram

- A myelogram is an *invasive* procedure that uses x-ray technology and contrast agents to evaluate the condition of the spine, spinal canal, spinal cord, spinal nerve roots, and vertebral discs.
- A lumbar puncture is used to draw CSF for testing. A contrast agent is then injected into the spinal canal, and an x-ray fluoroscope then records images produced by the contrast material.
- While regular x-rays of the spine only allow observation of the bones, the contrast agent used in a myelogram appears white on x-rays, allowing detailed observation of the spinal cord, nerves, and spinal canal.
- Myelograms can be used to evaluate and diagnose the following:
  - *Spinal cord tumors*
  - *Spinal abscesses*
  - *Herniated discs causing nerve root impingement*

Electroencephalography

- An electroencephalogram (EEG) is a *noninvasive* test that is able to record neural electrical patterns or brain waves.
- Most neurologic imaging scans have replaced the EEG in the determination of the exact location of anatomic pathology.
- The EEG is currently used as a pathophysiological tool able to identify *abnormal cerebral function that cannot be visualized through radiographic or magnetic imaging scans.*
- The EEG is most commonly used in the evaluation of the following:
  - *Seizures* (transient state)—while the EEG is useful for the evaluation of seizures, it is not useful for the evaluation of headache or dizziness.
  - *Epilepsy*
  - *Herpes simplex encephalitis* (evolving condition)
  - *Dementia* (global disorder)
- The EEG is able to measure the changing electrical potentials on the scalp in response to a controlled stimulus.
- Electrodes are placed on the scalp over specific brain regions (and attached with an adhesive substance). Electrodes are able to detect the small electrical signals that form brain waves and transmit these signals to the EEG machine.
- Limitations of the EEG include the following:
  - The test cannot detect brainstem activity well.
  - An EEG is not useful in the diagnosis of brain death.

Evoked Potentials

- Evoked potentials are used in the *assessment of several sensory organs* including vision, audition, and tactile sensation.

Visual System

- Visual evoked potentials are recorded over the occipital region in response to a controlled stimulus.
- This assessment can detect pre- and postchiasmal lesions and asymptomatic optic neuritis. Because pathology of the optic nerves is poorly detected by magnetic resonance imaging (MRI), visual evoked potentials have become an important adjunctive tool in the diagnosis of optic nerve lesions.

Auditory System

- Auditory evoked potentials are used in the evaluation of brainstem structures and auditory functions.
- Five specific waves are recorded from the vestibulocochlear nerve, the cochlear nucleus, the superior olivary complex, the lateral lemniscus, and the inferior colliculus.
- Auditory evoked potentials can detect physiologic lesions that are below the resolution limit of neurologic imaging techniques.

Somatosensory System

- Upper extremity somatosensory evoked potentials are recorded following stimulation of the median nerve.
- Lower extremity somatosensory evoked potentials are recorded following stimulation of the posterior tibial nerve.
- Somatosensory evoked potentials can aid in the diagnosis of multiple sclerosis, because multiple sclerosis plaques in the spinal cord are difficult to detect using MRI.

Electromyography and Nerve Conduction Studies

- Electromyography (EMG) and nerve conduction studies are valuable in the assessment and *diagnosis of peripheral nerve and muscular disorders.*
- EMG involves the recording of spontaneous, voluntary, and electrically stimulated muscle activity through small intramuscular needle electrodes.

- An EMG involves the insertion of very thin needles into specific muscles under evaluation. Each needle contains an electrode that is attached to a wire; the wire transmits signals to an oscilloscope, which records electrical signals in the form of wave patterns.
- Nerve conduction velocity (NCV) involves electrical stimulation of a peripheral nerve. The peripheral nerve's rate of transmission (or speed of conduction) and amplitude of response are then measured.
- NVC procedures involve the taping of small pads to the skin surface along the path of a nerve. These pads electrically stimulate the nerve under evaluation. Nerve signal conduction pathology is detected as the electrical current travels along the nerve and is picked up by the surface pads.
- EMG and NCV yield data that can determine the distinction among the following:
  - Primary motor neuron disease vs muscular disease
  - Demyelinating vs axonal neuropathy
  - Nerve root vs plexus disorders
- For example, these evaluations can determine the difference between motor neuron diseases (such as amyotrophic lateral sclerosis) and muscle disorders (such as myotonic dystrophy).
- EMG and NCV can also determine the presence and location of mononeuropathy. The most common entrapment syndrome best detected by an EMG is carpal tunnel syndrome.

Angiography

Cerebral Angiography

- Cerebral angiography is used to observe the blood vessels of the brain for narrowing or blockage.
- Angiography is an *invasive* procedure that involves *entering a catheter into the body to inject contrast material through the carotid arteries* (which supply the brain).
- Insertion of the catheter is usually made in the femoral artery. After insertion, the catheter is guided through the arterial system to the exact cerebral position under evaluation.
- Regular *x-rays are then used to image the contrast material* as it flows throughout the blood vessels.
- Cerebral angiography is most commonly performed after other tests (such as a CT scan) have detected pathology.
- The procedure provides a good visualization of the neurologic vascular system and is commonly used prior to neurosurgical procedures.
- Cerebral angiography is particularly useful in the detection of the following:
  - *Acute cerebral vascular accident*
  - *Cerebral anastomoses and occlusions*
  - *Aneurysm*
  - *Arteriovenous malformation (AVM)*
  - *Tumor*
  - *Arterial stenosis*

Magnetic Resonance Angiography

- Magnetic resonance angiography (MRA) uses *MRI technology* to detect and diagnose *cerebral vascular disease* and cardiac disease.
- MRA is a *noninvasive* method of evaluating cerebral vascular structures without the use of contrast material. Sometimes, however, contrast material is administered to enhance resolution.
- This technique is widely used to detect diseased intracranial arteries.
- For example, MRA can detect aneurysms that cannot be observed by conventional angiography.

Computerized Tomography Angiography

- Computerized tomographic angiography (CTA) is a *noninvasive* imaging method that uses *computerized tomography (CT)* to observe blood flow throughout the brain and body.
- CT combines the use of x-rays with computer analysis of images.
- X-rays are passed through the patient's body from different angles to produce cross-sectional images. These images are then assembled by computer into three-dimensional photos.

- CTA is *able to detect transient ischemic attacks (TIA)* in order to prevent a full CVA.
- Examining cerebral arteries using CTA may also help in the correct diagnosis of patients who report headache, dizziness, tinnitus, and/or syncope.

Magnetic Resonance Imaging

- Magnetic resonance imaging (MRI) uses magnets and radiowaves to detect subtle electromagnetic fields in the brain.
- Radiofrequency waves are directed at protons (the nuclei of hydrogen atoms). Protons emit radio signals that can be computer-processed to form an image.
- Images can be viewed in cross-sectional slices of the coronal, sagittal, and horizontal planes. Such cross-sectional slices can be as thin as one-quarter of an inch.
- MRI is a *noninvasive procedure* that does not require the injection of contrast material and does not expose the patient to ionizing radiation.
- It is valuable for providing images of the *heart*, *large blood vessels*, *soft tissues*, and *brain*.
- MRI is more sensitive than CT.
- Because MRI offers high-resolution images in multiple planes, it has become the preferred clinical and research tool for anatomic imaging.
- Diagnoses in which MRI is most useful include:
  - *Arteriosclerosis*
  - *Arteriovenous malformations*
  - *Vertebral disc disease*
  - *Spinal stenosis*
  - *Spinal metastases*
  - *Chiari malformation*
  - *Spina bifida*
  - *Spinal cord tumors*
  - *Demyelinating* and *degenerative brain disorders*
  - *Brain hemorrhages*, *infarcts*, and *space-occupying lesions*

Functional Magnetic Resonance Imaging

- Functional magnetic resonance (fMRI) is a *noninvasive* tool that combines the high resolution of conventional MRI with the opportunity to measure brain function and metabolic processes during neurologic activity.
- fMRI is able to detect small changes in the magnetic resonance signal resulting from an increase or decrease in blood oxygen level. Changes in blood oxygen level occur in response to the cellular activity of brain structures.
- This imaging technique allows researchers to *examine how brain structures function in various neurologic pathologies*—particularly neurobehavioral disorders (eg, schizophrenia, depression, autism).
- Researchers can use fMRI to determine which brain regions are responsible for functions such as cognition, language and speech, movement, and sensation.
- fMRI allows researchers to better understand how a normal, diseased, or injured brain is working.
- Additionally, neurosurgeons frequently use fMRI to map the brain for surgical procedures.

Computerized Tomography

- Computerized tomography (CT) involves the absorption of photons by tissues to generate data that, after computerized processing, are presented in a familiar gray scale format.
- In CT scans of the brain, numerous x-rays are passed through the skull and brain at various angles and then joined together to produce cross-sectional images of structures.
- CT is a *noninvasive* procedure; however, contrast dye may be injected intravenously to enhance resolution.
- While MRIs are more sensitive than CT scans, CTs are better than MRIs in the detection of fresh blood within cranial spaces. For this reason, it is the preferred method of imaging a *cerebrovascular accident (CVA)*.

- Emergency CT scans are used to rule out hemorrhage in the case of CVA. This knowledge is important when considering whether to administer anticoagulant (blood thinning) drugs—which are contraindicated in hemorrhagic stroke.
- In the subacute phase of CVA, CT scans are used to detect edema and infarction.
- CT scans are particularly effective in the detection of the following:
  - *Subarachnoid hemorrhages*
  - *Subdural collections of blood*
  - *Neoplasms*
  - *Vascular anomalies (eg, aneurysms and arteriovenous malformations)*

Magnetic Resonance Spectroscopy

- Magnetic resonance spectroscopy (MRS) is an imaging technique that applies computer software programming to the traditional MRI hardware.
- MRS is used to determine the molecular structure of a compound or to determine the compound's presence.
- It is based on the idea that nuclei of certain atoms have characteristic properties. As a result, signals can be received and displayed as an image.
- MRS is *noninvasive* and does not expose the patient to ionizing radiation.
- It is an important tool in the diagnosis and monitoring of the following:
  - *The progression of CVA*
  - *Ischemic injury*
  - *Intracranial tumors*
  - *Multiple sclerosis*
  - *Dementias*
  - *Encephalopathies*
- MRS is most commonly used as an adjunct to MRI when differential diagnoses cannot be otherwise made. For example, it is used to:
  - *Differentiate CVAs from neoplasms*
  - *Make clearer diagnoses when recurrent tumor, neoplasia, and necrosis cannot be otherwise diagnosed*

Positron Emission Tomography

- Positron emission tomography (PET) involves the measurement of positron emission from an injected radionuclide.
- A PET scan is an integration of two technologies—CT and radioactive tracers.
- The tracer is a radioactively labeled substance that is injected into the body and emits gamma rays. Gamma rays can be detected by PET in order to observe various metabolic processes in the brain and body. The tracers are similar to the contrast agents used in CT and MRI technologies.
- When the tracer is absorbed into the body, the patient is then scanned. PET is able to collect information emitted by the tracer and transforms it into two-dimensional cross-sectional images.
- The radioisotopes used in PET to label tracers can be detected by the scanner but pass through the body safely.
- Like fMRI, PET is able to *measure cellular changes resulting from increases or decreases in blood oxygen and glucose levels*. In other words, PET measures changes in regional cerebral blood flow when people perform specific activities (within the confines of the PET scanner).
- In addition to blood oxygen and glucose levels, PET can measure the following:
  - *Blood flow*
  - *Blood volume*
  - *Tissue acidity*
  - *Drug activity*
- PET is also the preferred tool to measure the *growth rate of malignant tumors*.
- Both PET and fMRI have enabled researchers to *examine how neuroanatomical structures function in neurobehavioral disorders*.

- Because PET requires a cyclotron to produce isotopes, its use is limited to major research and medical centers.
- Like fMRI, PET requires intravenous injection of radioactive material.

Single Photon Emission Computed Tomography

- Single photon emission computed tomography (SPECT) employs radiation detectors to determine the location of a tracer drug in the brain.
- Simplistically, SPECT is a form of nuclear imaging that is primarily used to view how blood flows through arteries and veins in the brain.
- Like PET, SPECT is based on the two technologies of CT and radioactive tracers.
- Just prior to the scan, patients are injected with a tracer chemical that is radioactively labeled and emits gamma rays. Gamma rays can be detected by the SPECT scanner.
- The computer then collects the information emitted by the gamma rays and transforms them into two-dimensional cross-sectional images.
- SPECT differs from PET in that the SPECT tracer remains in the bloodstream rather than being absorbed by surrounding tissues. This limits detection of images to regions of blood flow.
- SPECT uses a tracer with a long half-life, making possible studies that involve a prolonged series of scans over a 4-hour period.
- Like PET and fMRI, SPECT can be used to determine which brain regions are active during functional task performance. SPECT, however, has a poorer resolution than fMRI, making it less efficient in the examination of neurobehavioral disorders.
- While SPECT has less sensitivity than PET, it is less expensive and more available than PET.

Neurologic Indications for CT

- Acute cerebrovascular disease
- Acute head trauma
- Extracerebral tumors (particularly meningiomas)
- Intracranial calcification or osseous lesions
- Lumbar spine presurgical planning
- Subarachnoid hemorrhage

Neurologic Indications for MRI

- Cervical spine damage
- Demyelinating disorders
- Head trauma (after acute phase)
- Inflammatory disorders
- Intracranial tumors
- Osseous vertebral column metastases
- Posterior fossa and craniovertebral junction lesions
- Seizures
- Spinal cord lesions

SECTION 29

Neurotransmitters: The Neurochemical Basis of Human Behavior

- Neurotransmitters are chemicals that relay and modulate messages between neurons.
- Much of human behavior is mediated by the action of neurotransmitters in the brain. Researchers are also demonstrating that *behavioral pathology* is largely due to imbalances in one or more neurotransmitter systems. Physical diseases may also be due to specific neurotransmitter pathway disturbances (eg, Parkinson' disease).
- Neurotransmitters are chemicals that are stored in and released from the terminal boutons of neurons; they are released into the synaptic cleft.
- Once in the synaptic cleft, neurotransmitters can be destroyed by *antitransmitters*—a chemical substance that breaks down a neurotransmitter so that the postsynaptic neuron can repolarize in order to fire again.
- Neurotransmitters can also be reabsorbed by the presynaptic terminal boutons in a process called *reuptake*.
- Sometimes there is a decrease in the number of receptors for a neurotransmitter on the postsynaptic neuron due to long-term exposure to the neurotransmitter. This is called *down regulation*.
- Neurotransmitters can be classified into four major groups:
  1. *Amino acids (eg, glutamate [GLU], gamma aminobutyric acid [GABA], aspartic acid, glycine)*
  2. *Peptides (eg, vasopressin, somatostatin)*
  3. *Monoamines (norepinephrine [NE], dopamine, serotonin)*
  4. *Acetylcholine (ACh)*
- There are multiple neurotransmitter systems in the brain (and in the enteric nervous system).
- In the peripheral nervous system (PNS), only two neurotransmitters are used—ACh and norepinephrine.
- Neurotransmitters can be *excitatory* or *inhibitory* depending upon which receptor sites to which they bind. They work like the brain's brake and accelerator systems.

Major Neurotransmitters About Which Information Is Known

- Acetylcholine (ACh)
- Gamma aminobutyric acid (GABA)
- Glutamate (GLU)
- Dopamine (DA)
- Serotonin (5-HT)
- Norepinephrine (NE)
- Substance P
- Endorphins

Acetylcholine (ACh)

- ACh was the first neurotransmitter to be discovered.
- In the PNS, ACh is the major neurotransmitter that *controls muscle action*. There are comparatively few ACh receptors in the brain.
- This neurotransmitter most often has *excitatory effects*.

- In the PNS, ACh facilitates action at the *neuromuscular junction*.
- Excessive levels of ACh at the motor endplates of the neuromuscular junction can result in dyskinesia. *Dyskinesia* is a hyperkinetic motor disorder characterized by involuntary muscle contractions.
- Deficient levels of ACh at the neuromuscular junction can result in *paralysis*.
- In the CNS, ACh plays a role in regulating the autonomic nervous system (ANS)—such as in the regulation of heart rate.
- There are two main classes of ACh: a fast-acting receptor called *nicotinic*—because it is activated by the toxin in tobacco—and a slow-acting receptor named *muscarinic*—because it is activated by the toxin muscarine (found in poisonous mushrooms).
- *Acetylcholinesterase* is the *antitransmitter* that will stop the action of ACh at the neuromuscular junction. The effects of nerve agents (used in bioterrorism) such as *sarin gas* work because they inhibit acetylcholinesterase. This results in a painful continuous stimulation of muscles and glands.
- Certain insecticides similarly work because they inhibit acetylcholinesterase in insects.
- Some snake venoms contain toxins that can block nicotinic receptors and cause paralysis. For example, *curare* is a nicotinic blocking agent extracted from plants. Curare has been used as a poison (placed on arrowheads) by certain South American Indian groups.
- *Botulin* is also a poison that works as an ACh blocking agent causing paralysis. *Botox*, which is a botulin derivative, has become a popular cosmetic treatment to diminish facial wrinkles by temporarily paralyzing the responsible muscles.

Alzheimer's Disease

- A shortage of ACh in the brain has been implicated as a contributing factor in *Alzheimer's disease*. Certain drugs that inhibit acetylcholinesterase have been found to have some effectiveness in treatment of the disease.

Myasthenia Gravis

- *Myasthenia gravis* is a disease characterized by muscle weakness and fatigue. The disease occurs when the body inappropriately produces antibodies against ACh receptors. This causes ACh transmission to become impaired. Drugs that inhibit acetylcholinesterase have been found to be effective in the treatment of myasthenia gravis.

Gamma Aminobutyric Acid (GABA)

- GABA is the *major inhibitory neurotransmitter* of the brain.
- It essentially turns off the function of cells and acts like a brake on excitatory neurotransmitters that can cause anxiety.
- Without inhibitory neurotransmitters like GABA, brains cells would fire uncontrollably (as in epileptic seizures).
- GABA is most highly concentrated in the substantia nigra and globus pallidus. Other neural areas of high GABA concentration include the hypothalamus, periaqueductal gray matter (of the midbrain), and hippocampus.
- There are two known GABA receptors: GABA-A and GABA-B.
    - GABA deficiency is implicated in *anxiety disorders*, *insomnia*, and *epilepsy*.
    - GABA excess is implicated in *memory loss* and the *inability for new learning*.
- Preliminary research suggests that agents that can block GABA-B may improve learning and memory.

GABA and Anxiety Disorders

- People who experience anxiety and panic attacks may have an imbalance in the GABA system involving a depletion of this neurotransmitter.
- *Benzodiazepines* are a class of drugs (eg, Xanax, Valium, Ativan) used in the treatment of anxiety disorders.
- These drugs work by enhancing the effect of GABA on GABA-A receptors.
- Prolonged use of benzodiazepines results in *adaptation*—in other words, GABA receptors increase in number or sensitivity to the drug, making it less effective over time (referred to as *down regulation*).
- Larger doses of benzodiazepines are then needed to provide relief from anxiety. This is called *tolerance*.
- Stopping the use of benzodiazepines can result in diminished sensitivity of GABA receptors, causing heightened anxiety.

- In 2005, some dietary supplement companies began selling GABA as a sleep aid and antianxiety treatment; however, research demonstrating the effectiveness of such aids has not been extensively conducted as of this writing.

Glutamate (Glutamic Acid or GLU)

- Glutamate is the *most common neurotransmitter in the* CNS and may account for as much as half of all neurotransmitters in the brain.
- It is one of the *major excitatory neurotransmitters* of the CNS and is believed to have a significant role in *learning and memory*.
- Some researchers believe that glutamate can be used to enhance memory and new learning in *Alzheimer's disease*.
- Glutamate also serves as a *precursor* for the synthesis of the inhibitory neurotransmitter GABA.
- It is present in a wide variety of foods and is responsible for the taste sense of *umami* (ie, meaty or hearty). The sodium salt of glutamic acid—monosodium glutamate (MSG)—is a common food additive used to enhance flavor.

Epileptic Seizures

- *Overactivity of glutamate* may produce epileptic seizures.
- There is a delicate balance between excitation and inhibition in the brain.
- In epilepsy, the brain's chemistry may have titled too far toward excitation.
- Some researchers suggest that any disorder—Alzheimer's disease, schizophrenia, Parkinson's disease, epilepsy, CVA—may essentially involve an imbalance in one or more neurotransmitter systems.

Excitotoxicity

- An *excess of glutamate* can produce *neuronal damage* and *cell death*, or excitotoxicity. Frequently, trauma to the brain (eg, CVA, head injury) will trigger an excess release of glutamate resulting in the cell death of many more neurons than had occurred in the original trauma.
- For example, in CVA, researchers believe that it is not just oxygen deprivation that kills brain cells but also the release of too much glutamate.
- When the brain experiences a series of crises, glutamate is released 1000 times more than its normal level.
- This may serve as the organism's way of facilitating a painless death process when severe CNS trauma occurs.
- Physicians, in their attempt to save an individual's life, inadvertently stop the release of glutamate after it has already killed off an extensive amount of brain cells.
- Often, glutamate release is responsible for the greatest damage in TBIs and CVAs—rather than the initial injury.
- Researchers are attempting to develop drugs that can immediately prevent the release of glutamate after a severe brain injury.
- However, such drugs would have to be administered in the first hours after injury, and often patients are not transported to hospital emergency rooms in time.
- Another suggested strategy involves the development of drugs that could cause glutamate to bind with neighboring neurons without killing them.
- Essentially, researchers would be changing the way that glutamate works in the brain.

Dopamine (DA)

- The dopamine system has major effects on the *motor system*, *cognition*, and *motivation/reward*.
- Dopamine does not have an antitransmitter. Instead, it is removed from the receptor site by (a) reuptake into the presynaptic neuron, (b) enzymatic breakdown, or (c) diffusion out of the synaptic cleft.
- These processes are slower than breakdown by an antitransmitter, and thus, the effects of dopamine are longer lasting than ACh.
- Researchers have identified several *sources of dopamine in the* CNS:
  - Substantia nigra (in the midbrain)
  - Tegmentum (in the midbrain)

Four Dopaminergic Tracts in the Brain

- The *nigrostriatal tract* travels from the substantia nigra to the striatum and accounts for most of the brain's dopamine.
- The *tuberoinfundibular tract* travels from the arcuate nucleus of the hypothalamus to the pituitary stalk. This pathway controls the release of the hormone prolactin through D2 receptors.
- The *mesolimbic tract* travels from the ventral tegmental area to parts of the limbic system.
- The *mesocortical tract* travels from the ventral tegmental area to the neocortex—particularly the prefrontal area.

Parkinson's Disease

- Dopamine affects the basal ganglia and thus influences movement.
- *Loss of dopamine from the substantia nigra and the nigrostriatal pathway* is believed to be the primary cause of Parkinson's disease—causing paucity of movement, festinating gait, and masked face.
- Precursors to dopamine, such as L-dopa, can alleviate some of the symptoms of Parkinson's disease.

Schizophrenia

- *Too much dopamine* has been implicated in schizophrenia, causing hallucinations, delusions, disorganized thinking, and paucity of thought.
- Dopamine neurons in the mesolimbic pathway are associated with psychosis and schizophrenia.
- *Phenothiazines* are a class of antipsychotic drugs that block D2 dopamine receptors. These drugs can effectively *reduce psychotic symptoms*.
- Drugs such as *amphetamines* and *cocaine* greatly increase dopamine levels and can *cause psychosis*.
- Most modern antipsychotic medications are designed to block function of dopamine receptors to some degree.
- Conversely, *blockage of D2 dopamine receptors* has been associated with *depression relapse*. Such blockage can reduce the effectiveness of the class of antidepressant medications known as *selective serotonin reuptake inhibitors (SSRIs)*.
- Older classifications of drugs used to treat schizophrenia would often cause *Parkinsonian-like symptoms* called *tardive dyskinesia*. Tardive dyskinesia is a hyperkinetic motor disorder characterized by involuntary muscle contractions: lip smacking, repetitive tongue protrusion, and blepharospasm. *Clozapine* is a newer class of antipsychotic drug that reduces the affects of schizophrenia without inducing Parkinsonism.

The Reward System and Addiction

- Dopamine has a significant role in the brain's reward system and is commonly released in response to highly pleasurable experiences such as eating and sexual activity.
- Dopamine is also strongly involved in the process of addiction.
- When drugs such as heroin, amphetamines, and cocaine are ingested, *dopamine levels rise to unnaturally high levels in the brain*. This accounts for the high experienced in response to addictive substances.
- This unnatural boost of dopamine acts like an assault on the brain. To counteract the assault, the *brain begins to decrease its natural production of dopamine causing severe craving for the addictive substance*.

Cognitive Function

- Dopamine disorders in the frontal lobes are believed to be responsible for a *decline in cognitive functions such as memory, attention, and problem solving*.
- Some researchers suggest that diminished dopamine concentrations in the prefrontal cortex are a contributing factor in *attention deficit disorder (ADD)* and the *negative symptoms of depression and schizophrenia* (eg, social withdrawal, apathy, anhedonia).

Serotonin (5-HT)

- Serotonin is synthesized in serotonergic neurons in the CNS and gastrointestinal (GI) tract. It is synthesized from the amino acid *tryptophan*.
- In the CNS, the neurons of the raphe nuclei of the medulla are the principle source of serotonin release.
- There are more than 14 identified serotonin systems in the CNS.
- Serotonin has been implicated in sleep, *emotional control and equanimity*, *pain regulation*, *emesis (vomiting)*, *and carbohydrate feeding behaviors (or the binging behaviors that occur in certain eating disorders)*.

Regulation of Circadian Rhythms and Sleep-Wake Cycles

- The richest concentration of serotonin in the body is found in the *pineal gland*—although this gland does not use serotonin as a neurotransmitter. Rather the pineal gland uses serotonin to synthesize *melatonin*—a substance which is important in the regulation of circadian rhythms, diurnal patterns, and sleep-wake cycles.
- The *supra chiasmatic nucleus (SCN)* of the hypothalamus is extensively innervated by serotonergic input from the raphe nucleus and is responsible for regulating circadian rhythms.
- Low serotonin levels may disrupt SCN regulation of circadian rhythms seen in *seasonal affective disorder*.
- Depletion of serotonin may also be related to the disruption of sleep–wake cycles seen in *aging*.

Depression, Anger, Obsessive–Compulsive Disorder

- Low levels of serotonin are associated with *depression* and *suicidal behavior*.
- The *SSRIs* are a classification of antidepressants that increase the brain's levels of serotonin by blocking the reuptake of serotonin in the presynaptic neuron.
- *Kramer's Kindling Theory of Depression* is an attempt to describe the physiologic process of chronic, recurrent depression. The first three or four depressive episodes (which may be minor) can cause a permanent change in the serotonin system. With each successive depressive episode, the synaptic vessels produce and release less serotonin. Once the serotonin system becomes less responsive, it may never return to its normal activity level without intervention.
- Low levels of serotonin have also been associated with *aggression*, *anger*, and *violence*. Some researchers have suggested that impulsivity and aggression may be inversely correlated with the responsiveness of one's serotonin system. The less active or responsive a serotonin system is, the more impulsive and aggressive a person may behave. SSRIs have been used effectively to reduce levels of anger, irritability, and depression.
- In addition to depression and anger, too little serotonin has also been associated with *obsessive–compulsive disorder (OCD)*. The SSRIs have shown some effectiveness in the treatment of OCD.

Appetite and Eating Disorders

- Serotonin levels also play a role in *appetite* and *carbohydrate feeding behaviors*.
- Low serotonin levels have been shown to be related to increased appetite and carbohydrate craving. Studies on patients with *bulimia* have demonstrated that low serotonin levels are present in patients demonstrating *binging behaviors*.
- The antidepressant Prozac has been used effectively to treat patients with binging and compulsive eating disorders.

NOREPINEPHRINE (NE)

- Also called noradrenalin
- Most norepinephrine pathways in the brain have their origin in the *locus ceruleus* (a nucleus in the pons).
- Norepinephrine and ACh are the only two neurotransmitters used in the *peripheral nervous system (PNS)*.
- In the CNS, norepinephrine is released from the *medulla* (of the adrenal glands) as a *hormone* into the blood.
- Norepinephrine is also a *neurotransmitter* released from noradrenergic neurons during synaptic transmission.
- Norepinephrine does not have an antitransmitter. Rather, it is removed from the receptor site by active transport or by diffusion out of the synaptic cleft.
- These processes are slower than antitransmitter breakdown, and thus, the effects of norepinephrine can be long lasting.

Alertness and Attention

- Noradrenergic neurons that project from the locus ceruleus of the reticular activating system play a major role in *wakefulness/arousal*.
- Because of this connection to the *reticular activating system*, norepinephrine is important in the *active surveillance of one's surroundings* by increasing attendance to sensory information from the environment. It has thus become an important agent in pharmaceutical intervention for people with *attention deficit hyperactivity disorder (ADHD)* because it appears to enhance concentration and cognitive organization.

- Psychostimulant medications (such as Ritalin, Concerta, Dexedrine, and Adderall) are commonly prescribed to increase norepinephrine and dopamine levels in people with ADD/ADHD.
- Strattera is a selective norepinephrine reuptake inhibitor sometimes used in the treatment of ADHD that only affects norepinephrine rather than dopamine.
- As a result, Strattera has a lower potential for abuse and can remain active in the body for longer periods of time. Strattera has been shown to be effective for some patients with ADD/ADHD.

Stress Hormone

- As a stress hormone, norepinephrine is essential in *activating the sympathetic nervous system* to produce the *fight/flight response*, *increase heart rate*, *release energy from fat storage*, and *increase muscle preparedness*.
- Overactivity of the norepinephrine system produces *fear*, *anxiety*, and *panic* (through the action of cortical and limbic regions).
- *Beta blocking agents*—such as Propranolol—prevent norepinephrine from binding to beta receptors. This prevents sweating, rapid heart beat (tachycardia), and other sympathetic NS signs that may occur in stressful situations.
- Musicians, actors, and public speakers commonly use beta blockers before performances to reduce sympathetic NS signs and to enhance calmness.

Depression

- Low levels of norepinephrine have been implicated in *depression*.
- *Serotonin-norepinephrine reuptake inhibitors (SNRIs)* are a class of antidepressants that increase the amount of serotonin and norepinephrine available in the brain.
- There is recent evidence that SNRIs may increase dopamine levels as well.
- Like the SSRIs, the SNRIs have also shown effectiveness in the treatment of depression.

Substance P

- Substance P is classified as a neuropeptide that functions as a neurotransmitter and neuromodulator.
- In the CNS Substance P is associated with *mood regulation*, *anxiety*, *stress*, *neurogenesis*, *respiratory rhythm*, *neurotoxicity*, *vasodilation*, *nausea and emesis*, and *pain perception*.
- Substance P acts as a neurotransmitter in the nociceptive pathway and is involved in the *transmission of pain signals from peripheral receptors to the CNS*. The nociceptive pathway mediates the sensation of pain.
- Substance P is found in the dorsal horn of the spinal cord, substantia nigra, amygdala, hypothalamus, and the cerebral cortex.
- Some researchers have suggested that Substance P may play a role in *fibromyalgia*.
- The pain reliever capsaicin (the active ingredient in peppers) has been shown to reduce Substance P levels.

Opioid Peptides

- Opioid peptides include *endorphins*, *enkephalins*, and *dynorphins*.
- These peptides are neurotransmitters produced in the *pituitary gland* and *hypothalamus*. The highest concentration of opioid receptors is found in the sensory, limbic, and hypothalamic regions of the CNS. The amygdala and periaqueductal gray (of the midbrain) have particularly high concentrations of opioid receptors.
- The opioids resemble *opiates* (opium, morphine, heroin) in their ability to produce analgesia and feelings of well-being. They are sometimes referred to as the body's natural pain killers or natural morphine.
- The opioids have a major role in the perception of pleasure produced by the mesocorticolimbic system—the brain's reward system. Along with dopamine, the opioids may contribute to the mechanisms underlying addiction.
- The primary action of the opioids is the *inhibition of nociceptive or pain information*. Endorphins work conjointly with Substance P as a pain modulator. The analgesic capsaicin (the active ingredient in peppers) has been shown to stimulate the release of endorphins. Capsaicin in the form of a topical cream has been used effectively in the treatment of certain types of chronic pain.
- Some researchers have found that endorphin release may underlie the placebo effect in the treatment of pain.
- Other studies have found that *acupuncture* may trigger the production of endorphins and thus produce pain relief.

- In addition to modulating pain, opioid peptides also have a role in *cardiac*, *gastric*, and *vascular functions*.
- Low levels of opioid peptides have also been associated with *anxiety* and *panic*.
- There is evidence, too, that some opioid peptides play a role in *satiety* and *appetite control*.

Neurotransmitters Can Overlap in Function

- Several different neurotransmitters can regulate the same behaviors.
- This redundancy in the regulation of same functions is seen throughout the CNS and the PNS.
- This redundancy is likely the nervous system's way of creating a fool-proof system—akin to having back-up systems if something goes wrong. This mechanism may have developed as an evolutionary safe-guard for the survival of human beings.
- *Dopamine*, *serotonin*, and *opioid peptides* appear to regulate functions and disorders involving the experience of anger, depression, anxiety, and addictive behaviors.
- *Norepinephrine* also appears to regulate well being, anxiety, and arousal.
- GABA and *glutamate* have shared roles in memory and learning.
- ACh and *dopamine* have shared roles in the regulation of motor activity.
- Neurotransmitters involved in wakefulness and consciousness include *norepinephrine*, *dopamine*, ACh, *histamine*, and *glutamate*.
- Neuropeptides involved in wakefulness and consciousness include *corticotropin releasing hormone* (CRH), *thyrotropin releasing hormone* (TRH), and *vasoactive intestinal peptide* (VIP). Deficiency in one or more of these substances can produce *somnolence* (sleep).

Role of Neurotransmitters in the Treatment of *Depression*

- Depression is a serious clinical disorder that affects approximately 16% of the population at least once in their lives.
- The average age of onset occurs in the mid to late 20s and affects females more than males at a 2 to 1 ratio.
- Depression is often characterized by the following: anhedonia (loss of interest in once enjoyable activities), sadness, fear, hopelessness, helplessness, guilt, difficulty concentrating, decreased memory functions, loss of energy, psychomotor agitation, and changes in sleep patterns, appetite, and weight.
- The etiology of depression may have multiple contributing factors including heredity, physiology, medical conditions, and diet.
- Several neurotransmitters have been implicated in the cause of depression including serotonin, norepinephrine, and dopamine.

Selective Serotonin Reuptake Inhibitors (SSRIs)

- The SSRIs are a class of antidepressants currently considered as standard first line agents.
- Researchers have suggested that low levels of serotonin play a major role in the etiology of depression.
- The SSRIs may work, in part, by preventing the reabsorption of serotonin into nerve cells, thus increasing the amount available for use by the brain.
- SSRIs include fluoxetine (Prozac), paroxetine (Paxil), escitalopram (Lexapro), citalopram (Celexa), and sertraline (Zoloft).
- While these antidepressants work as effectively as older classes, they have fewer side effects and thus may be easier to tolerate.

(Continued)

Role of Neurotransmitters in the Treatment of Depression (Continued)

Norepinephrine Reuptake Inhibitors (NeRIs)

- Researchers have found that in some patients, low levels of norepinephrine may be a factor in the etiology of depression.
- Norepinephrine reuptake inhibitors (NeRIs) are a newer class of antidepressants that block the reabsorption of norepinephrine into the presynaptic neuron, thus increasing the amount available for use by the brain.
- Because norepinephrine is one of the neurochemicals believed to regulate concentration and motivation, the NeRIs are thought to have a positive effect on alertness and attention as well as mood.
- NeRIs include such drugs as reboxetine (Edronax).

Serotonin–Norepinephrine Reuptake Inhibitors (SNRIs)

- SNRIs are a newer class of antidepressant drug that work on both norepinephrine and serotonin. Recent research suggests that the SNRIs may also affect dopamine levels.
- They are typically used as second line agents when patients do not respond to SSRI treatment.
- Like the SSRIs, they have fewer side effects than the older classes of antidepressants but are about as effective.
- Examples include venlafaxine (Effexor) and duloxetine (Cymbalta).

Norepinephrine–Dopamine Reuptake Inhibitors (NDRIs)

- Bupropion (Wellbutrin) is an NDRI that is believed to inhibit the reuptake of dopamine.
- It also blocks the reuptake of serotonin and norepinephrine but to a much lesser extent when compared to the tricyclic drugs.
- Bupropion has been shown to be effective in some patients.
- Common side effects include increased restlessness, agitation, and insomnia, which may be intolerable to some patients.
- When administered in higher doses, bupropion may cause seizures and is thus contraindicated in patients with seizure disorders.
- Bupropion is also contraindicated in patients with anorexia nervosa, as this antidepressant may stimulate appetite loss.

Tricyclic Antidepressants

- The tricyclics are the oldest class of antidepressants and include drugs such as amitriptyline and desipramine.
- The tricyclics block the reuptake of norepinephrine and serotonin.
- While their effectiveness is equal or greater to the SSRIs, they are no longer commonly used because of their side effects—which are more severe than those of the SSRIs. An overdose of tricyclic drugs also has the potential to be lethal.
- Generally, tricyclic drugs are used when patients do not respond to the SSRIs, NeRIs, SNRIs, and NDRIs.

Monoamine Oxidase Inhibitors (MAOIs)

- MAOIs block the enzyme monoamine oxidase, which breaks down serotonin and norepinephrine.
- The MAOIs are as or more effective than the tricyclics but are rarely used because of their potentially fatal interactions with certain foods and drugs.
- MAOIs such as Nardil may be used when patients are not responsive to the SSRIs, NeRIs, SNRIs, NDRIs, or tricyclic drugs. Patients using an MAOI must remain on a restricted diet to avoid drug and food interactions.
- A new MAOI has recently been developed called moclobemide (Manerix). This drug is referred to as a reversible inhibitor of monoamine oxidase A (RIMA). Because it uses a very specific chemical pathway in the brain, it does not pose the threat of fatal drug/food interactions, and patients taking moclobemide do not have to remain on a restricted diet.

Role of Neurotransmitters in the Treatment of *Obsessive–Compulsive Disorder*

- OCD is a type of anxiety disorder involving obsessive thoughts and worries. To alleviate the obsessive worries, people engage in compulsive behaviors that become ritualistic (ie, performed repeatedly in the same way).
- Examples of OCD include a fear of dirt and germs (eg, compulsive hand washing rituals), checking disorders (eg, checking repeatedly to see if the door is locked), counting disorders (ie, compulsions to ritualistically count such things as floor tiles) and hoarding disorders (ie, fear of throwing anything away in case it may one day be needed).
- Approximately 2.5% of the population will experience some form of OCD at some point in their lives.
- The disorder occurs equally among males and females and all age groups.
- Researchers believe that the etiology of OCD involves genetics and an imbalance in certain neurotransmitter systems.
- Some researchers have suggested that OCD can sometimes be caused by a streptococcal infection occurring in childhood. Researchers found that a high rate of strep infections occurred in the 3-month period prior to the onset of OCD symptoms. A child who experienced several strep infections had nearly 14 times the average risk of developing OCD in the following year. It is suggested that antibodies recruited to defeat the infection attacked other body systems including the basal ganglia—a neural region implicated in OCD.
- PET scans of people diagnosed with OCD have shown malfunction in a neural circuit connecting the frontal lobes and the basal ganglia. The basal ganglia normally filter messages to and from the frontal lobes regarding body movements. In OCD, dysfunction of this circuit may lead to the obsessional thoughts and ritualistic compulsive behaviors seen in the disorder.
- Neurotransmitters that are found in the pathway between the frontal lobes and the basal ganglia include serotonin and dopamine.
- The role of these neurotransmitters, however, is unclear. The SSRIs have some effectiveness in the treatment of OCD, while drugs that increase the activity of dopamine worsen symptoms of the disorder.

Antidepressant Drug Treatment

- The primary choice for drug treatment of OCD is currently the selective serotonin reuptake inhibitors (SSRIs)—fluoxetine (Prozac), fluvoxamine (Luvox), sertraline (Zoloft), paroxetine (Paxil), escitalopram (Lexapro), or citalopram (Celexa).
- Venlafaxine (Effexor)—a SNRI—has become a recent alternative.
- Some researchers believe that clomipramine (Anafranil)—a tricyclic antidepressant—may be the most effective treatment currently available. Clomipramine blocks the reuptake of norepinephrine and serotonin from the presynaptic neuron.
- Generally, OCD requires higher doses of the above antidepressant drugs than are normally administered in the treatment of depression. The antidepressants also take longer to work when treating OCD than when treating depression—usually as long as 3 months (compared to 2 to 4 weeks in the treatment of depression).
- Approximately half of all treated patients experience some relief from symptoms but frequently relapse when drug treatment is stopped. If treatment is continued, approximately 80% experience a reduction of symptoms but not a complete cessation. In most cases, the disorder remains throughout life and worsens if left untreated.

Role of Neurotransmitters in the Treatment of *Anxiety Disorders*

- Anxiety disorders involve excessive fear, worries, and physiological distress that interfere with one's daily function.
- Physical symptoms of anxiety disorders can include shortness of breath, palpitations, chest pain, psychomotor agitation, restlessness, sweating, and choking sensations.
- Anxiety disorders include panic attack, agoraphobia, OCD, post-traumatic stress disorder, generalized anxiety disorder, acute stress disorder, social phobia, and specific phobia.
- Recent research suggests that anxiety disorders may be caused by abnormalities in the function of several neurotransmitter systems.

(Continued)

Role of Neurotransmitters in the Treatment of Anxiety Disorders (Continued)

- Primary neurotransmitters that have been shown to play a role in the etiology of anxiety disorders include serotonin, norepinephrine, GABA, and dopamine.
- Currently, the most effective pharmaceutical agents in the treatment of anxiety disorders affect the GABA and serotonin systems.
- Drugs that are primarily used to treat anxiety are called anxiolytics.

Benzodiazepines

- The benzodiazepines are a class of drugs that were developed in the 1950s and 1960s.
- Common benzodiazepines include alprazolam (Xanax), diazepam (Valium), lorazepam (Ativan), clonazepam (Klonopin), and triazolam (Halcion).
- Benzodiazepines are highly effective anxiolytics that work quickly and are well tolerated.
- The disadvantages of using benzodiazepines include drowsiness, impaired memory, and incoordination. In addition, physical dependence can develop after long-term use along with the occurrence of withdrawal symptoms upon discontinuation.

Selective Serotonin Reuptake Inhibitors (SSRIs)

- While the SSRIs were developed as antidepressants, they have been shown to have some efficacy in the treatment of anxiety disorders.
- Because they are well tolerated, have fewer side effects than older classes of antidepressants, and lack the potential for physical dependence (as seen with benzodiazepines), the SSRIs have become first line agents in the treatment of anxiety.
- The benzodiazepines are primarily recommended as adjunctive treatment for anxiety disorders; however, they are still used as first line agents when patients are unresponsive to or cannot tolerate the SSRIs.

Serotonin and Norepinephrine Reuptake Inhibitors (SNRIs)

- Venlafaxine (Effexor) is an SNRI that has recently been approved for the treatment of generalized anxiety disorder.
- There is evidence, too, that venlafaxine's effectiveness may extend to other forms of anxiety disorders.
- As with the SSRIs, side effects may be poorly tolerated by some patients. Common side effects include weight gain, sexual dysfunction, and agitation.

GABA Enhancing Pharmaceuticals

- Researchers have begun to generate evidence that the neurotransmitter GABA has a significant role in the etiology of anxiety and other mood disorders.
- A large amount of research data now suggests that drugs affecting the GABA system are highly effective in the treatment of anxiety.
- Such drugs include the benzodiazepines and certain anticonvulsant drugs including tiagabine, valproate, vigabatrin, and gabapentin. In the last several years, anticonvulsant drugs have been commonly used as mood stabilizers in the treatment of bipolar and unipolar depression.
- The above GABA enhancing drugs have demonstrated effectiveness in the treatment of anxiety. Tiagabine, topiramate, valproate, and carbamazepine (all anticonvulsant drugs with GABA enhancing properties) have shown effectiveness in the treatment of post-traumatic stress disorder (PTSD). One characteristic of PTSD is nighttime awakening from vivid and frightening dreams. Tiagabine in particular has been found to normalize sleep disturbances and reduce the frequency of flashbacks in PTSD.
- Tiagabine enhances CNS GABA transport through its unique ability to inhibit GABA reuptake. Currently, tiagabine is the only available selective GABA reuptake inhibitor (SBRI). Tiagabine increases available extracellular GABA by up to 200%.
- Gabapentin increases GABA primarily by enhancing release of the neurotransmitter from glia cells. Gabapentin has been shown to have anxiolytic effects similar to those of the benzodiazepine, diazepam—but without the memory impairment often seen in long-term benzodiazepine use. Gabapentin has also been shown to be effective in panic disorder, social phobia, OCD, and PTSD.
- Both tiagabine and gabapentin appear to have the same high effectiveness as the benzodiazepines but do not produce the physical dependence or withdrawal symptoms upon discontinuation that are characteristic of the benzodiazepines.

Role of Neurotransmitters in the Treatment of *Eating Disorders*

- Today, most researchers believe that the etiology of eating disorders is multifactorial and includes genetic, neurotransmitter, and hormonal abnormalities in addition to cultural ones.
- Anorexia nervosa is characterized by severe reduction in food intake (to the point of starvation), extreme weight loss, distorted body image, and obsession with weight and food.
- Bulimia nervosa is characterized by repeated episodes of uncontrollable binge eating followed by self-induced purging (eg, vomiting, laxative abuse, excessive exercise).
- More than 90% of patients diagnosed with an eating disorder are adolescent and young adult women. For many years this finding was interpreted as evidence that social and cultural expectations of thinness lay at the center of the disorders' etiology.
- In recent years, however, an increasing amount of research has pointed to genetic, neurochemical, and hormonal abnormalities as primary factors underlying eating disorder etiology.

Serotonin

- Abnormalities in the serotonergic pathways have been connected to the onset and persistence of eating disorders. A growing body of research indicates that increased serotonin levels in the brain may be responsible for anorexic behavior, while decreased serotonin levels may account for bulimic behaviors.
- A large amount of evidence has been generated showing that increased serotonin activity is associated with appetite suppression. Drugs that suppress appetite and are used in the treatment of obesity work by increasing serotonin levels in the brain. Researchers have suggested that increased serotonin levels may account for the food restriction that is a common characteristic of anorexia.
- Conversely, decreased levels of serotonin are related to increased appetite. High serotonin levels may account for the gnawing hunger and inability to feel satiated that is characteristic of bulimia.
- People with bulimia may have disturbances in their satiation response center. The brain's satiety center lies in the ventromedial hypothalamus. When this part of the brain is stimulated, feelings of satiety occur and eating stops. In contrast, the lateral hypothalamus governs hunger. When it is stimulated, feelings of hunger ensue and eating behaviors are triggered. Normally, these two hypothalamic systems work cooperatively to maintain normal weight and eating patterns. When pathology occurs in the system, eating disorders can emerge.
- Serotonin has been linked to feelings of well-being and satiation. After eating a carbohydrate-rich diet, the body converts these sugars to tryptophan—the precursor of serotonin. People with bulimia may engage in binge eating behaviors in an attempt to raise their own serotonin levels.
- Conversely, people with anorexia may have an overactive serotonergic system leading to abnormally high levels of brain serotonin. The food restriction that is commonly seen in anorexia may be a self-imposed attempt to lower serotonin levels. Very high levels of serotonin have been correlated with nervousness, irritation, and jitteriness. People with anorexia may be using food restriction as a way to control these uncomfortable feelings.
- Some researchers have also suggested that people with eating disorders may become addicted to fasting or binge eating. Fasting and binge eating both trigger opioid and dopamine release in the brain, just as other addictive substances and behaviors do. People with eating disorders may become addicted to the chemicals released in the brain's reward center during fasting and binge eating.

Comorbid Disorders: Depression and Obsessive-Compulsive Disorder

- There is a high comorbidity between depression and eating disorders. Because many people with eating disorders also have depression, researchers have proposed a physiological link between these two pathologies. Low levels of serotonin and norepinephrine have been found in both people with depression and people with eating disorders.
- SSRIs and SNRIs are two classes of antidepressant drug that have been found to be effective in the treatment of both depression and anorexia.
- Researchers have also found a high comorbidity between bulimia and OCD. A disproportionately high number of people with bulimia also have OCD. Similarly, people with OCD frequently possess some form of abnormal eating behaviors. Again, biochemical similarities between the two groups have been found to exist largely within the serotonin system.
- The hormone vasopressin has also been found to be abnormal in both people with eating disorders and people with OCD. Researchers have found that vasopressin levels are abnormally elevated in patients with OCD, anorexia, and bulimia. Vasopressin is normally released in response to either physical or emotional stress. Abnormally elevated levels of vasopressin may contribute to the obsessive behavior seen in some patients with eating disorders.

Role of Neurotransmitters in the Treatment of *Schizophrenia*

- Schizophrenia is a serious psychiatric disorder characterized by disorganized thinking and perception.
- Positive symptoms of schizophrenia (or the presence of abnormal behaviors) include hallucinations, delusions, disorganized speech and cognition, severely disorganized behavior, and catatonic behavior.
- Negative symptoms of schizophrenia (or the absence of normal behaviors) include social withdrawal, isolation, poor self-care, blunted mood and affect, and lack of spontaneous thinking.
- Schizophrenia affects 1 in 100 people and occurs in all cultures and socioeconomic groups. Males and females are affected equally, but the typical age of onset differs by gender (male age of onset occurs in the early 20s; female age of onset occurs in the late 20s).

Neural Regions Involved in Schizophrenia

- Brain abnormalities have been found in several consistent areas including the frontal lobe, temporal lobe, limbic system (specifically the cingulate gyrus, amygdala, and hippocampus), thalamus, and cerebellum. All of these areas have been found to have decreased volumes in people with schizophrenia.
- Additionally, many studies have found a general decrease in cortical mass as demonstrated by abnormally widened sulci, enlarged ventricular spaces, and decreased volumes of gray and white matter.
- Imaging scans have also demonstrated decreased cerebral blood flow and glucose metabolism in the frontal lobe.

Neurotransmitters Involved in Schizophrenia

- The dopamine hypothesis of schizophrenia suggests that the disease is caused by an overactive dopamine system. Excessive dopamine can be responsible for disruptions in motor, cognitive, and emotional function.
- Studies have shown that excessive amounts of dopamine can induce psychosis. Recent studies have demonstrated that reduced numbers of D1 dopamine receptors may increase dopamine levels in the brain.
- While dopamine abnormalities have been considered to be the primary factor in the etiology of schizophrenia, researchers have also begun to compile evidence that other neurotransmitters may also play significant roles. These include serotonin, glutamate, GABA, and acetylcholine.
- Researchers have proposed that the etiology of schizophrenia either involves an imbalance in a single neurotransmitter—such as dopamine—that pathologically affects other neurotransmitters or that the cause of schizophrenia involves an imbalance of multiple neurotransmitter systems.

Pharmaceutical Intervention

- Drug treatment for schizophrenia involves two main categories—first generation antipsychotic drugs and second generation antipsychotic drugs.
- First generation antipsychotic drugs include chlorpromazine, fluphenazine, and haloperidol. These drugs are dopamine antagonists, which work by stopping dopamine from binding to its receptor. First generation antipsychotics have a high risk of dangerous side effects including tardive dyskinesia (uncontrolled motor movements).
- Second generation antipsychotic drugs include clozapine, Olanzapine, risperidone, quetiapine, amisulpride, and ziprasidone. Second generation antipsychotics are dopamine and serotonin antagonists. These drugs work by preventing dopamine and serotonin from binding to their receptor sites. They have considerably fewer side effects than first generation drugs and are thus more widely prescribed.

SECTION 30

The Neurologic Substrates of Addiction

- Addiction is a *neurobiological phenomenon* that is rooted in *genetic factors*.
- The most recent research regarding the addiction process suggests the following:
  - *A person' risk for addiction is influenced by genetic, neurochemical, and environmental factors.*
  - *Once the brain becomes addicted to a substance, it may always remain in an addicted state unless pharmaceutical and psychosocial intervention are obtained.*
  - *An addicted brain remains in this state despite years of abstinence.*
- The course of addiction is characterized by *repeated remission and relapse*. Remission and relapse are largely regulated by neurological processes rather than indicative of weak moral character or poor volitional control.

The Addiction Experience

- Initially, the high produced by alcohol or other substances is experienced as a pleasurable sensation by the user.
- But very quickly, after repeated exposure to the substance, the amount that initially produced feelings of euphoria is no longer effective.
- People find that they require more of the substance to produce the sensation of well-being.
- The euphoria of the initial high is never obtained again. Instead, people require greater and greater amounts of the substance just to feel normal. Without the substance, they become depressed, irritable, and often physically ill.
- At this point, addiction takes hold, and people commonly lose control over their intended use—often ingesting far more than anticipated and for longer amounts of time.
- The brain has now become addicted to the substance and begins to elicit the severe cravings that drive people to relapse. Such neurochemically driven cravings compel people to seek and use substances despite clear feedback that the addiction is harming their health, finances, and relationships.

Neurobiology of the Addiction Process

The Brain's Reward System: The Mesolimbic Dopamine System

- The euphoria induced by substances of abuse occurs as a result of their effect on the brain's reward system.
- The *brain's reward system* is called the *mesocorticolimbic*, or *mesolimbic, system*. This system consists of a complex circuit of neurons that evolved to encourage people to repeat pleasurable behavior that supports survival. For example, eating, quenching thirst, and sexual behavior all produce surges of pleasurable neurochemicals in the mesolimbic system.
- Researchers have demonstrated that chronic substance use can cause changes in the structure and function of the mesolimbic system that can last for years after a person's last substance use experience.
- There are three primary changes in the brain's reward system:

1. *Tolerance*—when the substance is used chronically it no longer elicits the same pleasurable feelings.
2. *Cravings*—the addicted brain produces intense cravings for the substance that cause an escalation in substance use.
3. *Sensitization*—in a period of abstinence, the brain adapts to the lack of substance by becoming highly sensitized to the substance—so that ingestion of the substance after periods of abstinence is experienced with heightened pleasure. This often causes relapse.

- These changes in the mesolimbic system—in response to chronic substance use—trap people in a destructive spiral of escalating use, attempts at abstinence, and repeated relapse.

Neuroanatomical Structures of the Mesolimbic System

- The mesolimbic system is a set of interrelated neuroanatomical structures of the *cortex*, *midbrain*, and *limbic system*.
- This system is able to interpret whether behavior is rewarding or aversive and sends this information to other brain areas that function in decision making and action.
- The primary neural pathway of the mesolimbic system originates in the *ventral tegmental area* of the midbrain. This area sends projections to the *nucleus accumbens*, which is located deep beneath the frontal cortex.
- The primary neurotransmitter used by this system is *dopamine*.
- *Almost all drugs of abuse have the potential to become addictive because of their ability to increase dopamine levels in the brain's reward system.*
- Drugs of abuse activate the system with an intensity and persistence that far exceed that of nonaddicting substances, which can also trigger the reward system (such as eating and sex).
- The ventral tegmental area of the midbrain acts as a meter that measures reward. It sends signals to other brain regions with information about how pleasurable a specific substance or behavior is.
- The *amygdala* is a limbic system structure that also has a role in the interpretation of whether a behavior is pleasurable or aversive. When the amygdala is lesioned, individuals no longer link pleasure to its cause.
- The amygdala has connections with the *hippocampus*—another limbic system structure. The hippocampus records and stores the memories of emotionally laden events—such as the euphoria of a drug-induced state or the extreme irritability and physical illness associated with withdrawal.
- These subcortical structures send messages to the *frontal lobes* of the cortex, which processes this information on a conscious level and uses it to make decisions and initiate action.

Addiction Also Involves Distorted Learning and Memory

- In addition to the neurochemical addiction of the brain, the process of addiction involves a *distorted type of learning* and *memory*.
- Researchers have shown that addiction is associated with *alterations in the frontal lobe* that produce the following:
  - *An overvaluing of the reward*
  - *An undervaluing of the risks associated with obtaining the reward*
  - *A diminished ability to connect addictive behaviors with their negative consequences*

Priming

- Another cognitive distortion related to addiction is called priming.
- *Priming is a phenomenon in which people learn to associate the euphoria of a drug-induced state with the objects and people involved in the process of drug use.*
- For example, the sight of white lines on a mirror can produce anticipatory pleasure that activates the same areas of the brain as actual consumption of the substance.
- Any sensory stimuli associated with drug seeking and obtaining can cause priming and relapse—visual images of the substance, the paraphernalia used to ingest the substance, and the smell or taste of the substance.
- Researchers have shown that when cocaine addicts are shown videos of someone using cocaine, the structures of the mesolimbic system light up on functional magnetic resonance imaging (fMRI) and positron emission tomography (PET) scans. The same regions respond similarly when compulsive gamblers are shown images of slot machines. This finding suggests that the mesolimbic system responds similarly in nondrug addictions such as gambling, eating disorders, and compulsive shopping.

Dopamine

- *All drugs of abuse cause the nucleus accumbens to receive a flood of dopamine.* Repeatedly dosing the brain with addictive drugs is akin to a chemical assault that alters the structure of neurons in the brain's reward system.
- There is evidence that people with *low levels of dopamine D2 receptors are at greater risk for addiction and substance abuse.*
- People with low levels of D2 receptors may be less able to experience feelings of pleasure from activities that are commonly intrinsically rewarding—such as engaging in enjoyable hobbies or eating tasteful foods.
- Researchers have also shown that people with low D2 receptor levels are more likely to be obese or possess a binge eating disorder. Low levels of D2 receptors may compel people to overeat to feel satiated and a sense of gratification from food.
- Deficient levels of D2 receptors have also been linked to a range of compulsive behaviors including excessive hand washing and checking disorders (as seen in obsessive-compulsive disorder [OCD]).

Tolerance and Dependence

- Tolerance and dependence occur in the early stages of addiction. *Tolerance occurs when a drug makes the brain less responsive to that drug.* The reward response of the mesolimbic system becomes diminished, and people experience depressed mood and motivation.
- Consuming more of the substance is the easiest and quickest way for people to feel normal again.
- *Dependence involves the physical illness that occurs when the drug is not consumed.*
- Both tolerance and dependence occur because chronic drug use suppresses the brain's reward circuits.
- Chronic drug use causes neural receptors in the mesolimbic system to adapt by resisting the drug—thereby increasing the need for higher doses just to feel normal.
- The neurophysiologic mechanism that underlies tolerance and dependence involves a molecule known as CREB (cAMP [cyclic adenosine monophosphate] response element-binding protein).
- CREB is a protein that regulates the expression of genes. When drugs of abuse are consumed, dopamine levels increase in the nucleus accumbens causing the eventual activation of CREB.
- Chronic drug use causes sustained activation of CREB, which eventually dampens the responsiveness of the brain's reward system.

Sensitization

- *Sensitization is a heightened response to a substance of abuse after a period of abstinence.*
- After a period of abstinence from the substance—which could last for weeks, months, or even years—the *brain responds to re-exposure to the substance with increased sensitivity.*
- Sensitization is a neurobiological process that *can cause relapse* even after years of abstinence.
- Shortly after repeated exposure to a substance, CREB activity is high and tolerance to the substance increases. During this period, people need increasingly greater amounts of the drug to trigger the brain's reward pathways.
- However, if the person abstains from substance use, CREB activity declines. It is at this point that tolerance begins to lessen and sensitization becomes the dominant process, setting off the intense cravings that drive compulsive drug seeking.

Glutamate's Role in Sensitization

- Glutamate is a neurotransmitter important in learning and the formation of memories.
- Substances of abuse alter the sensitivity of glutamate in the ventral tegmental area and the nucleus accumbens for extended periods of time.
- *Researchers suggest that this state of prolonged heightened glutamate sensitivity strengthens the neural pathways of memories that link substance use with reward. This phenomenon appears to serve as the basis for the formation of memories that distort the substance abuse experience.*
- People tend to remember their experience as more positive and enjoyable than it actually was. They also tend to undervalue the risks associated with substance abuse and appear unable to connect their addiction to real life negative consequences.

Addiction to Compulsive Behaviors

- In addition to substances, people can develop addictions to compulsive behaviors.
- *Addiction is any repeated behavior that produces a chemical change in the brain's reward system, thus compelling the individual to repeat the addicting behavior.*
- Addiction commonly involves the following:
  - *Unsuccessful attempts to reduce, stop, or control one's use of the addictive substance or behavior*
  - *Neglect of major life role obligations and responsibilities*
  - *Increasingly greater time spent participating in addictive behaviors*
  - *Continued substance abuse or participation in behaviors despite health, financial, work, and family-related problems*
- The addiction literature extends the definition of addiction to compulsive behaviors that meet the above criteria. In addition to substance use, addiction also involves compulsive behaviors:
  - *Eating disorders (anorexia, bulimia, compulsive binge eating)*
  - *Work (workaholism)*
  - *Relationships/sex*
  - *Gambling*
  - *Shopping*
  - *Kleptomania*
- These repeated behaviors cause a chemical change in the brain that produces a sense of well-being.

Addictions Can Also Involve Obsessions

- *Obsessions are recurrent ideas that invade consciousness.*
- Obsessions are mediated by neural circuits that appear to *originate in the basal ganglia.*
- The abnormality in the neural pathway involves recurrent electrical brain waves that will not stop firing—in other words, the messages running in the pathway will not shut off.
- Researchers have long known that a link exists between obsessive thoughts and the need to act out compulsive behaviors. The basal ganglia appear to play an important role in this phenomenon.
- The basal ganglia are a part of the primitive survival centers in the brain.
- Obsessions involve ideas and behaviors that are often linked with survival and control, for example: food, money and possessions, and cleanliness and order.
- *OCD is a component of all addictions.*

- The etiology of OCD is believed to be rooted in the same dysfunction of the serotonin and dopamine systems that are believed to underlie the addiction process.
- OCD often involves obsessions with the following:
  - *Symmetry of objects*
  - *Checking behaviors*
  - *Cleanliness*
  - *Orderliness*
  - *Counting*
  - *Invasive thoughts about sex, religion, catastrophic events*
  - *Invasive music that runs through the mind repetitively*

Serotonin as an Obsession Blocker

- Serotonin helps to shut off the neural circuits that cause obsessive-compulsive behaviors.
- The selective serotonin reuptake inhibitors (SSRIs) increase serotonin levels in the brain and have shown effectiveness in the treatment of OCDs.

Addictive Behaviors Can Be Cross-Addicting

- People with addictions often leave rehabilitation programs and remain abstinent from their former addictive behaviors only to switch to another addictive substance or compulsive behavior.
- For example, a person with a gambling disorder may remain free of gambling behaviors after rehabilitation but may begin to shop compulsively.
- *Switching from one substance or compulsive behavior to another occurs because treatment did not address the neurochemical mechanisms maintaining the brain in an addicted state.*
- The addicted brain activates cravings that compel the person to use a similar substance or engage in a comparable compulsive behavior that will positively affect the brain's reward circuits.
- Individuals often recover from one addiction only to go to another.

Addictive Behaviors Are Heritable

- There is considerable evidence that addictive behaviors are genetically and neurochemically based.
- The same deficiency in the serotonin and dopamine systems that underlie addictive behaviors *tends to run in families.*
- While the neurophysiologic substrates of addiction may be similar in members of the same family, *the type of addiction that emerges in individual members may be different.*
- For example, it is common to observe that children of alcoholics often have OCD or eating disorders, and vice versa.
- As a result of cultural influences, environmental factors, and unique psychological make-up, people in the same biological family group may have similar deficiencies in the serotonin and dopamine systems that are expressed as distinctly different addictive behaviors.

Treating the Addicted Brain

- Pharmaceutical advances in addiction treatment have largely addressed physical dependence and withdrawal.
- In contrast, it has been more difficult to develop effective medications that address craving and relapse—the actual processes that maintain addiction.
- Interventions that do not address craving and relapse are unable to treat the brain's addicted state. This is the main factor accounting for the ineffectiveness of most addiction treatment—intervention may address dependence, withdrawal, and environmental stressors but leaves the brain in an addicted state.
- People having received intervention that leaves the brain in an addicted state have a high likelihood of relapse.

Three Primary Approaches to Pharmaceutical Treatment

- Medication that prevents a drug from reaching its neurologic target, thus rendering the drug's effects inert
- Medication that mimics a drug's action without causing the deleterious effects of the addictive substance—thereby reducing cravings
- Medication that stops the neurologic addiction process, thus eliminating cravings, sensitization, and relapse

Blocking Drug Targets

- The most common pharmaceutical approach to treat addiction involves blocking the substance from reaching its neurologic target.
- Naltrexone is a narcotic antagonist used to treat opiate and alcohol addiction.
- Naltrexone was designed to block opiate receptors in the brain, thereby preventing opiate action.
- It also triggers a withdrawal reaction in people who are physically dependent upon opiates.
- Studies have shown that naltrexone can reduce the risk of relapse in the first 3 months after withdrawal by 36%.
- However, because naltrexone does not stop cravings, many people who begin its use eventually stop taking the medication.
- Polysubstance abuse is also problematic. While naltrexone blocks the effects of opiates, it has no effect on other illicit drugs.

Mimicking Drug Action

- Medications that block drug targets do not address craving, and thus, relapse is likely.
- In contrast, medications that attempt to mimic a drug's action in the brain can alleviate cravings that can cause relapse.
- Methadone is a long-acting opioid receptor agonist used in the treatment of heroin addiction.
- Because of methadone's long half-life, people taking the medication have a low level of sustained opioid receptor activation. This prevents cravings and enables people to return to a more stable employment and family life.
- Methadone maintenance is controversial, as some believe that it legally reinforces opiate addiction.
- However, methadone maintenance has a significant amount of research supporting its effectiveness.
- Studies show that people on methadone maintenance are less depressed, more likely to maintain stable employment and family lives, less likely to commit crimes, and less likely to contract HIV or hepatitis.
- Methadone and other long-acting opioid agonists—such as levo-alpha acetyl methadol [LAAM]—are effective because they allow the brain to believe that it is receiving the addictive substance without causing the deleterious effects of that substance.

Halting the Addiction Process

- To truly end the addiction process and relapse, medication must both alleviate cravings and stop sensitization.
- While medications have been developed that can block drug targets and mimic drug action, it has been more difficult to design medication that addresses both craving and sensitization—the primary factors that maintain addiction.
- To date, while such medications are in development, they remain untested in humans.

SECTION 31

Learning and Memory

Early Theories About Memory

- Early theories describing memory were based on a computer model.
- Memories were compared to computer files that could be placed in storage and pulled up into consciousness when needed.
- Today, more is known about how memory works. Information has been gained through experimental research and research on patients with cerebral damage.

Memory Is Not a Factual-Based Record of Reality

- Our memories are not passive or literal recordings of reality.
- Instead, *our memories are recreations of specific events that over time become distorted and lose accuracy.*
- For example, members of one family commonly find that each person's memory of a specific event differs significantly from the way that other family members recall that event. Some members have forgotten the memory entirely while others adamantly recall the event's occurrence. Some family members insist that the sequence of events occurred differently from the way others recall it.
- People rarely recall all of the details in an event accurately.
- *We often recall occurrences that made general sense or fit our expectations of what should have happened but were not actually part of the original event.*
- False memories have become highly debated in recent years as reports of early childhood abuse are questioned. Some have asked if false memories are the product of psychological techniques that induce memories of events that never occurred.

Memories Are Encoded by Brain Networks

- A person's experience is encoded by *brain networks involving multiple cerebral structures*. There is no single anatomical structure that alone deals with memory functions.
- PET scans have enabled researchers to observe the brain in action while people remember specific events.
- Memories are encoded by *brain networks whose connections have already been shaped by previous encounters with the world.*
- This pre-existing structure powerfully influences how individuals encode and store new memories. It influences what an individual will remember and why.
- Memories that have a strong emotional significance appear to be more readily remembered. This may occur because the *hippocampus* and the *amygdala* (both limbic system emotional centers) play an important role in the *storage of long-term memories (LTMs)* that have a *strong emotional component*.

- But LTMs that have an emotional significance are also subject to distortion based on an individual's psychological interpretation of an event.
- *Human memory is thus predisposed to corruption by suggestive influences.* Memories more likely involve distortions of reality rather than being factual snapshots or video.

Types of Memory

Short-Term Memory

- Short-term memory (STM) is the ability to remember others and events *encountered less than 1 hour ago.*

Working Memory

- Working memory is a subcomponent of STM and involves *moment-to-moment awareness.*
- Working memory also plays a role in the *search and retrieval of archived information.*

Long-Term Memory

- LTM is the ability to remember one's past, familiar others, and events *encountered more than several hours ago.*

Recent Memory

- Recent memory is sometimes considered to be a component of LTM. Recent memory refers to the ability to recall events/information that *occurred hours to weeks ago.*

Remote Memory

- Remote memory is a component of LTM and refers to the ability to recall events/information that *occurred in the distant past* (eg, several years ago).

Semantic Memory

- Semantic memory involves the recall of *facts*—dates of people's birthdays, state capitols, the names of presidents.
- Semantic memory includes the *definitions of words* and how to use the *rules of grammar.*

Episodic Memory

- Episodic memory involves *significant events* that happened to an individual—the first day of school, one's wedding, the birth of a child.

Procedural Memory

- Procedural memory involves the recall of *steps involved in specific tasks.* For example, knowing the steps involved in fabricating a wrist cock-up splint or knowing the steps involved in changing a flat tire.

Explicit Memory

- Explicit memory involves *knowledge that an individual consciously knows he or she has acquired.* For example, an anatomist knows that he or she has acquired knowledge of the human body's musculoskeletal system.

Implicit Memory

- Implicit memory involves *knowledge that the individual does not consciously know he has acquired.* For example, someone with amnesia may not consciously know she has acquired knowledge of how to play the piano. But when seated at the piano, she can play fluently.

Source Memory

- Source memory involves the recall of *how information is learned.* For example, an anatomist may recall that he learned specific information about the human shoulder joint from a specific dissection class.
- Often, people remember facts but cannot recall how they came to learn those facts. An example involves the rehearsal of legal witnesses by lawyers. If lawyers over-rehearse a witness, the witness may confuse rehearsed information with the events that they actually observed.

Perceptual Memory

- Memory can also be categorized by sensation and perception:

Visual Memory

- Visual memory is the ability to accurately identify objects and people by recalling their unique visual characteristics.

Olfactory Memory

- Olfactory memory is the ability to accurately identify animate and inanimate objects by recalling their unique olfactory characteristics.

Gustatory Memory

- Gustatory memory is the ability to accurately identify objects by recalling their unique gustatory characteristics.

Somatosensory Memory

- Somatosensory memory is the ability to accurately identify animate and inanimate objects by recalling their unique somatosensory characteristics.

Auditory Memory

- Auditory memory is the ability to accurately identify objects, animals, and people by recalling their unique auditory characteristics.

Field Memories vs Observer Memories

- Field and observer memories refer to *a person's own perspective of him- or herself within the remembered event.*

Field Memory

- *In a field memory, people do not see themselves in the memory.* The memory is remembered as though the individual is seeing it through his or her own eyes.

Observer Memory

- *In an observer memory, the person has incorporated his or her own image into the remembered event.* The individual has altered the original scene—because the original perspective of an event is always viewed from a field perspective. The person is now able to visualize themselves within the remembered event.

Field and Observer Memories are a Factor of Time

- More *recent memories* are remembered from a *field perspective*, while *older memories* are remembered from an *observer perspective.*
- This is a phenomenon of time. People tend to incorporate themselves into the remembered event as time goes by.

Retrograde Amnesia vs Anterograde Amnesia

Retrograde Amnesia

- Retrograde amnesia is a common consequence of brain damage.
- This type of memory problem involves *the loss of one's entire personal past* and occurs after the injury or trauma.
- Memory of one's past is often recovered at some point after injury.
- LTM often returns to patients as their brain injury resolves. This may be due to *neurogenesis* (or the birth of new brain cells) that has been found to occur in the hippocampus throughout life, until old age (ie, the seventh decade of life).

- If retrograde amnesia resulted after an accident or traumatic event in which the individual lost consciousness, the accident is usually never remembered. It is suggested that this phenomenon occurs because the memory of the accident was never transferred from STM to LTM before the individual lost consciousness.

Anterograde Amnesia

- Anterograde amnesia is a memory dysfunction resulting from brain damage in which—after the injury or disease process—the person *cannot remember ongoing day-to-day events although memory of one's personal past remains intact.*
- Anterograde amnesia is a dysfunction of the encoding process—the individual *cannot transfer STMs into LTM storage.* In effect, the individual cannot develop any new LTMs.

The Encoding Process

- The encoding process involves the actual steps that the brain uses to turn an event into a stored memory that can be recalled.
- Items in STM can only be remembered for seconds unless they are encoded and stored.
- Encoding information involves *moving it from STM to LTM storage.* It also involves *associating novel information with information that has already been learned.*
- The encoding process must involve the formation of neural circuits that can be easily accessed in order to retrieve stored information. Otherwise, newly learned information could not be recalled.
- Being able to retrieve information involves a subcomponent of STM called *working memory.* Working memory is a function of the *prefrontal cortex* and consists of *moment-to-moment awareness.* Working memory also plays a role in the *search and retrieval of archived information.*
- LTM is the brain's archival system. It is a collaborative function of the hippocampus and other cortical areas that are only beginning to be understood.

How the Brain Stores Information

- Memories of single objects appear to be fragmented into pieces that are stored in different regions of the brain.
- For example, storage of the *names of tools* is a function of the *medial temporal lobe* and the *left premotor area.*
  - Humans name tools through the association of the *sounds* that they make. This is a function of *audition* and the *temporal lobe.*
  - We also name tools by their *actions.* This is a function of the *premotor area.*
- The storage for the *names of animals* is primarily a function of the *medial occipital lobe.*
  - Humans categorize animals primarily by their *appearance* (a function of the *occipital lobe*).
  - We also categorize animals by the *sounds* they make (a function of the *temporal lobe*).
  - The way the animal *feels to one's touch* is a function of the *parietal lobe.*
  - Storage for the *movement characteristics* of animals appears to be a function of the *premotor area.*
- When patients have *difficulty naming actions* (eg, swimming, driving), it usually indicates *prefrontal lobe damage.* The prefrontal lobe is where words of action are processed and where motor plans are stored.

Establishing a Durable Memory

- Incoming information must be encoded thoroughly, or deeply, by associating it with meaningful knowledge that already exists in memory storage.
- The human memory system is built so that individuals are more likely to remember what is most important to them and what has great emotional significance.
- The amygdala and hippocampus play an important role in encoding memories that have emotional significance.

Frontal Lobe Memory Functions

- The prefrontal lobe is the seat of STM and is largely responsible for the following:
    - Moment-to-moment awareness
    - Encoding STMs into LTM storage
    - Searching for and retrieving archived information
    - Recalling the source of information
- The premotor area specifically stores information about the motor actions of objects and people.

Parietal, Occipital, and Temporal Memory Functions

- The parietal, occipital, and temporal lobes participate in the storage of various attributes of objects and people.
- For example, the parietal lobe stores information about the somatosensory and visuospatial properties of an object or person. The occipital lobe stores information about the visual characteristics of an object or person.
- The temporal lobe stores information about the auditory properties of an object or person.

Hippocampus and Amygdala

- The hippocampus and amygdala are responsible for the storage and gating of LTMs that have great emotional significance.
- Some researchers have suggested that pathology in these structures may account for the symptoms of post-traumatic stress disorder in which people cannot stop flooding of images of the traumatic event.
- Researchers have also suggested that the amygdala may act as a gate, preventing the retrieval of specific LTMs that have become repressed.

Encoding and Mnemonic Devices

- Mnemonic devices are techniques that facilitate learning and the recall of learned material.

Method of Loci

- The method of loci is a mnemonic device involving the creation of a *visual map of one's house*.
- For example, if one wanted to remember to buy soda, potato chips, and soap, one would use the rooms of one's house and visualize the kitchen with spilled soda on the floor, the bedroom with potato chips scattered on the bed sheets, and the bathroom tub filled with soap bubbles.
- At the store, one would then take a mental walk around one's house and recall which object is in each room.
- The cognitive act of creating a visual image and linking it to a mental location is a form of *deep elaborative encoding*.

Acronyms

- Acronyms are another form of mnemonic device in which the first letter of each word is chained to form a meaningful sentence.
- For example, a common acronym used to remember the Classification of Living Systems (kingdom, phylum, class, order, family, genus, species) is "King Phyl came over for good spaghetti."
- A common acronym for remembering the order of the planets (Mercury, Venus, Earth, Mars, Jupiter, Saturn, Uranus, Neptune, Pluto) is "Mary's very extravagant mother just sold us ninety parrots."
- There are acronyms that students use to remember the cranial nerves and the bones of the skeleton.

Rhymes

- Rhymes are another form of mnemonic device that facilitate learning and memory.
- The familiar rhyme to remember the number of days of each month illustrates the effectiveness of using rhymes to enhance learning and memory (Thirty days has September, April, June, and November. All the rest have 31. February is great with 28. Leap year is fine with 29).

Damage to Frontal Lobe Regions and Encoding Problems

- The prefrontal cortex plays an important role in deep or elaborative encoding.
- Patients who have sustained damage to the frontal lobe regions often have encoding problems. In other words, *they cannot move information from STM to LTM storage.*
- Such patients fail to organize and categorize new information as it comes into STM. They have difficulty associating novel information with similar archived information in order to store and retrieve it.
- *The problem lies in the processes of working memory rather than LTM storage.* LTMs that were established prior to frontal lobe damage can easily be retrieved. This is why patients with frontal lobe damage can often recall events that occurred in their distant past, while they cannot remember events that occurred in the last 24 hours—because events that occurred recently can no longer be moved from STM to LTM storage; they cannot become LTMs.
- LTM may remain intact in brain injury because of the process of *neurogenesis*. Neurogenesis is the birth of new brain cells. Researchers have found that neurogenesis occurs in the hippocampus (the seat of LTM) throughout life.
- The *hippocampal region* becomes activated when people encode novel events.
- The brain activates the hippocampus in response to a novel stimulus to *determine if it has already developed associations to that stimulus.*
- When associations are found (or if none are found), the *prefrontal cortex* then becomes activated in preparation to *encode the novel stimulus.*
- Patients with frontal lobe damage often lose the ability to encode new memories.

Reconstruction of Memory

- The brain engages in an act of reconstruction during the retrieval process.
- Theorists once believed that engrams—or single neural pathways—existed for each stored memory. Today, researchers understand that the *reconstruction of a past memory involves a series of neural pathways and structures.*
- Retrieved memories are a temporary construction of information from several distinct brain regions—a reconstruction that has many anatomic contributors.
- We reconstruct our memories in a similar fashion to a jigsaw puzzle. Pieces of the puzzle are reconstructed until we have enough to remember an event. Fragments of the memory may be left out, just as pieces of the puzzle may be lost. Other pieces may be inaccurately borrowed from a similar memory and confused with the event we are attempting to recall.
- In the reconstruction of a memory, the perceptual regions of the brain that are concerned with sight and sound merge with the *posterior multimodal association area* (or convergence zone).
- The posterior multimodal association area contains codes that bind sensory fragments to one another and to pre-existing information.
- Researchers believe that the posterior multimodal association area contains a type of index that indicates the location of information stored in separate cortical areas.

- The index is initially needed to keep track of all sights, sounds, and thoughts that comprise an event until the memory can be held together by direct connections between the cortical regions themselves. Once direct connections between the cortical regions are established, the index is then no longer needed to recall a specific event.

The Consolidation Process

- LTM consolidation occurs, in part, because people talk and think about their past experiences.
- *The older the memory, the greater the opportunity for such past event rehearsal.*
- Thinking and talking about a past experience promotes the direct connections between cortical storage areas.
- *Once an experience has been repeatedly retrieved, it becomes consolidated and no longer depends upon the integrity of the posterior multimodal association area to act as an index.*
- Sleep also plays a role in the consolidation process—particularly *rapid eye movement (REM)* sleep.
- During sleep, the *hippocampus* plays back a recent experience to areas of the cortex where it will eventually be stored.

The Passage of Time and Decreasing Memory

- As time passes, humans encode and store new experiences that interfere with the ability to recall previous ones.
- *For about 1 year after the occurrence of an event, it can be readily accessed with virtually any cue.*
- As more time elapses, the *memory becomes blurry* and the *range of cues* that elicit a specific event *progressively narrows*.
- This means that when an individual suddenly and unexpectedly recovers a forgotten memory, it may be because she has stumbled upon a retrieval cue that easily elicits the faded memory.

Mirror Neurons and Learning

- Mirror neurons are specialized premotor cells designed to facilitate learning.
- The discovery that premotor neurons have a mirroring function was made approximately a decade ago by researchers who found that groups of specialized neurons fire whenever humans observe novel behaviors and then imitate such actions.
- The discovery of mirror neurons suggests that humans mentally rehearse all motor actions, language, and emotions prior to imitation.
- Mirror neurons may underlie the very basics of how we learn—from imitating facial expressions to replicating complex athletic movements. The presence of mirror neurons may shed light on how humans develop elaborate social interaction skills and social support systems.
- Mirror neurons have been found in the following neural regions:
    - Premotor cortex (motor planning and speech)
    - Parietal lobe (perception of oneself and one's relationship to the external environment; comprehension of language)
    - Temporal lobe and the insula (comprehension of emotions and the ability to feel empathy; expression of language)
- In addition to using mirror neurons to imitate and learn novel actions, it appears that humans also use mirror neurons to understand the intention behind others' actions. One type of mirror neuron fires in response to observing novel behaviors, language, emotions, facial expressions, and body gestures. A second type fires when we attempt to understand the intention behind the actions just observed. Understanding intention is a key component in the development of empathy and social bonding.
- Other studies have demonstrated that mirror neurons may enable humans to share others' emotions by mirroring their facial expressions and verbal tone. For example, when observing the emotion of disgust on others' faces, the same set of mirror neurons will be triggered in the viewer's insula, causing similar feelings. This process may serve as one neurologic basis of empathy.
- Some researchers have suggested that pathology in the mirror neuron system may, in part, underlie disorders such as autism in which the ability to learn and feel empathy by mirroring others is impaired. When researchers showed photos of people with distinctive facial expressions to two groups of adolescents (one group consisting of highly functioning adolescents with autism; the other group without autism), both groups could imitate the expressions and state which emotions they represented. But while the group without autism displayed heightened activity in their mirror neurons (in regions that corresponded to the emotions they observed), the adolescents without autism showed no mirror neuron activity. The group with autism stated that they had learned to cognitively understand the emotions represented by distinct facial expressions but felt no empathy in response to observing such emotions.
- Other studies have demonstrated that observing violent video games may reinforce the association among gain, power, and inflicting harm.

SECTION 32

The Neurologic Substrates of Emotion

- Research regarding how the brain processes emotion has been advanced in recent years by functional magnetic resonance (fMRI) and positron emission tomography (PET) scans. Imaging scans have enabled researchers to identify which brain regions become activated when people experience specific emotions.
- Much like memory, researchers used to believe that emotion was processed by one single cerebral system—largely the limbic system. Today, researchers believe that emotions are processed by multiple cerebral networks.

Major Neuroanatomical Structures Involved in Emotion

- The primary neuroanatomical structures involved in the processing of emotion include the following:
  - Prefrontal cortex (in both the left and right hemispheres)
  - Limbic system (including the amygdala, hippocampus, and cingulate gyrus)
- Secondary structures in the processing of emotion include the:
  - Thalamus
  - Anterior insular region (of the temporal lobe)
  - Septum pellucidum

Role of the Left vs the Right Hemisphere in Emotion

- The left and right hemispheres have different roles in the processing of emotion.
- Both hemispheres act in a mutual cross-modulatory relationship—a relationship much like reciprocal inhibition in which *each hemisphere keeps the other's innate emotional response in check.*
- The two hemispheres are specialized for the processing of different kinds of emotions—just as they are specialized for different kinds of cognitive, sensory, perceptual, and motor skills.

Prefrontal Areas of the Left vs the Right Hemisphere

- Imaging scans have revealed that activation of the *left prefrontal area* coincides with *positive emotions* and a sense of well-being.
- Activation of the *right prefrontal area* coincides with emotions of *agitation*, *nervousness*, *distress*, *anxiety*, *sadness*, and *depression.*
- Stimulation of the *left prefrontal area* through *repeated transcranial magnetic stimulation (rTMS)* appears to *enhance one's mood.*
- Stimulation of the *right prefrontal area* by rTMS produces depressed moods in subjects.

Right and Left Hemispheres May Divide Negative and Positive Emotions

- When a person reports feelings of *sadness*, *depression*, and *anxiety*, his or her *right prefrontal lobe* shows increased electrical activity on SPECT scans.
- Emotional states of *happiness* and *enjoyment* increase the electrical activity in the *left prefrontal lobe.*

- Infants who are prone to distress when separated from their mothers show increased activity in the right prefrontal lobe. People who identify themselves as pessimists also show increased activity in the right prefrontal lobe.
- People who were at one time depressed (but who do not report feelings of depression in the present) show decreased left prefrontal lobe activity when compared to people who never experienced depression. This finding supports *Kramer's Kindling Theory of Depression*—each depressive episode produces permanent changes in the brain's structure and chemical balance. Such changes tend to make recurrence of depressive episodes more likely.

Two Syndromes Related to the Site of Brain Injury

Left Prefrontal Lobe Damage

- When brain damage occurs to the left hemisphere, the intact right hemisphere assumes control of emotional regulation.
- Left brain damage appears to release the right hemisphere's predisposition towards anxiousness and depression.
- Patients with left prefrontal lobe damage tend to be *emotionally labile*, *depressed*, and *despondent*. This is sometimes referred to as *catastrophic reaction*. Catastrophic reaction tends to occur during the acute stages of brain damage.

Right Prefrontal Lobe Damage

- Right prefrontal lobe damage often leaves patients with an indifference to their impairment (referred to as *anosognosia*). Patients commonly report states of *euphoria* and *well-being*.
- This may occur because right hemisphere brain damage allows the left hemisphere to assume control of emotional regulation.
- The left hemisphere appears to have an innate emotional bias toward experiences of *optimism* and *well-being*.
- This may account for the optimistic denial of disability and a disowning of impaired appendages that is clinically observed in the acute stages of right brain damage.

Left Hemisphere Regulates the Emotional Responses of Both Hemispheres

- In the absence of pathology it appears that the left hemisphere may regulate the emotional responses of both hemispheres.
- When *pathology occurs to both hemispheres*, however, patients tend to become *depressed*—similar to cases of unilateral left hemisphere damage.
- In bilateral hemisphere damage, it is as though the co-existing right hemisphere injury had not occurred. The left hemisphere is unable to alleviate the right brain's negative emotional response and patients tend to become depressed.
- It appears that the *left hemisphere* may play a role in *balancing* a person's *emotional equanimity*—so that, under normal circumstances, the right brain's propensity toward depression does not take over.
- In patients with chronic episodic depression, the left brain's ability to mitigate right brain emotional negativity appears to become decreased with each depressive episode.

Therapeutic Significance

- If depressed patients exhibit heightened activity in the right hemisphere, one way to enhance mood may be to provide activity that stimulates left brain function.
- This may account for the effectiveness of *cognitive therapy* with depressed patients. Research has shown that cognitive therapy—concomitantly administered with drug therapy—has been found to be more effective than drug therapy alone.
- Cognitive therapy challenges patients to assess their cognitive mindset in order to determine if their thought patterns are reality-based. If one's thought patterns are found to be nonreality-based, the therapist challenges the patient to make needed mental set changes. These activities require left brain cognitive functions.
- *Occupational therapy* that provides depressed patients with *activity that stimulates left brain function* may also be an effective treatment for depression (if administered conjointly with drug therapy).
- It is also likely that *workaholism* has the same effect as stimulating left brain functions. People who are workaholics may use work as an attempt to self-medicate an underlying depression.

Orbitofrontal vs Dorsolateral Frontal Lobe Lesions

- There are two further emotional syndromes related to the site of brain injury/disease.

Orbitofrontal Lobe Lesions

- Patients with *orbitofrontal lobe lesions* exhibit *impulsiveness* and *disinhibition*.
- The orbitofrontal region mediates the brain's executive functions. When the orbitofrontal region is lesioned, the *executive functions become impaired*.
- Patients exhibit poor judgment, risk-taking behaviors, rowdiness, and socially inappropriate behaviors.
- Patients may also more easily become irritated, agitated, aggressive, and angry.

Dorsolateral Frontal Lobe Lesions

- Patients with *dorsolateral frontal lobe lesions* exhibit *decreased drive and motivation*, *lethargy*, and a *flat affect*.
- Such patients rarely initiate activity or display emotion. It is as though their emotional center has become blunted.

The Amygdala's Role in Emotion

- The amygdala has several roles in the processing of emotion.
- The amygdala is part of the *mesolimbic dopamine system*, or the *pleasure center* of the brain.
- Patients with lesions to the amygdala do not experience pleasure from people and activities that once provided them with an internal sense of happiness (referred to as *anhedonia*).
- Patients who have *bilateral lesions to the amygdala* no longer experience the emotion of fear.
- Such patients have difficulty recognizing dangerous situations and often engage in risk-taking behavior as a result.
- The amygdala works cooperatively with the anterior cingulate gyrus, prefrontal lobe regions, and the right hemisphere to assess whether situations are dangerous.
- This circuit generates a response that is transmitted along two pathways for feedback about the internal and external environment.
  - One pathway travels to the *sympathetic nervous system* to check the body's sympathetic response to a stimulus. Messages from the sympathetic nervous system are then sent to the *secondary somatosensory area (SS2)*.
  - In the second pathway, SS2 sends the information to the *thalamus* and the *prefrontal lobe* for cognitive decisions regarding the appropriate response to the stimulus in question. Example:
    - An individual hears the roar of a lion.
    - Her sympathetic nervous system becomes activated.
    - The sympathetic nervous system information travels to SS2.
    - SS2 then sends this information back to the thalamus and prefrontal cortex.
    - The prefrontal cortex, observing that the lion is in a confined zoo area, overrides the amygdala's initial fear response.

Patients with Lesioned Amygdalae Cannot Recognize Fear in Others

- When presented with a chart of faces with different emotions, patients with lesioned amygdalae are unable to recognize the emotion of fear on the observed faces. However, they can identify expressions of happiness, sadness, and anger.
- They are also unable to produce the facial expression of fear when asked to do so.

The Amygdala May Play a Role in the Recognition of Social and Emotional Cues

- There is evidence that individuals with autism have impairment within the limbic structures—particularly the amygdala.

- The *inability to interpret social cues* and generate an appropriate emotional response (eg, empathy, sadness, or happiness for others) is commonly impaired in people with autism.
- The anterior insular region, the amygdala, and the hippocampus appear to be involved in *attaching emotional significance to thoughts and memories*. If any of these brain regions are lesioned, the individual has difficulty identifying important and meaningful experiences.
- Such individuals also have difficulty recognizing meaningful events for others and tend to react without emotion in response to another's discussion of emotionally significant experiences. This, too, is a common characteristic of autism.

The Amygdala and Post-Traumatic Stress Disorder

- In PET scans of people who experience *post-traumatic stress disorder (PTSD)*, the *amygdala becomes highly activated*—perhaps overactive.
- People with PTSD report feeling that the traumatic event is vividly occurring all over again—as though the amygdala and other limbic system structures are flooding the visual, auditory, and somatosensory cortices with the painful memories.
- Another phenomenon of PTSD is that the language areas of the brain shut down during the PTSD experience. People experiencing PTSD episodes report that they cannot transfer their emotions regarding the event into words (referred to as *alexithymia*).
- Without the ability to use language to describe their emotions, they cannot integrate the right hemisphere's emotional response to the event with the left hemisphere's ability to cognitively analyze the event.
- The traumatic event continues to be experienced as an overwhelming somatic experience that has a life of its own and that invades one's consciousness without modulation from the left hemisphere.
- Treatment for PTSD has not been significantly effective. This may be related to the finding that emotional memories involving extreme fear are permanently encoded into the limbic structures. While such memories can be suppressed, they may never be fully erased.
- The goal of treatment is to mitigate flooding experiences and help the individual to use cognitive and emotional strategies to subdue PTSD when it occurs.

The Limbic System's Role in Anxiety, Panic Attacks, Phobias, and Obsessive Compulsive Disorder

- PET scans of individuals with anxiety, panic attacks, phobias, and obsessive–compulsive disorder (OCD) reveal heightened activity in limbic system structures—particularly the anterior cingulate gyrus, amygdala, and anterior temporal cortex (the location of the parahippocampal gyrus). Other structures include the insula and the posterior orbitofrontal cortex.
- This circuit is sometimes called the *worry circuit*.
- The locus ceruleus (part of the brainstem that rouses the body to action by secreting norepinephrine) also becomes activated during anxiety, panic attacks, phobias, and OCD.
- There is evidence that pathways leading from the amygdala to the cortex are far greater in number than pathways traveling from the cortex back to the amygdala.
- This suggests that the amygdala structures are designed to alert the cortex to threatening situations and danger.
- Such cerebral architecture may serve as the brain's attempt to ensure that danger is consciously recognized and addressed.
- In a modern world, however, in which threats to one's survival may be ongoing and without clear end (eg, fear of losing employment), the amygdala's alerting system may go into overdrive and become dysfunctional.
- In such situations, the lesser number of pathways leading from the cortex to the amygdala may mean that the conscious mind is unable to shut off the amygdala's fear signals through cognitive self-talk.
- What was once an adaptive cytoarchitectural design ensuring safety may now become a maladaptive strategy.

The Septal Area and Anger

- Another brain region that appears to mediate anger is the septal area—a region in both cerebral hemispheres comprising the subcallosal area and the septum pellucidum. The area has olfactory, hypothalamic, and hippocampal connections.
- The role of the septal area is to provide a means of communication between the limbic structures and the diencephalon (particularly the thalamus and hypothalamus).
- If the septal area is stimulated with an electrode in a cat, the cat lashes out with rage.
- PET scans reveal heightened activity in the septal area when people experience anger.
- A lesion to the septal area in animals produces rage and hyper-emotionality.
- In humans, destruction of the septal area gives rise to emotional overreactions to stimuli in the environment.

Differences in the Neurologic Substrates of Emotion Between Males and Females

- The brains of males and females generate specific emotions using different patterns of brain activity.
- When females feel sadness, they experience greater activity in the anterior limbic system than do males.
- According to some research studies, females also tend to experience a more profound sadness than do males. This tends to support the research finding that females experience twice the risk of depression than men.
- Because females activate both hemispheres during the experience of significant emotions (as observed in PET scans), they may be more able to describe their emotions using words—a process requiring the use of both left and right hemisphere language areas. The ability to integrate left and right language areas facilitates the association between words and emotions. As a result, females may be more aware of their emotions than males.
- Males do not as readily integrate both hemispheres during the experience of significant emotions. This may predispose males to greater difficulty attaching words to their emotions.

Depression

- Depression affects approximately 16% of the population at least once throughout the lifespan.
- The mean age of onset is in the late 20s.
- Twice as many females as males report experiencing depression, although this disparity appears to be decreasing in recent decades.
- Male/female differences in incident rate disappears roughly after age 55 (after most females have experienced menopause).
- The World Health Organization (WHO) has reported that major depression is currently the leading cause of disability in the United States and is expected to become the second leading cause of disability worldwide by 2020.
- Primary symptoms of depression follow:
    - Feelings of extreme sadness
    - Anhedonia or an inability to experience pleasure from previously enjoyed activities
    - Changes in appetite (loss of or increased appetite)
    - Changes in sleep patterns (insomnia, hypersomnia, or disrupted sleep)
    - Psychomotor agitation or retardation
    - Mental and/or physical fatigue and loss of energy
    - Feelings of guilt, helplessness, and/or hopelessness
    - Decreased concentration and a general slowing of mental functions
    - Recurrent thoughts of death, suicidal ideation with or without specific plans, suicide attempt
- Anxiety and debilitating fear commonly accompany depression.
- Neurologic substrates of depression have been identified as follows:
    - Decreased activation of the prefrontal cortex (primarily in the left hemisphere)
    - Decreased activation of the anterior cingulate gyrus
    - Increased activity of the amygdala and hippocampus (Researchers have suggested that increased activity of the amygdala and hippocampus may represent an alerting system for the cortex. In other words, in depressive states the amygdala and hippocampus may become highly sensitive to negative stimuli. These structures initiate messages to the cortex that facilitate the cognitive processes of worrying and ruminating over negative thoughts.)
- Additionally, neurotransmitter systems that are believed to be involved in states of depression include serotonin, norepinephrine, and dopamine. Such neurochemicals may become imbalanced causing a dysregulation of emotions and mood.

SECTION 33

The Aging Brain

Epidemiology

- In developed nations, the leading cause of senile dementia (loss of memory and reason in the elderly) is *Alzheimer's disease (AD)*.
- Other diseases of high incidence in the elderly include *Parkinson's disease* and *multiple cerebrovascular accidents (CVAs)*.

Positron Emission Tomography Scans of Aged Brains

- Positron emission tomography (PET) scans of aged brains reveal the following:
  - *Cortical atrophy*
  - *Broadening of the sulci*
  - *Decrease in the size of the gyri*
  - *Widening of ventricular cavities*
- Brain weight and volume also decrease with age. Approximately 5% to 10% of brain volume is lost between the ages of 30 and 90.

Age-Related Damage

- Age-related damage depends upon the following:
  - *Age of onset when damage occurred*
  - *Type and extent of physical brain alterations*
  - *Past and current medical history*
  - *Genetic history*
  - *Use of brain structures throughout the lifespan*
- Most age-related structural and chemical changes become apparent in *late middle life*—the *50s and 60s*.
- Some age-related changes become pronounced after age 70.

Age-Related Damage in the Diencephalon and Brainstem

- Certain cells and brain areas are more susceptible to age-related damage than others.
- Scientists used to believe that as people aged, their overall number of neurons decreased. Recent research involving neurologic imaging suggests that *most of the brain's neurons remain intact throughout life* unless pathology or disease exists.
- Areas of the brain where *neurons do appear to be lost* include the *substantia nigra* and *locus ceruleus* (both located in the brainstem). Such neuronal loss in these areas may account for the slowing of motor skills seen in the elderly. *Parkinson's disease* can destroy *70% or more* of the neurons in the substantia nigra and locus ceruleus.
- In the healthy elderly, only 20% to 40% of neurons are lost from the substantia nigra and locus ceruleus.

Age-Related Damage in the Limbic System

- *The hippocampus can lose 5% of neurons each decade in the second half of life.* Twenty percent of hippocampal neurons can be lost as a result of the normal aging process.
- *Large neurons tend to be more affected than short neurons.* It has been found that large neurons in the hippocampus and cortex have a greater tendency than shorter neurons to shrink.
- *Cell bodies and axons also degenerate in certain acetylcholine-secreting neurons.* These types of neurons project from the forebrain to the hippocampus.
- Some neuronal changes may represent attempts by surviving neurons to compensate for loss or shrinkage of other neurons and their projections.
- Studies have found an *increase in the net growth of dendrites* in the hippocampus and some areas of the cortex occurring in the *40s and 50s*.
- In late life (the *80s and 90s*), the *net growth of dendrites* in the hippocampus *decreases*.
- This suggests that certain areas of the brain are capable of dynamically remodeling their neuronal connections.
- This may also suggest that therapy can augment this type of age-related neuronal plasticity.

Age-Related Changes in Neurons, Cell Bodies, and Extracellular Spaces

- In the normal aging process, the cytoplasm of cells in the hippocampus and the cortex (both critical for learning and memory) begin to fill with protein filaments known as *neurofibrillary tangles*.
- An abundance of these tangles was once believed to contribute to *AD*.
- Recent research instead suggests that the formation of neurofibrillary tangles occurs in normal aging rather than from a disease process.
- Additionally, the extracellular spaces between neurons in the hippocampus and certain cortical areas accumulate moderate numbers of deposits known as *plaques*.
- Plaques, or amyloid bodies, develop very slowly in the presence of inflammation in the brain.
- While such amyloid proteins are now thought to form as a normal part of aging, research also suggests that amyloid proteins develop in brain regions that are particularly affected by AD—for example, the hippocampus.
- Once formed, amyloid proteins appear to interfere with the hippocampus' ability to encode short-term memories into long-term memory storage.

DNA Damage and the Aging Process

- Predominant theories suggest that *the body ages as a result of DNA damage in the cells.*
- The enzymatic mechanisms that are designed to repair faulty DNA in cellular structures become less efficient as we age, particularly in late life (70s+).
- The DNA in mitochondria (the energy provider of the cells) slowly become defective.

Oxidation of Cells and the Aging Process

- The levels of *oxidized proteins in human cells also progressively increase with age.*
- Cells from young adults with progeria (a syndrome of premature aging) contain levels of oxidized proteins that approach those found in healthy 80-year-old persons.
- There is evidence to suggest that such oxidation of proteins may lead to a loss of mental function.

Degeneration of Myelin and the Aging Process

- Age-related changes occur in the *myelin that sheathes and insulates axons.*
- Alterations of myelin can have a measurable effect on the speed and efficiency with which neurons propagate electrical impulses.
- Functionally, this may translate into decreased speed of movement and thought.

THE HEALTHY ELDERLY

- In the healthy elderly, the extent of anatomic and physiologic alterations tends to be modest.
- The degree of neuronal loss, the accumulation of plaques, and the deficiency of enzymatic activity range from 5% to 30% above those of young adults.
- Such gradual declines often appear to have little functional effect on mental faculties.
- PET scans of the healthy elderly have shown that the brains of healthy people in their 80s were almost as active as people in their 20s.
- The brain appears to have considerable physiologic reserve and the ability to tolerate small losses of neuronal function.
- Epidemiological studies show that, as a group, almost 90% of all people older than 65 are free of dementia. Fewer than 5% of people aged 65 to 75 years exhibit symptoms of dementia.
- In people aged 75 to 84, 20% are found to exhibit some symptoms of dementia. The percentage then jumps to 50% in people older than 85.
- Studies have found that when people in their 70s and 80s remain in good health, they show only a *subtle decline in performance on tests of memory, perception, and language*.
- The *speed of processing*, however, significantly declines.
- While the healthy elderly may be unable to retrieve certain details of events or information (eg, dates, places, names) as quickly, they are nevertheless often able to recall the information minutes or hours later.
- Given enough time and a stress-reduced environment, most healthy elderly score about as well as young or middle-aged adults on tests of mental performance.
- The more complex a task is (eg, a multiple step math problem), however, the more likely a healthy elder will perform less well than a young adult.
- While one may not learn or remember quite as rapidly during healthy late life, the ability to learn and remember declines only minimally.

Age-Related Dementia and Alzheimer's Disease

- Approximately 4 million people in the United States are diagnosed with AD. The World Health Organization (WHO) projects this number to increase four times by 2040.
- AD is a degenerative disease characterized by a progressive loss of memory and cognitive function.
- Related symptoms include depression, disorientation, decreased concentration, difficulty communicating, loss of bowel and bladder, personality change, and severe mood swings.
- One of the strongest indicators of AD and age-related dementia involves difficulty carrying out daily activities that were once easily mastered—for example, difficulty understanding how to dress or carry out other self-care tasks, difficulty remembering the route to a familiar store, or difficulty remembering how to prepare a simple meal.
- Death from AD or age-related dementia usually occurs within 5 to 10 years after diagnosis.
- There are currently no reliable or accurate diagnostic procedures to identify AD and age-related dementia. Once dementia is diagnosed, however, brain scans can be helpful to monitor the progression of the disease by showing changes in blood glucose metabolism.
- The average age of onset is in the eighth decade of life; however, the disease can occur as early as the 40s, 50s, or 60s.
- AD and dementia are genetically based. Approximately 50% of people who have a parent or sibling with AD will also develop the disease by age 87.

Protecting the Aging Brain

- Researchers suggest that remaining physically fit and mentally active may lessen age-related cognitive deficits. Older people who consistently exercise and read perform better on cognitive tests than do sedentary individuals of the same age and health status.
- Physicians also must be cautious about the type and mix of medications they prescribe for elderly patients. One research study suggested that the average person aged 65 and older takes approximately 8 to 10 different prescription or over-the-counter medications. People over 60 are particularly sensitive to *benzodiazepines* (sedatives like Valium) and other depressants and stimulants of the central nervous system (CNS). Compared with young adults, older people experience a greater decline in reasoning while such drugs are in their system. The elderly are affected longer and react to lower doses more strongly. Drug interactions can also have serious side effects that impair mental function.
- The same is true of *anesthesia* used during surgical procedures. General anesthesia has a greater effect and lasts longer in the elderly, causing disorientation and confusion.
- Health care professionals must also be aware that *nutritional deficiencies* can mask as dementia. A common cause of memory loss is an insufficient supply of nutrients to the brain. Deficiencies in amino acids and B vitamins (particularly B6 and B12) can produce dementia-like symptoms—which can be reversed with nutritional supplements.

The Benefits of Cognitively Stimulating and Leisure Activities

- There is growing evidence that cognitively stimulating activity reduces the risk of AD and other forms of age-related dementia.
- Some of the strongest research supporting the protective factors of mentally stimulating activities have emerged from the Nun Studies—a body of research demonstrating that nuns who remained intellectually active throughout life and displayed high linguistic and verbal reasoning skills were significantly less likely to develop AD in later life. Maintaining intellectually stimulating activities into old age and mastering verbal reasoning were significantly associated with higher brain weight, less cerebral atrophy, and reduced neurofibrillary pathology.
- Similarly, researchers have found that participation in mentally stimulating leisure and social activities can significantly decrease one's risk for AD and related dementias. Although the exact mechanism of this protection is unknown, some researchers have suggested that synaptic complexity and neuronal reserve—which both result from participation in intellectually stimulating activities—may play a role.
- Synaptic complexity occurs when neurons build greater connections with each other as the brain learns new skills and associates newly learned skills with mastered ones having existing neural pathways. Neuronal reserve involves the preservation of functioning neurons that are silent or non-active.
- In the event of brain damage, silent neurons can take over the functions of damaged neurons. Neuronal reserve allows the brain to remain plastic and flexibly adapt to change—whether from the natural aging process, illness, or accident. The more a person builds new neural connections and preserves existing neurons, the less likely it is that neurodegeneration will occur.
- Conversely, participation in mentally passive activities has been found to significantly increase one's risk for the development of AD and related dementias. Several studies have shown that television viewing actually heightens one's risk for dementia. In one study, the risk for developing AD increased 1.3 times for each daily hour of television viewing a person engaged in. In contrast, for each hour spent engaged in intellectual activities per day, the risk of developing AD decreased by 16%.
- Gerontologists suggest that the participation in intellectually and socially stimulating activities throughout the lifespan provides protection against age-related dementias. Intellectually stimulating activities include reading, completing crossword puzzles, playing a musical instrument, engaging in crafts and fine arts, writing, playing cards and board games, participating in needlework, and completing handy work and home repairs. Researchers suggest that in addition to these activities, people can participate in intellectually stimulating activities by learning new skills and hobbies, taking educational courses, participating in book clubs, and traveling to new places.

SECTION 34

Sex Differences in Male and Female Brains

Research Regarding Sex-Based Brain Differences

- Findings regarding the differences between male and female brains have largely come from:
  - Positron emission tomography (PET), single photon emission computed tomography (SPECT), and magnetic resonance imaging (MRI) scans
  - Research regarding noninjured and brain-injured patients
  - Research regarding sex-based hormonal differences in fetal development
- It is important—as with any research that is in the early stages of development—that findings are not misinterpreted and that false generalizations are not extrapolated from the data.

Overlap Between the Male and Female Brain

- Despite male and female brain differences found through research, there is still greater overlap of brain similarity than dissimilarity.
- In both males and females, there is a substantial range of abilities within each sex.
- While males as a group may perform better on tasks of spatial skills than females, there are females who perform equally or better than men on spatial skill tasks, and vice versa.

The Effects of Sex Hormones on Brain Organization In Utero

- Differing patterns of ability between males and females likely reflect different *hormonal influences upon fetal development of each sex.*
- The action of *estrogen* (female hormone) and *androgens* (male hormones—chief of which is testosterone) establish sexual differentiation in utero.
- All mammal embryos—including human—have the potential to develop into a male or female in the first weeks of life.
- Humans have 46 chromosomes: XX (female) and XY (male)
- If a Y chromosome is present, testes or male gonads form. This is the critical first step toward becoming male.

Female Default System

- If gonads do not produce male hormones—or if the hormones cannot act on the fetal tissue—the XY embryo defaults into the form of a female organism. This is called the *female default system.*
- There are many abnormalities that can occur regarding the release of specific hormones necessary to differentiate the unisex embryo into a male. The female default system is an evolutionary safeguard to maintain the life of the fetus.
- If testes are formed, they produce substances that facilitate the development of a male. One such substance, testosterone, causes masculinization by promoting the male or Wolffian ducts to convert to the external appearance of a scrotum and penis. Wolffian ducts in the male embryo develop into the vas deferens and seminal vesicles.
- The Mullerian ducts cause the female gonads to regress physically. In a female embryo, the Mullerian ducts develop into the uterus, fallopian tubes, and vagina.

Adam Plan

- Like the female default system, there is an evolutionary safeguard that maintains fetal life if a prenatal anomaly prevents the XX organism from developing into a female. This is called the *Adam plan,* and it involves a *defeminization process of the female template*. The Adam plan stops the embryonic development of the female sex anatomy and instead, facilitates the masculinization of embryonic structures through Mullerian inhibiting hormone, dihydrotestosterone, and testosterone.

Gender-Based Brain Differences: Established In Utero by Sex Hormones

- Gender-based brain differences begin in the womb. The hormones responsible for transforming embryonic tissue into male or female anatomy act on brain development early in the gestational period.
- The most current research regarding brain development suggests that the human brain is permanently and irreversibly transformed into male or female by the 18th week of pregnancy.

Abnormalities that Occur In Utero

Turner Syndrome

- Chromosomal anomaly of 45 XO. The female is missing one sex chromosome.
- Causes the absence of ovaries. The female is infertile.
- Other signs and symptoms include short stature and webbed neck.
- Females require sex hormone treatment in puberty to induce secondary sex characteristics and menstruation.

Klinefelter's Syndrome

- Chromosomal anomaly in males. The male has an extra sex chromosome: 47 XXY.
- Signs and symptoms include small penis, small testes, sterility (no sperm production), tall and thin body type, and male gynecomastia (development of male breasts). Secondary male sex characteristics are often weakly developed and do not respond well to hormonal therapy.

Dihydrotestosterone Deficiency Syndrome

- Dihydrotestosterone (DHT) deficiency syndrome; also called deficiency of 5-alpha reductase.
- DHT is a powerful androgen converted from testosterone (through the enzymatic action of 5-alpha reductase).
- This directs the differentiation of the penis and scrotum (in utero).
- DHT is converted from testosterone. But in DHT deficiency syndrome, DHT cannot be produced.
- DHT is necessary for the development of male external sexual anatomy.
- The fetus possesses the normal 46 XY chromosomes.
- However, the external anatomy remains female because testosterone cannot convert into DHT, which directs the differentiation of the penis and scrotum.
- The testes develop in the second month of pregnancy but do not descend.
- Mullerian inhibiting hormone is present and negates the development of the fallopian tubes, uterus, and vagina.
- At puberty, the common testosterone surge (occurring at this stage of life) may trigger the transformation of a clitoral-like phallus into a small penis.
- Weak secondary male sexual characteristics may develop at this time.
- It appears that the little girl, at puberty, turns into a male.
- This anomaly has been reported commonly in the Dominican Republic.

Testicular Feminizing Syndrome

- Also called androgen insensitivity syndrome
- The embryo possesses the normal 46 XY chromosomes.
- However, the cells of the embryo are not able to respond to testosterone.
- The internal anatomy is male but is incomplete.
- The female default plan is attempted by the organism but cannot be carried out fully. Because of the presence of testosterone, the development of internal female anatomy is prevented—the development of ovaries cannot occur and the individual is infertile.
- The external anatomy appears as a normal female.

Retained Mullerian Syndrome

- This disorder results in a type of pseudo-hermaphrodite.
- Internal female sex organs are present.
- External male sex organs are also present.

In Addition to Differentiating Male/Female Sex Organs, Sex Hormones May Also Organize Male/Female Behavior in Animals and Humans

Animal Studies

- If a rodent with functional male genitals is deprived of androgens immediately after birth, male sexual behavior—such as mounting—will be reduced.
- Instead, female behaviors—such as lordosis (arching of the back to receive intercourse)—will be increased.
- Similarly, if androgens are administered to female rodents directly after birth, the female rodent displays more male sexual behavior and less female behavior.

Congenital Adrenal Hyperplasia in Humans

- Congenital adrenal hyperplasia (CAH) occurs as a result of a genetic anomaly that causes large amounts of adrenal androgens to be produced in the XX embryo.
- These are females who were *exposed to an excess of androgens in the prenatal stage.*
- Although the consequent masculinization of genitals can be surgically corrected in early life, and drug therapy can stop the overproduction of androgens, the effects of prenatal exposure on the brain cannot be reversed.
- Females who were exposed to prenatal androgens grow up to be self-reported tomboys and prefer typically masculine toys more than their unaffected biological sisters.
- Researchers also found that the CAH females performed better than their unaffected biological sisters on spatial manipulation tasks.

Differences in the Anatomy of the Male vs Female Brain

- Anatomical differences between male and female human brains exist.
- Males generally have about 4% more brain cells and approximately 100 grams more brain tissue than females.
- Females have more dendritic connections between cells.
- While the male brain is somewhat larger than the female brain, specific regions of each brain have been found to have differing sizes.

Frontal Cortex

- Certain regions of the frontal cortex have been found to be larger in females than in males. One such area is the region where *higher level executive functions* are mediated (ie, skills including judgment, problem solving, and regulating raw emotions).

Parietal Cortex

- The *inferior parietal lobule* is a bilateral region in the parietal lobes that has been linked to *higher level, abstract mathematical skills*. This region was shown to be enlarged in the brain of Albert Einstein and in other mathematicians and physicists.
- The inferior parietal lobule (IPL) has been found to be significantly larger in males than in females.
- The right IPL has been correlated with the ability to manipulate objects in space.
- The left IPL appears to be associated with the perception of time and speed and the ability to rotate three-dimensional figures.
- Research regarding gender-based skills have shown that men are generally more adept at rotating objects in space, abstract mathematical reasoning skills, route navigation, and targeted motor skills (such as intercepting projectiles like a ball tossed in the air).
- Researchers have also shown that the brain regions responsible for the above skills appear to mature approximately 4 years earlier in boys than in girls.

Corpus Callosum

- PET scans have shown that the corpus callosum is more extensive in women than in men. Women have 11% more neurons in the corpus callosum, thus creating more neural fibers that connect the two hemispheres.
- This may suggest that women are more able to integrate both hemispheres in activities than men.
- In *language skills*, women activate both hemispheres to interpret the literal meaning of the words (left brain function) as well as the emotional tone of the words (right brain function).
- Men activate only the left hemisphere in language skills and have greater difficulty interpreting the subtle emotional tone of a conversation.

- The incidence of *language disorders (aphasia)* is higher in men than in women after damage to the left hemisphere.
- Research has also shown that females typically recover from aphasia after cerebrovascular accident more readily than do men. This may relate to a female's greater ability to integrate both hemispheres during language tasks.

Broca's and Wernicke's Area

- *Broca's area* is located in the left frontal lobe (beneath the premotor area) and has been associated with *expressive communication.*
- *Wernicke's area* is located in the left parietal lobe in the superior temporal gyrus and has been associated with *receptive communication.*
- Both areas have been found to be significantly larger in women than in men—which may account for women's greater performance in tests of language-based skills. Women have approximately 23% more volume in Broca's area and 13% more volume in Wernicke's area than men.
- Researchers have also found that the brain regions involved in language and fine motor skills appear to mature 6 years earlier in girls than in boys.

The Amygdala and Limbic System: Emotion

- Research has found that men generally possess larger amygdala structures than women.
- Women, conversely, have more dendritic connections between the amygdala and the cortex.
- The amygdala is a structure in the limbic system that is believed to be responsible for the feelings of anger and fear.
- *The connections between the amygdala and the cortex may be used to alert the cortex to the threat of danger.*
- Researchers have suggested that because women possess greater connections between the amygdala and cortex, they may be better at modulating their anger and using symbolic signals to deescalate a potentially threatening situation.
- Because of this anatomical difference, men may have greater difficulty modulating anger and may respond with more overt displays of aggression.
- PET scans have also shown that men exhibit greater brain activity in the more ancient regions of the limbic system—the regions that mediate action in response to emotion.
- Women exhibit greater activity in the phylogenetically newer and more complex regions of the limbic system that mediate symbolic action in response to emotion.
- One researcher explained that, if a dog is angry and jumps and bites, that is an active response to the emotion of anger. If the dog bears its teeth and growls, that is a symbolic response to the emotion of anger.
- The implication may be that women are more predisposed to a symbolic display of their emotions, while men react with action to their emotions.
- Women are also more adept at recognizing symbolic expressions of emotions in others than are men. Men seem to need an overt reaction that clearly demonstrates emotion—otherwise they miss emotions that are symbolically or subtly displayed.
- Studies have shown that women are more sensitive at recognizing subtle facial expressions than are men (a limbic system function).
- In studies, both men and women were equally adept at recognizing happiness in the photos of faces of both men and women.
- But women were more adept at determining if a man or woman was sad. A woman's face had to be tearful for men to recognize the emotion of sadness. Subtle expressions of sadness were not recognized.
- Researchers have also found that emotional activity appears to be localized in the amygdala. At 6 years of age, the connections to the cortex have not been fully established, and thus, the lack of connections between the amygdala and cortex may account for a child's difficulty regulating emotions and attaching words to feelings.
- Such connections between the amygdala and cortex begin to more fully develop in adolescence—although predominately in females.

Nature vs Nurture

- Researchers have suggested that the effects of sex hormones on brain organization occur so early in development that from birth, the environment is acting on differently wired male and female brains.

- One study found that 3-year-old boys perform better at physical targeting skills (guiding or intercepting projectiles) than girls of the same age.
- Other studies have shown that the extent of experience playing sports does not account for sex differences in targeting skills exhibited by young adults.
- Some researchers have argued that sex differences in spatial rotation performance are present before puberty.

Skills that Males Perform Better Than Females

- Skills that adult males perform better than adult females include the following. It should be noted that this finding considers men as a group, without considering differences occurring in both sexes.
  - Spatial tasks requiring the individual to imagine rotating an object or visually manipulating it in some way
  - Mathematical reasoning
  - Navigating through a route
  - Target-directed motor skills (guiding or intercepting projectiles)

Skills that Females Perform Better Than Males

- Skills that adult females perform better than adult males include the following. It should be noted that this finding considers women as a group, without considering differences occurring in both sexes.
  - Rapidly identifying matching items (called perceptual speed)
  - Verbal fluency (including the ability to find words that begin with a specific letter or fulfill some other constraint)
  - Arithmetic calculation
  - Recalling landmarks from a route
  - Precision fine motor skills

Route Navigation

- In studies, men learned a route in fewer trials than women and made fewer errors.
- Once learned, however, women remembered more of the landmarks than did men.
- Women appear to use landmarks as a strategy to orient themselves.
- Men appear to navigate using the spatial relationships of routes as depicted on maps.

The Recall of Objects and Their Locations

- In studies that tested the ability of men and women to recall objects and their locations in a confined space (a room or tabletop), it was found that women were better able to remember whether an item had been displaced or not.
- Women were also better able at replacing objects in their original locations after they had been moved.

Possible Evolutionary Significance

- Some theorists suggest that the above sex-based differences may have evolutionary roots.
- Such theorists suggest that for thousands of years during which human brain characteristics evolved, humans lived in relatively small hunter–gatherer groups.
- The division of labor between the sexes in such societies was probably distinct—as it is in existing hunter–gatherer societies.
- Men were believed to be responsible for large game hunting requiring long-distance travel.
- Women were believed to gather food near the campsite, prepare food and clothing, and care for children.
- Theorists argue that such specializations placed different evolutionary selection pressures on men and women.
- Men would require long-distance route-finding abilities to recognize a geographic area from varying locations. They would also need targeting skills for hunting.
- Women would require short-range navigation using landmarks, fine motor capabilities carried out within a circumscribed space, and perceptual discrimination skills sensitive to small changes in the environment or in children's appearance and behavior.

The Empathizing vs Systematizing Brain

- Simon Baron-Cohen (Director of the Autism Research Centre at Cambridge University) developed the idea that male and female brains are specialized for either empathizing or systematizing.
- Baron-Cohen suggests that the female brain is predominantly hard-wired for empathy, identifying the meanings of facial expression, decoding nonverbal communication, detecting subtle nuances of tone and gesture, analyzing contextual social cues, and responding in emotionally appropriate ways that foster cooperation and well-being of the group as a whole.
- The male brain is predominantly hard-wired for constructing rule-based analyses of the external world—particularly of inanimate objects and events. They seek to understand and build systems—whether mechanical, governmental, or gaming systems.
- The greater spatial skill of males, and the greater language and emotional skills of females, seem to serve the basic differences of empathizing and systematizing.
- Baron-Cohen suggests that this neural difference appears in children's preferences for toys (mechanical trucks versus humanlike dolls), styles of interaction (ordering versus negotiating), and methods of navigation (women appear to personalize space through landmarks; men more commonly visualize space as a geometric system).

Joining and Hosting Groups

- To better examine empathizing versus systematizing skills, researchers have developed tests that assess the differences between the sexes in various scenarios.
- In one scenario, children are tested to determine the differences between girls and boys in the way each joins a new group. This has been investigated by introducing a new child to a group that has already formed.
- As a newcomer to an established play group, female children tend to stand back and observe for a while before joining. They evaluate what is happening in the group and attempt to fit into the ongoing activity.
- Male children, as newcomers, are more likely to attempt to take over the game, try to change its rules, and direct everyone's attention onto them.
- Another scenario involves investigating differences between males and females as they attempt to host a group. Female children are more attentive to the newcomer and offer ways through which the newcomer can become part of the group activities.
- Male children, as hosts, more commonly ignore the newcomer's attempt to join in and are more likely to continue the activities in which they were already engaged.

Eye Contact and Language

- Sex differences can be found as early as birth. Researchers have shown that 1-day-old female infants look longer at a face while 1-day-old males look longer at a suspended mechanical mobile.
- The amount of eye contact that children make is in part determined by prenatal testosterone.
- At 1-year-old, male toddlers show a stronger preference for watching video games of cars. One-year-old female toddlers show a greater preference for watching videos of human faces demonstrating a range of emotions.
- Testosterone level during fetal development also influences language skills. The higher the prenatal testosterone level, the smaller a child's vocabulary was measured to be at 18 and 24 months.

Asperger's Syndrome

- Baron-Cohen suggests also that autism—particularly Asperger's syndrome—may reflect a brain that has been too highly systematized by gestational hormones in utero.
- Asperger's syndrome is a form of autism in which intelligence is average or higher but social communication is highly impaired. Asperger's syndrome predominantly affects males at a 10:1 ratio.
- Baron-Cohen's theory is similar to one that Geschwind and Galaburda had proposed years earlier—that the male hormone testosterone slows the growth of the brain's left hemisphere and accelerates growth on the right. Some have suggested that when too much testosterone tilts this maturational process to one extreme, a range of problems can potentially occur, including learning disability, attention-deficit disorder (ADD) and attention-deficit hyperactivity disorder (ADHD), and autism. All of these disorders disproportionately affect males.
- Lack of eye contact and poor language skills are early signs of autism. Other characteristics include difficulty developing social relationships, difficulty interpreting and responding to social cues, narrow interests, and a strong adherence to routines.
- Baron-Cohen suggests that Asperger's syndrome may be indicative of an extreme male brain—particularly talented at systemizing and deficient in the skill of empathy.

References

Section 1: Directional Terminology

Gray H, et al. *Gray's Anatomy: The Anatomical Basis of Clinical Practice*. 39th ed. New York: Elsevier Churchill Livingstone; 2005.

Netter FH, Hansen JT. *Atlas of Human Anatomy*. 3rd ed. Teterboro, NJ: Icon Learning Systems; 2003.

Stedman, TL. *Stedman's Medical Dictionary for the Health Professions and Nursing*. 5th ed. Philadelphia: Lippincott, Williams, and Wilkins; 2005.

Section 2: Division of the Nervous System

Agur AMR, Grant JCB. *Grant's Atlas of Anatomy*. 11th ed. Philadelphia: Lippincott, Williams, and Wilkins; 2005.

Gray H, et al. *Gray's Anatomy: The Anatomical Basis of Clinical Practice*. 39th ed. New York: Elsevier Churchill Livingstone; 2005.

Netter FH, Hansen JT. *Atlas of Human Anatomy*. 3rd ed. Teterboro, NJ: Icon Learning Systems; 2003.

Rohen JW, Yokochi C, Lutjen–Drecoll E. *Color Atlas of Anatomy: A Photographic Study of the Human Body*. 5th ed. Philadelphia: Lippincott, Williamson, and Wilkins; 2002.

Snell RS. *Clinical Anatomy*. Philadelphia: Lippincott, Williams, and Wilkins; 2004.

Stedman TL. *Stedman's Medical Dictionary for the Health Professions and Nursing*. 5th ed. Philadelphia: Lippincott, Williams, and Wilkins; 2005.

Section 3: Gross Cerebral Structures

Agur AMR, Grant JCB. *Grant's Atlas of Anatomy*. 11th ed. Philadelphia: Lippincott, Williams, and Wilkins; 2005.

Amaral DG. The amygdala, social behavior, and danger detection. *Ann N Y Acad Sci*. 2003;1000:337–347.

Bannerman DM, et al. Double dissociation of function within the hippocampus: Spatial memory and hyponeophagia. *Behav Neurosci*. 2002;116:884–901.

Basso MA, et al. Cortical function: A view from the thalamus. *Neuron*. 2005;45:485–488.

Bllom JS, Hynd GW. The role of the corpus callosum in interhemispheric transfer of information: Excitation or inhibition? *Neuropsychol Rev*. 2005;15:59–71.

Daskalakis ZJ, et al. Exploring the connectivity between the cerebellum and motor cortex in humans. *J Physiol*. 2004;557:689–700.

Deadwyler SA, Hampson RE. Differential but complementary mnemonic functions of the hippocampus and subiculum. *Neuron*. 2004;42:465–476.

Gray H, et al. *Gray's Anatomy: The Anatomical Basis of Clinical Practice*. 39th ed. New York: Elsevier Churchill Livingstone; 2005.

Harrington D, et al. Does the representation of time depend on the cerebellum: Effect of cerebellar stroke. *Brain*. 2004;127:1–14.

Hoshi E, et al. The cerebellum communicates with the basal ganglia. *Nat Neurosci*. 2005;8:1291–1493.

Horn AK. The reticular formation. *Prog Brain Res*. 2005;151:127–155.

Joels M, et al. Effects of chronic stress on structure and cell function in rat hippocampus and hypothalamus. *Stress*. 2004;7:221–231.

Jonas R, et al. Cerebral hemispherectomy: Hospital course, seizure, developmental, language, and motor outcomes. *Neurology*. 2004;62:1712–1721.

Lundy RF, Norgen R. Activity in the hypothalamus, amygdala, and cortex generates bilateral and convergent modulation of pontine gustatory neurons. *J Neurophysiol*. 2004;91:1143–1157.

Malberg JE. Implications of adult hippocampal neurogenesis in antidepressant action. *Rev Psychiatr Neurosci.* 2004;29:196–205.

Malberg JE, Schecter, LE. Increasing hippocampal neurogenesis: a novel mechanism for antidepressant drugs. *Curr Pharm Design.* 2005;11:145–155.

Miller EK, Wallis JD. *The Prefrontal Cortex and Executive Brain Functions.* San Diego: Academic Press; 2003.

Morgane PJ, et al. A review of systems and networks of the limbic forebrain/limbic midbrain. *Prog Neurobiol.* 2005;75:143–160.

Neary NM, et al. Appetite regulation: from the gut to the hypothalamus. *Clin Endocrinol.* 2004;60:153–160.

Netter FH, Hansen JT. *Atlas of Human Anatomy.* 3rd ed. Teterboro, NJ: Icon Learning Systems; 2003.

Rohen JW, et al. *Color Atlas of Anatomy: A Photographic Study of the Human Body.* 5th ed. Philadelphia: Lippincott, Williamson, and Wilkins; 2002.

Saab C, Willis WD. The cerebellum: organization, functions and its role in nociception. *Brain Res Rev.* 2003;42:85–95.

Schmahmann JD. Disorders of the cerebellum: ataxia, dysmetria of thought, and the cerebellar cognitive affective syndrome. *J Neuropsychiatr Clin Neurosci.* 2004;16:367–378.

Schwabe A, et al. The role of the human cerebellum in short- and long-term habituation of postural responses. *Gait Posture.* 2004;19(1):16–23.

Sherman SM, Guillery RW. *Exploring the Thalamus and Its Role in Cortical Function.* Cambridge, MA: MIT Press; 2003.

Skuse D, et al. The amygdala and development of the social brain. *Ann N Y Acad Sci.* 2003;1008:91–101.

Steriade M. Local gating of information processing through the thalamus. *Neuron.* 2004;41:493–494.

Snell RS. *Clinical Anatomy.* Philadelphia: Lippincott, Williams, and Wilkins; 2004.

Strauss, MM. fMRI of sensitization to angry faces. *NeuroImage.* 2005;26:389–413.

Swaab DF. *The Human Hypothalamus: Basic and Clinical Aspects.* Boston: Elsevier; 2004.

Thach WT, et al. Role of the cerebellum in the control and adaptation of gait in health and disease. *Prog Brain Res.* 2004;143:353–366.

Theil A, et al. Essential language functions of the right hemisphere in brain tumor patients. *Ann Neurol.* 2004;57:128–131.

Tomaz C, et al. Integrative function of the amygdala in emotional memory storage. *Intern Congress Ser.* 2003;1250:335–346.

Wells S, Murphy D. Transgenic studies on the regulation of the anterior pituitary gland function by the hypothalamus. *Frontiers Neuroendocrinol.* 2003;24:11–26.

Winstanley CA, et al. Contrasting roles of basolateral amygdala and orbitofrontal cortex in impulse choice. *J Neurosci.* 2004;24:4718–4722.

Section 4: Ventricular System

Bateman GA, et al. The pathophysiology of the aqueduct stroke volume in normal pressure hydrocephalus: Can co-morbidity with other forms of dementia be excluded? *Neuroradiology.* 2005;47:741–748.

Blomsterwall E, et al. Postural disturbance in patients with normal pressure hydrocephalus. *Acta Neurologica Scandinavica.* 2000;102:284–291.

Chen Cl, et al. Intraventricular cavernous hemangioma at the foramen of Monro: case report and literature review. *Clin Neurol Neurosurg.* 2006;108:604–609.

D'Angelo VA, et al. Lateral ventricle tumors: surgical strategies according to tumor origin and development—a series of 72 cases. *Neurosurgery.* 2005;56(1):36–45.

Dastgir G, et al. Unilateral hydrocephalus due to foramen of Monro stenosis. *Minimally Invasive Neurosurg.* 2006;49:184–186.

Di Rocco, et al. Lateral ventricle choroid plexus papilloma extending into the third ventricle. *Images Pediatr Neurosurg.* 2004;40:14–316.

Duinkerke A, et al. Cognitive recovery in idiopathic normal pressure hydrocephalus after shunt. *Cogn Behav Neurol.* 2004;17:179–184.

Graft Cl, Pollack GM. Drug treatment at the blood–brain barrier and the choroid plexus. *Curr Drug Metab.* 2004;5:95–108.

Gray H, et al. *Gray's Anatomy: The Anatomical Basis of Clinical Practice.* 39th ed. New York: Elsevier Churchill Livingstone; 2005.

Jean WC, et al. Subtonsillar approach to the foramen of Luschka: an anatomic and clinical study. *Neurosurgery.* 2003;52:860–866.

Kolarcik C, Bowser R. Plasma and cerebrospinal fluid-based protein biomarkers for motor neuron disease. *Mol Diagn Ther.* 2006;10:281–292.

Meier U, et al. Predictors of outcome in patients with normal-pressure hydrocephalus. *Eur Neurol.* 2004;51:59–67.

Moody DM. The blood–brain barrier and blood–cerebral spinal fluid barrier. *Semin Cardiothorac Vascular Anesth.* 2006;10:128–131.

Netter FH, Hansen JT. *Atlas of Human Anatomy.* 3rd ed. Teterboro, NJ: Icon Learning Systems; 2003.

Owler BK, Pickard JD. Normal pressure hydrocephalus and cerebral blood flow: a review. *Acta Neurologica Scandinavica.* 2001;104:325–342.

Ramtahal J. Arachnoid cyst mimicking normal pressure hydrocephalus. A case report and review of the literature. *J Neurol Sci.* 2006;50:79–81.

Serot JM, et al. Choroid plexus, aging of the brain, and Alzheimer's disease. *Frontiers Biosci.* 2003;8:S515–S521.

Silverberg GD, et al. Alzheimer's disease, normal-pressure hydrocephalus, and senescent changes in CSF circulatory physiology: a hypothesis. *Lancet Neurol.* 2003;2:506–511.

Sorteberg A. A prospective study on the clinical effect of surgical treatment of normal pressure hydrocephalus: The value of hydrodynamic evaluation. *Brit J Neurosurg.* 2004;18:149–157.

Section 5: The Cranium

Agur AMR, Grant JCB. *Grant's Atlas of Anatomy.* 11th ed. Philadelphia: Lippincott, Williams, and Wilkins; 2005.

Gray H, et al. *Gray's Anatomy: The Anatomical Basis of Clinical Practice.* 39th ed. New York: Elsevier Churchill Livingstone; 2005.

Netter FH, Hansen JT. *Atlas of Human Anatomy.* 3rd ed. Teterboro, NJ: Icon Learning Systems; 2003.

Rohen JW, et al. *Color Atlas of Anatomy: A Photographic Study of the Human Body.* 5th ed. Philadelphia: Lippincott, Williamson, and Wilkins; 2002.

Snell RS. *Clinical Anatomy.* Philadelphia: Lippincott, Williams, and Wilkins; 2004.

Stedman, TL. *Stedman's Medical Dictionary for the Health Professions and Nursing.* 5th ed. Philadelphia: Lippincott, Williams, and Wilkins; 2005.

Section 6: The Meninges

Al-Sarraj S, et al. Subdural hematoma (SDH): assessment of macrophage reactivity within the dura mater and underlying hematoma. *Clin Neuropathol.* 2004;23:62–75.

Aimedicu P, Grebe R. Tensile strength of cranial pia mater: preliminary results. *J Neurosurg.* 2004;100:111–114.

Ballabh P, et al. The blood–brain barrier: an overview. Structure, regulation, and clinical implications. *Neurobiol Dis.* 2004;16:1–13.

Bayer SA, Altman J. *Atlas of Human Central Nervous System Development.* Boca Raton, FL: CRC Press; 2004.

Chronic subdural hematoma. Bleeding on the brain. *Mayo Clin Health Lett.* 2006;24(8):4–5.

Erol FS, et al. Irrigation vs. closed drainage in the treatment of chronic subdural hematoma. *J Clin Neurosci.* 2005;12:261–263.

Graff CL, Pollack GM. Drug transport at the blood–brain barrier and the choroid plexus. *Curr Drug Metab.* 2004;5:95–108.

Greenstein B, Greenstein A. *Color Atlas of Neuroscience: Neuroanatomy and Neurophysiology.* New York: Thieme; 2000.

Gregory D, et al. *Basic and Clinical Anatomy of the Spine, Spinal Cord, and ANS.* 2nd ed. St. Louis, MO: Mosby; 2005.

Haines DE. *Neuroanatomy: An Atlas of Structures, Sections, and Systems*. Philadelphia: Lippincott, Williams, and Wilkins; 2003.

Neuwelt E. Mechanisms of disease: the blood–brain barrier. *Neurosurgery*. 2004;54:131–142.

Ozawa H, et al. Mechanical properties and function of the spinal pia mater. *J Neurosurg–Spine*. 2004;1:122–127.

Reina MA, et al. The origin of the spinal subdural space: ultra structure findings. *Anesth Analg*. 2002;94:991–995.

Singh S, et al. Cerebral venous sinus thrombosis presenting as subdural hematoma. *Aust Radiol*. 2005;49:101–103.

Stilling M, et al. Subdural intracranial pressure, cerebral perfusion pressure, and degree of cerebral swelling in supra– and infratentorial space-occupying lesions in children. *Acta Neurochirurgica*. 2005;95(Suppl):133–136.

Weller RO. Microscopic morphology and histology of the human meninges. *Morphologie*. 2005;89(284):22–34.

Section 7: Spinal Cord Anatomy

Arts MP, de Jong TH. Thoracic meningocele, meningomyelocele or myelocystocele? Diagnostic difficulties, consequent implications and treatment. *Pediatr Neurosurg*. 2004;40:75–79.

Banerjee A. *Clinical Physiology: An Examination Primer*. Cambridge, England: Cambridge University Press; 2005.

Bayer SA, Altman J. *Atlas of Human Central Nervous System Development*. Boca Raton, FL: CRC Press; 2004.

Bergman RA, Afifi AK. *Functional Neuroanatomy: Text and Atlas*. 2nd ed. New York: McGraw–Hill; 2005.

Ganong WF. *Review of Medical Physiology*. 22nd ed. New York: McGraw–Hill; 2005.

Greenstein B, Greenstein A. *Color Atlas of Neuroscience: Neuroanatomy and Neurophysiology*. New York: Thieme; 2000.

Gregory D, et al. *Basic and Clinical Anatomy of the Spine, Spinal Cord, and ANS*. 2nd ed. St. Louis, MO: Mosby; 2005.

Haines DE. *Neuroanatomy: An Atlas of Structures, Sections, and Systems*. Philadelphia: Lippincott, Williams, and Wilkins; 2003.

Hendelman WJ. *Atlas of Functional Neuroanatomy*. Boca Raton; FL: CRC Press; 2000.

Kumar R, et al. Occurrence of splint cord malformation in meningomyelocele: complex spina bifida. *Pediatr Neurosurg*. 2002;36:119–127.

Lamin S, Bhattacharya JJ. Vascular anatomy of the spinal cord and cord ischemia. *Pract Neurol*. 2003;3:92–95.

Mimura T, et al. A case of Arnold Chiari syndrome with flaccid paralysis and huge syringomyelia. *Spinal Cord*. 2004;42:541–544.

Mitchell L, et al. Spina bifida. *Lancet*. 2004;364:1885–1895.

Randy JJ. The anatomic and physiologic basis of local, referred and radiating lumbosacral pain syndromes related to disease of the spine. *J Neuroradiol*. 2004;31:163–180.

Sheerin F. Spinal cord injury: anatomy and physiology of the spinal cord. *Emerg Nurse*. 2004;12:30–36.

Section 8: The Cranial Nerves

Benatar M, Edlow J. The spectrum of cranial neuropathy in patients with Bell's palsy. *Arch Intern Med*. 2004;164:2383–2385.

Bennetto L, et al. Trigeminal neuralgia and its management. *Brit Med J*. 2007;334:201–205.

Brewer AA, et al. Visual field maps and stimulus selectivity in human ventral occipital cortex. *Nat Neurosci*. 2005;8:1102–1109.

Dalton P. Olfaction and anosmia in rhinosinusitis. *Curr Allergy Asthma Rep*. 2004;4:230–236.

Eaton DA, et al. Recovery of facial nerve function after repair or grafting: our experience with 24 patients. *Am J Otolaryngol*. 2007;28(1):37–41.

Kaido T, et al. Traumatic oculomotor nerve palsy. *J Clin Neurosci*. 2006;13:852–855.

Kasse CA, et al. Clinical data and prognosis in 1521 cases of Bell's palsy. *Int Congress Ser*. 2003;1240:641–647.

Kitagawa J, et al., Pharyngeal branch of the glossopharyngeal nerve plays a major role in reflex swallowing from the pharynx. *Am J Physiol—Regul Integr Comp Physiol*. 2002;282:R1342–R1347.

Kusher BJ. Perspective on strabismus. *Arch Ophthalmol*. 2006;124:1321–1326.

Leff A. A historical review of the representation of the visual field in primary visual cortex with special reference to the neural mechanisms underlying macular sparing. *Brain Lang*. 2004;88:268–278.

Lewis SL. *Field Guide to the Neurologic Examination*. Philadelphia: Lippincott, Williams, and Wilkins; 2004.

Monkhouse S. *Cranial Nerves: Functional Anatomy*. Cambridge, England: Cambridge University Press; 2006.

Murphy GJ, et al. Sensory neuron signaling to the brain: properties of transmitter release from olfactory nerve terminals. *J Neurosci*. 2004;24:3023–3030.

Nuti D, et al. Acute vestibular neuritis: prognosis based upon bedside clinical tests (thrusts and heaves). *Ann N Y Acad Sci*. 2005;1039:359–367.

Papanas N, et al. Simultaneous, painless, homolateral oculomotor and trochlear nerve palsies in a patient with type 2 diabetes mellitus. Neuropathy or brainstem infarction? *Acta Diabetologica*. 2006;43(1):19–21.

Piercy J. Bell's palsy. *Brit Med J*. 2005;330:1374.

Schwartzman RJ, Schwartzman D. *Neurologic Examination*. Malden, MA: Blackwell; 2006.

Sherman SC, Saadatmand B. Pontine hemorrhage and isolated abducens nerve palsy. *Am J Emerg Med*. 2007;25:104–105.

Spickler EM, Govila L. The vestibulocochlear nerve. *Sem Ultrasound CT MR*. 2002;23:218–237.

Straetmans J, et al. Horner's syndrome as a complication of acute otitis media. *B–ENT*. 2006;2:181–184.

Susarla SM, et al. Functional sensory recovery after trigeminal nerve repair. *J Oral Maxillofac Surg*. 2007;65(1):60–65.

Trip SA, et al. Retinal nerve fiber layer axonal loss and visual dysfunction in optic neuritis. *Ann Neurol*. 2005;58:383–391.

Witt RL, et al. Spinal accessory nerve monitoring with clinical outcome measures. *Ear Nose Throat J*. 2006;85:540–544.

Zafeiriou DI, Pavlov E. Images in clinical medicine. Hypoglossal–nerve palsy. *N Engl J Med*. 2004;350:e4.

SECTION 9: SENSORY RECEPTORS

Banerjee A. *Clinical Physiology: An Examination Primer*. Cambridge, England: Cambridge University Press; 2005.

Bayer SA, Altman J. *Atlas of Human Central Nervous System Development*. Boca Raton, FL: CRC Press; 2004.

Beenhakker MP, et al. Proprioceptor regulation of motor circuit activity by presynaptic inhibition of a modulatory projection neuron. *J Neurosci*. 2005;25:8794–8806.

Bergman RA, Afifi AK. *Functional Neuroanatomy: Text and Atlas*. 2nd ed. New York: McGraw–Hill; 2005.

Cahusac PM. Senok SS. Metabolic glutamate receptor antagonists selectively enhance responses of slowly adapting type I mechanoreceptors. *Synapse*. 2006;59:235–242.

Collins DF, et al. Cutaneous receptors contribute to kinesthesia at the index finger, elbow, and knee. *J Neurophysiol*. 2005;94:1699–1706.

Ganong WF. *Review of Medical Physiology*. 22nd ed. New York: McGraw–Hill; 2005.

Greenstein B, Greenstein A. *Color Atlas of Neuroscience: Neuroanatomy and Neurophysiology*. New York: Thieme; 2000.

Haines DE. *Neuroanatomy: An Atlas of Structures, Sections, and Systems*. Philadelphia: Lippincott, Williams, and Wilkins; 2003.

Hendelman WJ. *Atlas of Functional Neuroanatomy*. Boca Raton; FL: CRC Press; 2000.

Kaufman MP. Mechanoreceptors and central command. *Am J Physiol—Heart Circ Physiol*. 2007;292:H117–118.

Kiernan JA. *Barr's the Human Nervous System: An Anatomical Viewpoint*. 8th ed. Philadelphia: Lippincott, Williams, and Wilkins; 2004.

Kozyreva TV. Central and peripheral thermoreceptors. Comparative analysis of the effects of prolonged adaptation to cold and noradrenaline. *Neurosci Behav Physiol*. 2007;37:191–198.

Martin JH. *Neuroanatomy: Text and Atlas*. New York: McGraw–Hill; 2003.

Moreira TS, et al. Central chemoreceptors and sympathetic vasomotor outflow. *J Physiol*. 2006;577:369–386.

Najafipour H, Ketabchi F. The receptors and role of angiotensin II in knee joint blood flow regulation and role of nitric oxide in modulation of their function. *Microcirculation*. 2003;10:383–390.

Nijsen M, et al. Divergent role for CRF1 and CRF2 receptors in the modulation of visceral pain. *Neurogastroenterol Motility*. 2005;17:423–432.

Oh EC, et al. From the cover: transformation of cone precursors to functional rod photoreceptors by bZIP transcription factor NRL. *Proc Nat Acad Sci U S A*. 2007;104:1679–1684.

Pearson KG. Spinal cord injury reveals unexpected function of cutaneous receptors. *J Neurophysiol.* 2003;90:3583–3584.

Pintelon I, et al. Sensory receptors in the visceral pleura: neurochemical coding and live staining in whole mounts. *Am J Respir Cell Mol Biol.* 2006; [Epub ahead of print].

Snell RS. *Clinical Anatomy*. Philadelphia, PA: Lippincott Williams and Wilkins; 2004.

Sommer MA, Wurtz RH. What the brain stem tells the frontal cortex. Oculomotor signals sent from superior colliculus to frontal eye field via mediodorsal thalamus. *J Neurophysiol.* 2003;91:1381–1402.

Section 10: Neurons and Action Potentials

Banerjee A. *Clinical Physiology: An Examination Primer*. Cambridge, England: Cambridge University Press; 2005.

Barnett MH, Prineas JW. Relapsing and remitting multiple sclerosis: pathology of the newly forming lesion. *Ann Neurol.* 2004;55:458–468.

Bechtl W. *Discovering Cell Mechanisms: The Creation of Modern Cell Biology*. Cambridge, England: Cambridge University Press; 2005.

Clark BA, et al. The site of action potential initiation in cerebellar Purkinje neurons. *Nat Neurosci.* 2005;8:137–139.

Davies RW, Morris BJ, eds. *Molecular Biology of the Neuron*. 2nd ed. Oxford, England: Oxford University Press; 2004.

Eisen A. *Clinical Neurophysiology of Motor Neuron Diseases*. Boston, MA: Elsevier; 2004.

Eshong Y, et al. Gliomedin mediates Schwann cell–axon interaction and the molecular assembly of the nodes of Ranvier. *Neuron.* 2005;47:215–229.

Fischer LR, et al. Amyotrophic lateral sclerosis is a distal axonopathy: evidence in mice and man. *Exp Neurol.* 2004;185:232–240.

Ganong WF. *Review of Medical Physiology*. 22nd ed. New York: McGraw–Hill; 2005.

Hafler DA. Multiple sclerosis. *J Clin Investigat.* 2004;113:788–794.

Hughes BW, et al. Pathophysiology of myasthenia gravis. *Semin Neurol.* 2004;24(1):21–30.

Kiernan JA. *Barr's the Human Nervous System: An Anatomical Viewpoint*. 8th ed. Philadelphia: Lippincott, Williams, and Wilkins; 2004.

Matthews GG. *Cellular Physiology of Nerve and Muscle*. 4th ed. Malden, MA: Blackwell; 2003.

Pehar M, et al. Astrocytic production of nerve factor in motor neuron apoptosis: implications for amyotrophic lateral sclerosis. *J Neurochem.* 2004;89:464–473.

Rudolph M, et al. A method to estimate synaptic conductances from membrane potential fluctuations. *J Neurophysiol.* 2004;91:2884–2896.

Shiraishi H, et al. Acetylcholine receptors loss and post synaptic damage in Mu SK antibody–positive myasthenia gravis. *Ann Neurol.* 2005;57:289–293.

Snell RS. *Clinical Anatomy*. Philadelphia: Lippincott, Williams, and Wilkins; 2004.

Thakur MK, Prasad S. *Molecular and Cellular Neurobiology*. Oxford, England: Alpha Science International; 2005.

Uchoa LC. *Basic Histology*: Text and Atlas. New York: McGraw–Hill; 2005.

Section 11: Special Sense Receptors

Aiba T, et al. Magnetic resonance imaging for diagnosis of congenital anosmia. *Acta Otolaryngologica Supplementum.* 2004;554:50–54.

Ayres AJ. *Sensory Integration and Praxis Tests*. Los Angeles, CA: Western Psychological Services; 1989.

Benloucif S, et al. Stability of melatonin and temperature as circadian phase markers and their relation to sleep times in humans. *J Biol Rhythms.* 2005;20:178–188.

Bonanni E, et al. Quantitative EEG analysis in post-traumatic anosmia. *Brain Res Bull.* 2006;71:69–75.

Bowers D, Kremers J. *The Primate Visual System*. New York: John Wiley and Sons; 2005.

Brechman A, Scheich H. Hemispheric shifts of sound representation in auditory cortex with conceptual listening. *Cereb Cortex.* 2005;15:578–587.

Brown RL, Robinson PR. Melanopsin—shedding light on the elusive circadian photopigment. *Chronobiol Int.* 2004;21:189–204.

Chang W, et al. The development of semicircular canals in the inner ear: role of FGFs in sensory cristae. *Development.* 2004;131:4201–4211.

Dornhoffer J, et al. Stimulation of the semicircular canals via the rotary chairs as a means to test pharmacologic countermeasures for space motion sickness. *Otol Neurol.* 2004;25:740–745.

DuBois GE. Unraveling the biochemistry of sweet and umami tastes. *Proc Nat Acad Sci U S A.* 2004;101:13972–13973.

Duwel P, et al. Subclassification of vestibular disorders by means of statistical analysis in caloric labyrinth testing. *Acta Otolaryngologica.* 2004;124:595–602.

Gaillard I, et al. Olfactory receptors. *Cell Mol Life Sci.* 2004;61:456–469.

Ganong WF. *Review of Medical Physiology.* 22nd ed. New York: McGraw–Hill; 2005.

Gottfried JA, et al. Remembrance of odors past: human olfactory cortex in cross–modal recognition memory. *Neuron.* 2004;42:687–695.

Hashizume Y, et al. A 65–year–old man with headaches and left homonymous hemianopsia. *Neuropathol.* 2004;24:350–353.

Hawkes C. Olfaction in neurodegenerative disorder. *Movement Disord.* 2003;18:364–372.

Kastners S, et al. Functional imaging of the human lateral geniculate nucleus and pulvinar. *J Neurophysiol.* 2004;91:438–448.

Kehagias DT. A case of headache and disordered vision: cavernous hemangioma of the optic chiasm. *Eur Radiol.* 2003;13:2552–2553.

Lee AG, Brazis PW. Localizing forms of nystagmus: symptoms, diagnosis, and treatment. *Curr Neurol Neurosci Rep.* 2006;6:414–420.

Lopez C, et al. Torsional optokinetic nystagmus after unilateral vestibular loss: asymmetry and compensation. *Brain.* 2005;128:1511–1524.

Nelken I. Processing of complex stimuli and natural scenes in the auditory cortex. *Curr Opin Neurobiol.* 2004;14:474–480.

Rowe F. *Visual Fields Via the Visual Pathway.* Malden, MA: Blackwell; 2006.

Saeed P, et al. Optic nerve sheath meningiomas. *Ophthalmology.* 2003;110:2019–2030.

Savic–Berglund I. Imaging of olfaction and gustation. *Nutr Rev.* 2004;62:S205–207.

Shomstein S, Yantis S. Control of attention shifts between vision and audition in human cortex. *J Neurosci.* 2004;24:10702–10706.

Simon SA, et al. The neural mechanisms of gustation: a distributed processing code. *Nat Rev Neurosci.* 2006;7:890–901.

Thesen T, et al. Neuroimaging of multisensory processing in vision, audition, touch, and olfaction. *Cogn Proc.* 2004;5:84–93.

Weinberger NM. Specific long-term memory traces in primary auditory cortex. *Nat Rev Neurosci.* 2004;5:279–290.

Wichman W, Muller–Forell W. Anatomy of the visual system. *Eur J Radiol.* 2004;49(1):8–30.

Wooding S, et al. Natural selection and molecular evolution in PTC, a bitter–taste receptor gene. *Am J Human Genet.* 2004;74:637–646.

Zou Z, et al. Odor maps in the olfactory cortex. *Proc Nat Acad Sci U S A.* 2005;102:7724–7729.

Section 12: Vestibular System

Alpini D, et al. Aging and vestibular system: specific tests and role of melatonin in cognitive involvement. *Arch Gerontol Geriatr.* 2004;38:13–25.

Banfield GK, et al. Does vestibular habituation still have a place in the treatment of benign paroxysmal positional vertigo? *J Laryngol Otol.* 2000;114:501–505.

Barmack NH. Central vestibular system: vestibular nuclei and posterior cerebellum. *Brain Res Bull.* 2003;60:511–541.

Baselli G, et al. Assessment of inertial and gravitational inputs to the vestibular system. *J Biomech.* 2001;34:821–826.

Cullen KE, Roy JE. Signal processing in the vestibular system during active versus passive head movements. *J Neurophysiol.* 2004;91:1919–1933.

Della–Santina CC, et al. Orientation of human semicircular canals measured by three–dimensional multiplanar CT reconstruction. *J Assoc Res Otolaryngol.* 2005;6:191–206.

Deutschlander A, et al. Sensory system interactions during simultaneous vestibular and visual stimulation in PET. *Hum Brain Mapping.* 2002;16:92–103.

Dieterich M, Brandt T. Vestibular system: anatomy and functional magnetic resonance imaging. *Neuroimag Clin N Am.* 2001;11:263–273.

Dozza M, et al. Audio-biofeedback improves balance in patients with bilateral vestibular loss. *Arch Phys Med Rehab.* 2005;86:1401–1403.

Fitzpatrick RC, Day BL. Probing the human vestibular system with galvanic stimulation. *J Appl Physiol.* 2004;96:2301–2316.

Godemann F, et al. What accounts for vertigo one year after neuritis vestibularis—anxiety or a dysfunctional vestibular organ? *J Psychiatr Res.* 2005;39:529–534.

Golz A. The effects of noise on the vestibular system. *Am J Otolaryngol.* 2001;22:190–196.

Graham BP, Dutia MB. Cellular basis of vestibular compensation: analysis and modeling of the role of the commissural inhibitory system. *Exp Brain Res.* 2001;137:387–396.

Guerraz M, Day BL. Expectation and the vestibular control of balance. *J Cogn Neurosci.* 2005;17:463–469.

Gupta VK. Motion sickness is linked to nystagmus-related trigeminal brain stem input: a new hypothesis. *Med Hypothesis.* 2005;64:1177–1181.

Horii A, et al. Vestibular function and vasopressin. *Acta Otolaryngol.* 2004;553:50–53.

Ko JS, et al. Precision speed control of PMSM for stimulation of the vestibular system using rotatory chair. *Mechatronics.* 2003;13:465–475.

Lackner JR, DiZio P. Vestibular, proprioceptive, and haptic contributions to spatial orientation. *Ann Rev Psychol.* 2005;56:115–47.

Lai CH, Chan YS. Development of the vestibular system. *Neuroembryology.* 2002;1:61–71.

Munchau A, Bronstein AM. Role of the vestibular system in the pathophysiology of spasmodic torticollis. *J Neurol Neurosurg Psychiatry.* 2001;71:285–288.

Nuti D, et al. Acute vestibular neuritis: prognosis based upon bedside clinical tests (thrusts and heaves). *Ann N Y Acad Sci.* 2005;1039:359–367.

Pierrot–Deseilligny C, Milea D. Vertical nystagmus: clinical facts and hypotheses. *Brain.* 2005;128:1237–1246.

Rabbitt, et al. Hair-cell versus afferent adaptation in the semicircular canals. *J Neurophysiol.* 2005;93:424–436.

Sheykholeslami K, et al. Vestibular function in auditory neuropathy. *Acta Otolaryngol.* 2000;120:849–854.

Van Der Torn M, Van Dijk JE. Testing the central vestibular functions: a clinical survey. *Clin Otolaryngol.* 2000;25:298–304.

Yates BJ, Bronstein AM. The effects of vestibular system lesions on autonomic regulation: observations, mechanisms, and clinical implications. *J Vestibular Res.* 2005;15:119–129.

Section 13: Autonomic Nervous System

Carney RM, et al. Depression, the autonomic nervous system, and coronary heart disease. *Psychosom Med.* 2005;67: S29–S33.

Chen J, et al. Back to the future: the role of the autonomic nervous system in atrial fibrillation. *Pacing Clin Electrophysiol.* 2006;29:413–421.

Fleshner M. Physical activity and stress resistance: sympathetic nervous system adaptations prevent stress-induced immunosuppression. *Exerc Sport Sci Rev.* 2005;33:120–126.

Grimm DR, et al. Autonomic nervous system function among individuals with acute musculoskeletal injury. *J Manipulative Physiol Ther.* 2005;28(1):44–51.

Janig W. *Integrative Action of the Autonomic Nervous System: Neurobiology of Homeostasis.* Cambridge, England: Cambridge University Press; 2006.

Kawashima T. The autonomic nervous system of the human heart with special reference to its origin, course, and peripheral distribution. *Anat Embryol.* 2005;209:425–438.

Kiernan JA. *Barr's The Human Nervous System: An Anatomical Viewpoint.* 8th ed. Philadelphia: Lippincott, Williams, and Wilkins; 2004.

Kingsley RE. The autonomic nervous system. In: *Concise Text of Neuroscience*. 2nd ed. Philadelphia: Lippincott, Williams, & Wilkins; 2000:471–487.

Krout KE, et al. CNS neurons with links to both mood-related cortex and sympathetic nervous system. *Brain Res.* 2005;1050:199–202.

Leung RS, et al. Respiratory modulation of the autonomic nervous system during Cheyne–Stokes respiration. *Can J Physiol Pharmacol.* 2006;84(1):61–66.

Lindmark S, et al. Dysregulation of the autonomic nervous system can be a link between visceral adiposity and insulin resistance. *Obes Res.* 2005;13:717–728.

Lychkova AE. Balance of activity of sympathetic, parasympathetic, and serotoninergic divisions of the autonomic nervous system in rabbits. *Bull Exp Biol Med.* 2005;140:486–488.

Makarenko NV, et al. Reactions of the autonomic nervous system with different characteristics of higher nervous activity in the situation of examination stress. *Hum Physiol.* 2006;32:368–370.

Molina PE. Neurobiology of the stress response: contribution of the sympathetic nervous system to the neuroimmune axis in traumatic injury. *Shock.* 2005;24(1):3–10.

Neumann SA, et al. White–coat hypertension and autonomic nervous system dysregulation. *Am J Hypertens.* 2005;18:584–588.

Sloan RP, et al. Socioeconomic status and health: is parasympathetic nervous system activity an intervening mechanism? *Int J Epidemiol.* 2005;34: 309–315.

Stauss HM, Persson PB. Cardiovascular variability and the autonomic nervous system. *J Hypertens.* 2006;24:1902–1905.

van Orshoven NP, et al. Subtle involvement of the parasympathetic nervous system in patients with irritable bowel syndrome. *Clin Auton Res.* 2006;16(1):33–39.

Woo MA, et al. Functional abnormalities in brain areas that mediate autonomic nervous system control in advanced heart failure. *J Cardiac Failure.* 2005;11:437–446.

Section 14: Enteric Nervous System

Burns AJ. Migration of neural crest–derived enteric nervous system precursor cells to and within the gastrointestinal tract. *Int J Dev Biol.* 2005;49:143–150.

Burns AJ, Le Douarin NM. Enteric nervous system development: analysis of the selective developmental potentialities of vagal and sacral neural crest cells using quail–chick chimeras. *Anat Rec.* 2001;262(1): 16–28.

De Giorgio R, et al. Primary enteric neuropathies underlying gastrointestinal motor dysfunction. *Scand J Gastroenterol.* 2000;35:114–122.

De Giorgio R, et al. Inflammatory neuropathies of the enteric nervous system. Gastroenterology. 2004;126:1872–1883.

Ekblad E, Mei Q, Sundler F. Innervation of the gastric mucosa. *Microsc Res Technique.* 2000;48:241–257.

Furness JB. *The Enteric Nervous System.* Malden, MA: Blackwell; 2006.

Galligan JJ. Pharmacology of synaptic transmission in the enteric nervous system. *Curr Opin Pharmacol.* 2002;2:623–629.

Genton L, Kudsk KA. Interactions between the enteric nervous system and the immune system: role of neuropeptides and nutrition. *Am J Surg.* 2003;186:253–258.

Gershon MD. Nerves, reflexes, and the enteric nervous system: pathogenesis of the irritable bowel syndrome. *J Clin Gastroenterol.* 2005;39:S184–193.

Grundy D, Schemann M. Enteric nervous system. *Curr Opin Gastroenterol.* 2005;21:176–182.

Hansen MB. The enteric nervous system I: organization and classification. *Pharmacol Toxicol.* 2003;92:105–113.

Hansen MB. The enteric nervous system II: gastrointestinal functions. *Pharmacol Toxicol.* 2003;92:249–257.

Holzer P. Opioids and opioid receptors in the enteric nervous system: from a problem in opioid analgesia to a new prokinetic therapy in humans. *Neurosci Lett.* 2004;361:192–195.

Kirchgessner AL. Glutamate in the enteric nervous system. *Curr Opin Pharmacol.* 2001;1:591–596.

Lomax AE, et al. Plasticity of the enteric nervous system during intestinal inflammation. *Neurogastroenterol Motil.* 2005;17(1):4–15.

Newgreen D, Young HM. Enteric nervous system: development and developmental disturbances—part 2. *Pediatr Dev Pathol.* 2002;5:329–349.

Schemann M, Neunlist M. The human enteric nervous system. *Neurogastroenterol Motil.* 2004;16(Suppl 1):55–59.

Sternini C. Receptors and transmission in the brain–gut axis: potential for novel therapies. III. Mu–opioid receptors in the enteric nervous system. *Am J Physiol/Gastrointest Liver Physiol.* 2001;281(1):G8–15.

Tomita R, et al. Regulation of the enteric nervous system in the colon of patients with slow transit constipation. *Hepatogastroenterology.* 2002;49:1540–1544.

Wade PR. Aging and neural control of the GI tract. I. Age-related changes in the enteric nervous system. *Am J Physiol/Gastrointest Liver Physiol.* 2002;283:G489–495.

Wood JD, Galligan JJ. Function of opioids in the enteric nervous system. *Neurogastroenterol Motil.* 2004;16(Suppl 2):17–28.

Section 15: Pain

Ainsworth L, et al. Transcutaneous electrical nerve stimulation (TENS) reduces chronic hyperalgesia induced by muscle inflammation. *Pain.* 2006;120:182–187.

Bailey CP, Connor M. Opioids: cellular mechanisms of tolerance and physical dependence. *Curr Opin Pharmacol.* 2005;5(1):60–68.

Bender T, et al. Hydrotherapy, balneotherapy, and spa treatment in pain management. *Rheumatol Int.* 2005;25:220–224.

Bennett JS, et al. The use of nonsteroidal anti-inflammatory drugs (NSAIDs): a science advisory from the American Heart Association. *Circulation.* 2005;111:1713–1716.

Bowsher D, Haggett C. Paradoxical burning sensation produced by cold stimulation in patients with neuropathic pain. *Pain.* 2005;117:230.

Brodner G, et al. Acute pain management: analysis, implications and consequences after prospective experience with 6349 surgical patients. *Eur J Anaesthesiol.* 2000;17:566–575.

Chan AT, et al. Nonsteroidal antiinflammatory drugs, acetaminophen, and the risk of cardiovascular events. *Circulation.* 2006;113:1578–1587.

Cohen SP, et al. Lumbar discography: a comprehensive review of outcome studies, diagnostic accuracy, and principles. *Reg Anesth Pain Med.* 2005;30:163–183.

Dworkin RH, et al. Core outcome measures for chronic pain clinical trials: IMMPACT recommendations. *Pain.* 2005;113:9–19.

Dziedzic K, et al. Effectiveness of manual therapy or pulsed shortwave diathermy in addition to advice and exercise for neck disorders: a pragmatic randomized controlled trial in physical therapy clinics. *Arthritis Rheum.* 2005;52:214–222.

Gibson W, et al. Referred pain and hyperalgesia in human tendon and muscle belly tissue. *Pain.* 2006;120:113–123.

Hadzic A, et al. Peripheral nerve blocks result in superior recovery compared with general anesthesia in outpatient knee arthroscopy. *Anesth Analg.* 2005;100:976–981.

Harden R, et al. Biofeedback in the treatment of phantom limb pain: a time-series analysis. *Appl Psychophysiol Biofeedback.* 2005;30:83–93.

Hav EM, et al. Comparison of physical treatments versus a brief pain-management programme for back pain in primary care: a randomized clinical trial in physiotherapy practice. *Lancet.* 2005;365:2024–2030.

Herpfer I, Lieb K. Substance P receptor antagonists in psychiatry: rationale for development and therapeutic potential. *CNS Drugs.* 2005;19: 275–293.

Hsieh YL. Effects of ultrasound and diclofenac phonophoresis on inflammatory pain relief: suppression of inducible nitric oxide synthase in arthritic rats. *Phys Ther.* 2006;86(1):39–49.

Kelly R, et al. Effect of fluidotherapy on superficial radial nerve conduction and skin temperature. *J Orthop Sports Phy Ther.* 2005;35(1):16–23.

Khadilkar A, et al. Transcutaneous electrical nerve stimulation (TENS) for chronic low–back pain. *Cochrane Database Syst Rev.* 2005;3:CD003008.

Lin EL, et al. Cervical epidural steroid injections for symptomatic disc herniations. *J Spinal Disord Techniques.* 2006;19:183–186.

Malt EA, et al. Factors explaining variance in perceived pain in women with fibromyalgia. *BMC Musculoskeletal Disord.* 2002;3:12.

Manheimer E, et al. Meta-analysis: acupuncture for low back pain. *Ann Intern Med.* 2005;142:651–663.

Mason L, et al. Topical NSAIDs for chronic musculoskeletal pain: systematic review and meta-analysis. *BMC Musculoskeletal Disord.* 2004;5(1):28.

Melzack R, Wall PD. *The Challenges of Pain.* 2nd ed. New York: Penguin; 1996.

Melzack R, Wall PD. Pain mechanisms: a new theory. *Science.* 1965;150(3699):971–979.

Melzack R, Wall PD. *The Textbook of Pain.* 3rd ed. Edinburgh: Churchill Livingstone; 1994.

Ng LCL, Sell P. Predictive value of the duration of sciatica for lumbar discectomy: a prospective cohort study. *J Bone Joint Surg.* 2004;86:546–549.

Ong CK, et al. The efficacy of preemptive analgesia for acute postoperative pain management: a meta-analysis. *Anesth Analg.* 2005;100:757–773.

Shim JK, et al. Ultrasound-guided lumbar medial-branch block: a clinical study with fluoroscopy control. *Reg Anesth Pain Med.* 2006;31:451–454.

Sprott H, et al. Acetaminophen may act through beta endorphin. *Ann Rheum Disord.* 2005;64:1522.

Watanabe K, et al. Lumbar spinous process-splitting laminectomy for lumbar canal stenosis. *J Neurosurg–Spine.* 2005;3:405–408.

SECTION 16: PERIPHERAL NERVE INJURY AND REGENERATION

Boyd JG, Gordon T. Neurotrophic factors and their receptors in axonal regeneration and functional recovery after peripheral nerve injury. *Mol Neurobiol.* 2003;27:277–324.

Braga-Silva J, et al. Endoscopic exploration of a brachial plexus injury. *J Reconstr Microsurg.* 2006;22:539–541.

Bryan DJ, et al. Influence of glial growth factor and Schwann cells in a bioresorbable guidance channel on peripheral nerve regeneration. *Tissue Eng.* 2000;6:129–138.

Conlon CF, Rempel DM. Upper extremity mononeuropathy among engineers. *J Occup Environ Med.* 2005;47:1276–1284.

Dolu H, et al. Evaluation of central neuropathy in type 2 diabetes mellitus by multimodal evoked potentials. *Acta Neurologica Belgica.* 2003;103:206–211.

Fukaya K, et al. Oxidized galectin-1 stimulates the migration of Schwann cells from both proximal and distal stumps of transected nerves and promotes axonal regeneration after peripheral nerve injury. *J Neuropathy Exp Neurol.* 2003;62:162–172.

Hiraga A, et al. Recovery patterns and long term prognosis for axonal Guillain-Barré syndrome. *J Neurol Neurosurg Psychiatry.* 2005;76:719–722.

Kasalova Z, et al. Relationship between peripheral diabetic neuropathy and microvascular reactivity in patients with type 1 and type 2 diabetes mellitus—neuropathy and microcirculation in diabetes. *Exp Clin Endocrinol Diabetes.* 2006;114:52–57.

Liu T, et al. Depletion of macrophages reduces axonal degeneration and hyperalgesia following nerve injury. *Pain.* 2000;86:25–32.

Modic MT, et al. Acute low back pain and radiculopathy: MR imaging findings and their prognostic role and effect on outcome. *Radiology.* 2005;237:597–604.

Mueller M, et al. Macrophage response to peripheral nerve injury: the quantitative contribution of resident and hematogenous macrophages. *Lab Invest.* 2003;83:175–185.

Newswanger DL, Warren CR. Guillain-Barré syndrome. *Am Fam Physician.* 2004;69:2405–2410.

Perlas A, et al. The sensitivity of motor response to nerve stimulation and paresthesia for nerve localization as evidenced by ultrasound. *Reg Anesth Pain Manage.* 2006;31:445–450.

Rempel DM, Diao E. Entrapment neuropathies: pathophysiology and pathogenesis. *J Electromyogr Kinesiol.* 2004;14:71–75.

Schmidt H, et al. Autonomic dysfunction predicts mortality in patients with multiple organ dysfunction syndrome of different age groups. *Crit Care Med.* 2005;33:1994–2002.

Singleton JR, et al. Polyneuropathy with impaired glucose tolerance: implications for diagnosis and therapy. *Curr Treatment Options Neurol.* 2005;7:33–42.

Souayah N, Sander HW. Lumbosacral magnetic root stimulation in lumbar plexopathy. *Am J Phys Med Rehabil.* 2006;85:858–861.

Terenghi G. Peripheral nerve regeneration and neurotrophic factors. *J Anat.* 1999;194(1):1–14.

Wake DJ, et al. Ulnar mononeuropathy in diabetes mellitus. *Brit J Diabetes Vascular Dis.* 2005;5:171–172.

Williams FH, et al. Neuromuscular rehabilitation and electrodiagnosis. 1. Mononeuropathy. *Arch Phys Med Rehabil.* 2005;86(3 Suppl 1): S3–10.

Zhang J, et al. Ciliary neurotrophic factor for acceleration of peripheral nerve regeneration: an experimental study. *J Reconstr Microsurg.* 2004;20:323–327.

Section 17: Phantom Limb Phenomenon

Anke K, et al. Reorganization of Motor and somatosensory cortex in upper extremity amputees with phantom limb pain. *J Neurosci.* 2001;21:3609–3618.

Chainay H, et al. Foot, face and hand representation in the human supplementary motor area. *NeuroReport.* 2004;15:765–768.

Ehde DM, et al. Chronic phantom sensations, phantom pain, residual limb pain, and other regional pain after lower limb amputation. *Arch Phys Med Rehabil.* 2000;81:1039–1044.

Farah MJ. Why does the somatosensory homunculus have hands next to face and feet next to genitals? A *Hypothesis. Neural Comput.* 1999;10:1983–1985.

Farne A, et al. Face or hand, not both: Perceptual correlates of reafferentation in a former amputee. *Curr Biol.* 2002;12:1342–1346.

Flor H. Phantom-limb pain: characteristics, causes, and treatment. *Lancet Neurol.* 2002;1:182–184.

Flor H, et al. Phantom limb pain: a case of maladaptive CNS plasticity? *Nat Rev Neurosci.* 2006;7:873–881.

Greenlee JDW, et al. Functional connection between inferior frontal gyrus and orofacial motor cortex in human. *J Neurophysiol.* 2004;92:1153–1164.

Hamada Y, Suzuki R. Hand posture modulates neuronal interaction in the primary somatosensory cortex of humans. *Clin Neurophysiol.* 2003;114:1689–1696.

Kiefer RT, et al. Continuous brachial plexus analgesia and NMDA-receptor blockade in early phantom limb pain: a report of two cases. *Pain Med.* 2002;3:156–160.

Miyamoto JJ, et al. The representation of the human oral area in the somatosensory cortex: A functional MRI study. *Cereb Cortex.* 2005;16:669–675.

Nguyen BT, et al. Face representation in the human primary somatosensory cortex. *Neurosci Res.* 2004;227–232.

Prantl L, et al. Surgical treatment of chronic phantom limb sensation and limb pain after lower limb amputation. *Plast Reconstr Surg.* 2006;118:1562–1572.

Ramachandran VS, Rogers–Ramachandran D. Phantom limbs and neural plasticity. *Arch Neurol.* 2000;57:317–320.

Ramachandran VS, Rogers–Ramachandran DC. Perceptual correlates of massive cortical reorganization. *Science.* 1992;258:1159–1160.

Sacks O. *The Man Who Mistook His Wife for a Hat and Other Clinical Tales.* New York: Harper Perennial; 1990.

Tinazzi M, et al. Plastic interactions between hand and face cortical representations in patients with trigeminal neuralgia: a somatosensory–evoked potentials study. *Neuroscience.* 2004;127:769–776.

Van Deusen J. *Body Image and Perceptual Dysfunction in Adults.* Philadelphia: Saunders; 1993.

Wheaton KJ, et al. Viewing the motion of human body parts activates different regions of premotor, temporal, and parietal cortex. *NeuroImage.* 2004;22:277–288.

Woodhouse A. Phantom limb sensation. *Clin Exp Pharmacology and Physiology.* 2005;32:132–134.

SECTION 18: SPINAL CORD TRACTS

Boyle R. Vestibulospinal control of reflex and voluntary head movement. *Ann N Y Acad Sci*. 2001;942:364–380.

Cruccu G, et al. Conduction velocity of the human spinothalamic tract as assessed by laser evoked potentials. *NeuroReport*. 2000;11:3029–3032.

Doherty JG, et al. Prevalence of upper motor neuron versus lower motor neuron lesions in complete lower thoracic and lumbar spinal cord injuries. *J Spinal Cord Med*. 2002;25:289–292.

Hankey GJ, Wardlaw JM. *Clinical Neurology*. New York: Demos Medical, 2002.

Henderson RD, Pender MP. Lower motor neuron weakness after diving-related decompression. *Neurol*. 2006;66:451–452.

Holondny AI, et al. Tumor involvement of the corticospinal tract: diffusion magnetic resonance tractography with intraoperative correlation. *J Neurosurg*. 2001;95:1082.

Ivanhoe CB, Reistetter TA. Spasticity: the misunderstood part of the upper motor neuron syndrome. *Am J Phys Med Rehabil*. 2004;83(Suppl 10):3–9.

Kao CD, et al. MR findings of decerebrate rigidity with preservation of consciousness. *Am J Neuroradiol*. 2006;27:1074–1075.

Marshall RS, et al. Evolution of cortical activation during recovery from corticospinal tract infarction. *Stroke*. 2000;31:656–661.

Palecek J. The role of dorsal columns pathway in visceral pain. *Physiol Res*. 2004;53(Suppl 1):125–130.

Palacek J, et al. Fos expression in spinothalamic and postsynaptic dorsal column neurons following noxious visceral and cutaneous stimuli. *Pain*. 2003;104:249–257.

Saade NE, et al. The role of the dorsal columns in neuropathic behavior evidence for plasticity and non-specificity. *Neuroscience*. 2002;115:403–413.

van Den Berg-Vos RM, et al. The spectrum of lower motor neuron syndromes. *J Neurol*. 2003;250:1279–1292.

Weaver DF. A clinical examination technique for mild upper motor neuron paresis of the arm. *Neurology*. 2000;54:531–532.

Welgampola MS, Colebatch JG. Vestibulospinal reflexes: quantitative effects of sensory feedback and postural task. *Exp Brain Res*. 2001;139:345–353.

Willis WD. Long-term potentiation in spinothalamic neurons. *Brain Res*. 2002;40:202–214.

SECTION 19: SPINAL CORD INJURY AND DISEASE

Baleriaux DL, Neurgroschl C. Spinal and spinal cord infection. *Eur Radiol*. 2004;14:E72–83.

Benevento B, Sipski ML. Neurogenic bladder, neurogenic bowel, and sexual dysfunction in people with spinal cord injury. *Phys Ther*. 2002;82:601–612.

Bode RK, Heinemann AW. Course of functional improvement after stroke, spinal cord injury, and traumatic brain injury. *Arch Phys Med Rehabil*. 2002;83:100–106.

Boot CRL, et al. Body temperature responses in spinal cord injured individuals during exercise in the cold and heat. *Int J Sports Med*. 2006;27:599–604.

Carlson GD, et al. Sustained spinal cord compression: part I. Time-dependent effect on long-term pathophysiology. *J Bone Joint Surg*. 2003;85:86–94.

Degenerative and regenerative mechanisms governing spinal cord injury. *Neurobiol Dis*. 2004;15:415–436.

Ditunno JF, et al. Spinal shock revisited: a four-phase model. *Spinal Cord*. 2004;42:383–395.

Fowler CJ, O'Malley KJ. Investigation and management of neurogenic bladder dysfunction. *J Neurol Neurosurg Psychiatry*. 2003;74(Suppl 4):27–31.

Gill D, et al. Acute Brown-Séquard syndrome. *Arch Neurol*. 2004;61:131.

Guest J, et al. Traumatic central cord syndrome: results of surgical management. *J Neurosurg*. 2002;97(1):25–32.

Helkowski WM, et al. Autonomic dysreflexia: incidence in persons with neurologically complete and incomplete tetraplegia. *J Spinal Cord Med*. 2003;26:244–247.

Kelley MD, et al. Early effects of spinal cord transaction on skeletal muscle properties. *Appl Physiol Nutr Metabol*. 2006;31:398–406.

Krassiovkov AV, et al. Autonomic dysreflexia in acute spinal cord injury: an under–recognized clinical entity. *J Neurotrauma*. 2003;20:707–716.

Marsh DR, Weaver LC. Autonomic dysreflexia, induced by noxious or innocuous stimulation, does not depend on changes in dorsal horn Substance P. *J Neurotrauma*. 2004;21:817–828.

Poonoose PM, et al. Missed and mismanaged injuries of the spinal cord. *J Trauma—Inj Infect Crit Care*. 2002;53:314–320.

Sandberg A, Stalberg E. Reflexes in prior polio and their relation to weakness and anterior horn cell loss. *J Electromyography Kinesiol*. 2006;16:611–620.

Vora SK, Lyons RW. The medical Kipling—syphilis, tabes dorsales, and Romberg's test. *Emerging Infect Dis*. 2004;10:1160–1162.

Weaver LC. What causes autonomic dysreflexia after spinal cord injury? *Clin Auton Res*. 2002;12:424–426.

Weaver LC, et al. Central mechanisms for autonomic dysreflexia after spinal cord injury. *Prog Brain Res*. 2002;137:83–95.

Section 20: Proprioception

Cordo PJ, et al. Position sensitivity of human muscle spindles: single afferent and population representations. *J Neurophysiol*. 2002;87:1186–1195.

Chalmers G. Do Golgi tendon organs really inhibit muscle activity at high force levels to save muscles from injury, and adapt with strength training? *Sports Biomechanics*. 2002;1:239–249.

Dietz V. Proprioception and locomotor disorders. *Nat Rev Neurosci*. 2002;3:781–790.

Friemert B, et al. Benefits of active motion for joint position sense. *Knee Surg Sports Traumatol Arthroscopy*. 2006;86:197–203.

Gandevia SC, et al. Proprioception: peripheral inputs and perceptual interactions. *Adv Exp Med Biol*. 2002;508:61–68.

Gilman S. Joint position sense and vibration sense: anatomical organization and assessment. *J Neurol Neurosurg Psychiatry*. 2002,73:473–477.

Holmes NP, et al. Reaching with alien limbs: visual exposure to prosthetic hands in a mirror biases proprioception without accompanying illusions of ownership. *Perception Psychophysics*. 2006;68:685–701.

Kerr GK, Worringham CJ. Velocity perception and proprioception. *Adv Exp Med Biol*. 2002;508:79–86.

Laskowski ER, et al. Proprioception. *Phys Med Rehabil Clin N Am*. 2000;11:323–340.

Lee HM, et al. Evaluation of shoulder proprioception following muscle fatigue. *Clin Biomech*. 2003;18:843–847.

Lin SI. Motor function and joint position sense in relation to gait performance in chronic stroke patients. *Arch Phys Med Rehabil*. 2005;86:197–203.

Naito E, et al. Dominance of the right hemisphere and role of area 2 in human kinesthesia. *J Neurophysiol*. 2005;93:1020–1034.

Proske U. What is the role of muscle receptors in proprioception? *Muscle Nerve*. 2005;31:780–787.

Proske U. Kinesthesia: the role of muscle receptors. *Muscle Nerve*. 2006;34:545–558.

Proske U, Gregory JE. Signaling properties of muscle spindles and tendon organs. *Adv Exp Med Biol*. 2002;508:5–12.

Riemann BL, Lephart SM. The sensorimotor system, part II: the role of proprioception in motor control and functional joint stability. *J Athletic Training*. 2002;37:80–84.

Sacks O. *The Man Who Mistook His Wife for a Hat and Other Clinical Tales*. New York: Harper Perennial; 1990.

Sadato N, Naito E. Emulation of kinesthesia during motor imagery. *Behav Brain Sci*. 2004;27:412–413.

Treffort N, et al. The structure and response properties of Golgi tendon organs in control and hypodynamia–hypokinesia rats. *Exp Neurol*. 2005;195:313–321.

Tunik E, et al. Arm–trunk coordination in the absence of proprioception. *Exp Brain Res*. 2003;153:343–355.

Section 21: Disorders of Muscle Tone

Beres-Jones JA, et al. Clonus after human spinal cord injury cannot be attributed solely to recurrent muscle–tendon stretch. *Exp Brain Res*. 2003,149:222–236.

Conn PJ, et al. Metabolic glutamate receptors in the basal ganglia motor circuit. *Nat Rev Neurosci.* 2005;6:787–798.

Damiano DL, et al. What does the Ashworth scale really measure and are instrumented measures more valid and precise? *Dev Med Child Neurol.* 2002;44:112–118.

Gelber DA. *Clinical Evaluation and Management of Spasticity*. Totowa, NJ: Humana Press; 2002.

Ghiglione P, et al. Cogwheel rigidity. *Arch Neurol.* 2005;62:828–830.

Krakauer JW. Arm function after stroke: from physiology to recovery. *Semin Neurol.* 2005;25:384–395.

Lehericy S, et al. Motor control in basal ganglia circuits using fMRI and brain atlas approaches. *Cereb Cortex.* 2006;16:149–161.

Mettaffie JG, et al. Subcortical loops through the basal ganglia. *Trends Neurosci.* 2005;28:403–407.

Pandyan AD, et al. Spasticity: clinical perceptions, neurological realities and meaningful measurement. *Disability Rehabil.* 2005;27(1–2):2–6.

Platz T, et al. Clinical scales for the assessment of spasticity, associated phenomena, and function: A systematic review of the literature. *Disability Rehabil.* 2005;27:7–18.

Sheean G. The pathophysiology of spasticity. *Eur J Neurol.* 2002;9(1):3–9.

Thompson AJ, et al. Clinical management of spasticity. *J Neurol Neurosurg Psychiatry.* 2005;76:459–463.

Voerman GE, et al. Neurophysiological methods for the assessment of spasticity: the Hoffman reflex, the tendon reflex, and the stretch reflex. *Disability Rehabil.* 2005;27(1–2):33–68.

Wallace DM, et al. Motor unit behavior during clonus. *J Appl Physiol.* 2005;99:2166–2172.

Wood DE, et al. Biomechanical approaches applied to the lower and upper limb for the measurement of spasticity: a systematic review of the literature. *Disability Rehabil.* 2005;27(1–2):19–33.

Xia R, Rymer W. The role of shortening reaction in mediating rigidity in Parkinson's disease. *Exp Brain Res.* 2004;156:524–528.

Section 22: Motor Functions and Dysfunctions of the Central Nervous System: Cortex, Basal Ganglia, Cerebellum

Albin RL. Neurobiology of basal ganglia and Tourette syndrome: striatal and dopamine function. *Adv Neurol.* 2006;99:99–106.

Bekkering H, et al. Goal–directed imitation in patients with ideomotor apraxia. *Cogn Neuropsychol.* 2005;22:419–432.

Carr J, Shepard R. *Movement Science: Foundations for Physical Therapy Rehabilitation.* 2nd ed. Gaithersburg, MD: Aspen; 2000.

Christine CW, Aminoff MJ. Clinical differentiation of Parkinsonian syndromes: prognostic and therapeutic relevance. *Am J Med.* 2004;117:412–418.

Chiu CS, et al. GABA transporter deficiency causes tremor, ataxia, nervousness, and increased GABA–induced tonic conductance in cerebellum. *J Neurosci.* 2005;25:3234–3245.

Chung GH. Functional heterogeneity of the supplementary motor area. *Am J Neurol.* 2005;26:1819–1823.

Daskalakis ZJ, et al. Exploring the connectivity between the cerebellum and motor cortex in humans. *J Physiol.* 2004;557:689–700.

Devanne H, et al. The comparable size and overlapping nature of upper limb distal and proximal muscle representations in the human motor cortex. *Eur J Neurosci.* 2006;23:2467–2476.

Habas C, Cabanis EA. Cortical areas functionally linked with the cerebellar second homunculus during out–of–phase bimanual movements. *Neuroradiology.* 2006;48:273–279.

Harrington D, et al. Does the representation of time depend on the cerebellum: effect of cerebellar stroke. *Brain.* 2004;127:1–14.

Himle MB, Woods DW. An experimental evaluation of tic suppression and tic rebound effect. *Behav Res Ther.* 2005;43:1443–1451.

Hoekstra PJ, et al. Is Tourette's syndrome an autoimmune disease? Mol Psychiatry. 2002;7:437–445.

Hoshi E, et al. The cerebellum communicates with the basal ganglia. *Nat Neurosci.* 2005;8:1291–1493.

Kamada K et al. Functional identification of the primary motor area by corticospinal tractography. *Neurosurgery.* 2005;23:56:98–109.

Leckman JF. Phenomenology of tics and natural history of tic disorders. *Brain Dev.* 2003;25:S24–S28.

Liss B, Roeper J. Correlating function and gene expression of individual basal ganglia neurons. *Trends Neurosci.* 2004;27:475–481.

Margolese HC, et al. Tardive dyskinesia in the era of typical and atypical antipsychotics. Part 2: incidence and management strategies in patients with schizophrenia. *Can J Psychiatry.* 2005;50:703–714.

Margolis AM, et al. Interhemispheric connectivity and executive functioning in adults with Tourette syndrome. *Neuropsychology.* 2006;20:66–76.

Miller TM, Johnston SC. Should the Babinski sign be part of the routine neurologic examination? *Neurology.* 2005;65:1165–1168.

Montgomery EB. Dynamically coupled, high-frequency restraint, non-linear oscillators embedded in scale-free basal ganglia–thalamic–cortical networks mediating function and deep brain stimulation effects. *Nonlinear Stud.* 2004;11:385–421.

Nakanisha S, et al. Role of synaptic integration of dopaminergic and cholinergic transmissions in basal ganglia function. *Int Congress Ser.* 2003;1250:487–492.

Ouchi Y, et al. Microglia activation and dopamine terminal loss in early Parkinson's disease. *Ann Neurol.* 2005;57:161–162.

Pearce JMS. Dystonia. *Eur Neurol.* 2005;53:151–152.

Pilarczyk M, et al. Hemiballismus following general anesthesia. *Neurologic Sci.* 2003;24:299–300.

Popescu A, Lippa CF. Parkinsonian syndromes: Parkinson's disease dementia, dementia with Lewy bodies and progressive supranuclear palsy. *Clin Neurosci Res.* 2004; 3:461–468.

Sacks O. *Awakenings.* New York: Harper Perennial; 1990.

Sacks O. *The Man Who Mistook His Wife for a Hat and Other Clinical Tales.* New York: Harper Perennial; 1990.

Schmahmann, JD. Disorders of the cerebellum: ataxia, dysmetria of thought, and the cerebellar cognitive affective syndrome. *J Neuropsychiatry Clin Neurosci.* 2004; 16:367–378.

Schwabe A, et al. The role of the human cerebellum in short- and long-term habituation of postural responses. *Gait Posture.* 2004;19(1):16–23.

Singer HS, et al. Infection: a stimulus for tic disorders. *Pediatr Neurol.* 2000;22:380–383.

Thach WT, et al. Role of the cerebellum in the control and adaptation of gait in health and disease. *Prog Brain Res.* 2004;143:353–366.

Thobois S, et al. PET and SPECT functional imaging studies in Parkinsonian syndromes: from the lesion to its consequences. *NeuroImage.* 2004;23:1–16.

van Boxtel, et al. Prevalence of primitive reflexes and the relationship with cognitive change in healthy adults: a report from the Maastrict Aging Study. *J Neurol.* 2006;253:935–941.

Vidailhet M, et al. Bilateral deep–brain stimulation of the globus pallidus in primary generalized dystonia. *N Engl J Med.* 2005;352:459–467.

Yasoshima Y, et al. Subthalamic neurons coordinate basal ganglia function through differential neural pathways. *J Neurosci.* 2005;25:7743–7753.

Zhuang P, et al. Neuronal activity in the basal ganglia and thalamus in patients with dystonia. *Clin Neurophysiol.* 2004;115:2542–2557.

Section 23: Sensory Functions and Dysfunctions of the Central Nervous System

Antal A, et al. Transcranial direct current stimulation and the visual cortex. *Brain Res Bull.* 2006;68:459–463.

Bender D, Wang X. The neuronal representation of pitch in primate auditory cortex. *Nature.* 2005;436:1161–1165.

Born RT, Bradley DC. Structure and function of visual area MT. *Annu Rev Neurosci.* 2005;28:157–189.

Chung T, et al. Diffusion weighted MR imaging of acute Wernicke's encephalopathy. *Eur J Radiol.* 2003;45:256–258.

Corchs S, Gustavo D. Feature-based attention in human visual cortex: simulation of fMRI data. *NeuroImage.* 2004;21:36–45.

Cramer SC, et al. A pilot study of somatotopic mapping after cortical infarct. *Stroke.* 2000;31:668–671.

Downer J, et al. A multimodal cortical network for the detection of changes in the sensory environment. *Nat Neurosci.* 2000;3:277–283.

Fiebach CJ, et al. Revisiting the role of Broca's area in sentence processing: syntactic integration versus syntactic working memory. *Hum Brain Mapping.* 2005;24:79–91.

Friederici AD. Broca's area and the ventral premotor cortex in language functional differentiation and specificity. *Cortex.* 2006;42:472–475.

Harris J, et al. The cortical distribution of sensory memories. *Neuron.* 2001;30:315–318.

Johnson SC, et al. Mapping language comprehension: convergent auditory and visual activation of Wernicke's area with fMRI. *NeuroImage.* 2001;13:S547

Joyce CA, et al. Early selection of diagnostic facial information in the human visual cortex. *Vis Res.* 2006;46:800–813.

Kayser, et al. Integration of touch and sound in auditory cortex. *Neuron.* 2005;48:373–384.

Kell CA, et al. The sensory cortical representation of the human penis. Revisiting somatotopy in the male homunculus. *J Neurosci.* 2005;25:5984–5987.

Koechlin E, Jubault T. Broca's area and the hierarchical organization of human behavior. *Neuron.* 2006;50:963–974.

Kohn A, Whitsel BL. Sensory cortical dynamics. *Behav Brain Res.* 2002;135:119–126.

Mante V, Carandini M. Mapping of stimulus energy in primary visual cortex. *J Neurophysiol.* 2005;98:788–798.

Martinez LM, et al. Receptive field structure varies with layer in the primary visual cortex. *Nat Neurosci.* 2005;8:372–379.

Mukamel R, et al. Coupling between neuronal firing, field potentials, and fMRI in human auditory cortex. *Science.* 2005;309:951–954.

Murray SO, et al. Perceptual grouping and the interactions between visual cortical areas. Neural Networks. 2004;17:695–705.

Nikolaev AR, et al. Correlation of brain rhythms between frontal and left temporal (Wernicke's) cortical areas during verbal thinking. *Neurosci Lett.* 2001;298:107–110.

Pekkola J, et al. Primary auditory cortex activation by visual speech: an fMRI study at 3 T. *NeuroReport.* 2005;16:125–128.

Sacks O. *A Leg to Stand On.* New York: Harper Perennial; 1993.

Sacks O. *An Anthropologist on Mars.* New York: Vintage Books; 1995.

Sacks O. *The Man Who Mistook His Wife for a Hat and Other Clinical Tales.* New York: Harper Perennial; 1990.

Schummers J, et al. Local networks in visual cortex and their influence on neuronal responses and dynamics. *J Physiol.* 2004;98:429–441.

Serre T, et al. Object recognition with features inspired by visual cortex. *Comput Vis Pattern Recogn.* 2005;2:994–1000.

Simos PG, et al. Reproducibility of measures of neurophysiological activity in Wernicke's area: a magnetic source imaging study. *Clin Neurophysiol.* 2005;116:2381–2391.

Tettamanti M, Weniger D. Broca's area: a supramodal hierarchical procession? *Cortex.* 2006;42:491–494.

Section 24: Thalamus and Brainstem Sensory and Motor Roles: Function and Dysfunction

Basso MA, et al. Cortical function: a view from the thalamus. *Neuron.* 2005;45:485–488.

Bernat JL. The concept and practice of brain death. *Prog Brain Res.* 2005;150:369–379.

Bernat JL. Chronic disorders of consciousness. Lancet. 2006;367:1181–1192.

Bien MY, et al. Instability of spontaneous breathing patterns in patients with persistent vegetative state. *Respir Physiol Neurobiol.* 2005;145:163–175.

Bonfiglio L, et al. Spontaneous blinking behaviour in persistent vegetative and minimally conscious states: Relationships with evolution and outcome. *Brain Res Bull.* 2005;68:163–170.

Carpentier A, et al. Early morphologic and spectroscopic magnetic resonance in severe traumatic brain injuries can detect "invisible brain stem damage" and predict "vegetative states." *J Neurotrauma.* 2006;23:674–685.

Cruccu G, et al. Brainstem reflex circuits revisited. *Brain.* 2005;128:386–394.

de Beaufort I. Patients in a persistent vegetative state—a Dutch perspective. *N Engl J Med*. 2005;352:2373–2375.

Escudero D, et al. The Bispectral Index Scale: its use in the detection of brain death. *Transplantation Proc*. 2005;37:3661–3663.

Faran S, et al. Late recovery from permanent traumatic vegetative state heralded by event-related potentials. *J Neurol Neurosurg Psychiatry*. 2006;77:998–1000.

Fischer C, et al. Improved prediction of awakening or nonawakening from severe anoxic coma using tree-based classification analysis. *Crit Care Med*. 2006;34:1520–1524.

Gilbert DL, et al. Altered mesolimbic and thalamic dopamine in Tourette syndrome. *Neurol*. 2006;67:1695–1697.

Hayashi M, et al. Gamma knife surgery of the pituitary: new treatment for thalamic pain syndrome. *J Neurosurg*. 2005;102:S38–S41.

Horn AK. The reticular formation. *Prog Brain Res*. 2005;151:127–155.

Koltzenburg M, Scadding J. Neuropathic pain. *Curr Opin Neurol*. 2001;14:641–647.

Kotchoubey B, et al. Information processing in severe disorders of consciousness: vegetative state and minimally conscious state. *Clin Neurophysiol*. 2005;116:2441–2453.

Laureys S, et al. Cortical processing of noxious somatosensory stimuli in the persistent vegetative state. *NeuroImage*. 2002;17:732–741.

Laureys S, et al. Brain functions in coma, vegetative state, and related disorders. *Lancet Neurol*. 2004;3:537–546.

Lee PH, et al. Thalamic infarct presenting with thalamic astasia. *Eur J Neurol*. 2005;12:317–319.

Linek V, et al. Dysexecutive syndrome following anterior thalamic ischemia in the dominant hemisphere. *J Neurological Sci*. 2005;229–230:117–120.

Marchetti C, et al. Crossed right hemisphere syndrome following left thalamic stroke. *J Neurol*. 2005;252:403–411.

Owen AM, et al. Detecting awareness in the vegetative state. *Science*. 2006;313:1402.

Roehrs T, Roth T. Sleep and pain: interaction of two vital functions. *Semin Neurol*. 2005;25:106–116.

Saposnik G, et al. Spontaneous and reflex movements in 107 patients with brain death. *Am J Med*. 2005;118:311–314.

Shemie SD, et al. Diagnosis of brain death in children. *Lancet Neurol*. 2007;6:87–92.

Sherman SM, Guillery RW. *Exploring the Thalamus and Its Role in Cortical Function*. Cambridge, MA: MIT Press; 2003.

Smith E, Delargy M. Locked-in syndrome. *Brit Med J*. 2005;330:406–409.

Steriade M. Local gating of information processing through the thalamus. *Neuron*. 2004;41: 493–494.

Tatlisumak T, Forss N. Brain death confirmed with CT angiography. *Eur J Neurol*. 2007;14:42–43.

Wijdicks EF. Minimally conscious state vs. persistent vegetative state: the case of Terry (Wallis) vs. the case of Terri (Schiavo). *Mayo Clin Proc*. 2006;81:1155–1158.

Section 25: Right vs Left Brain Functions and Disorders

Agnew JA, et al. Left hemisphere specialization for the control of voluntary movement rate. *NeuroImage*. 2004;22:289–303.

Beis JM, et al. Right spatial neglect after left hemisphere stroke. *Am Acad Neurol*. 2004;63:1600–1605.

Buklina S. The corpus callosum, interhemisphere interactions, and the functions of the right hemisphere of the brain. *Neurosci Behav Physiol*. 2005;35:473–480.

D'Arcy RCN, et al. The fan effect in fMRI: left hemisphere specialization in verbal working memory. *NeuroReport*. 2004;15:1851–1855.

Fiebach CJ, Friederici AD. Processing concrete words: fMRI evidence against a specific right-hemisphere involvement. *Neuropsychologia*. 2004;42:62–70.

Geschwind N. *Cerebral Lateralization: Biological Mechanisms, Associations, and Pathology*. Cambridge, MA: MIT Press; 1987.

Geschwind N, Galaburda AM, eds. *Cerebral Dominance: The Biological Foundations*. Cambridge, MA: MIT Press; 1988.

Grafman J. Conceptualizing functional neuroplasticity. *J Commun Disord*. 2000;33:345–356.

Kangas M, Tate RL. The significance of clumsy gestures in apraxia following a left hemisphere stroke. *Neuropsychol Rehab*. 2006;16:38–65.

Kelly SD, Goldsmith LH. Gesture and right hemisphere involvement in evaluating lecture material. *Gesture*. 2004;4(1):25–42.

Lattner S, et al. Voice perception: sex, pitch, and the right hemisphere. *Hum Brain Mapping*. 2004;24(1):11–20.

Lu L, et al. Modification of hippocampal neurogenesis and neuroplasticity by social environments. *Exp Neurol*. 2003;183:600–609.

Marshall N, et al. The role of the right hemisphere in processing nonsalient metaphorical meanings: application of principal components analysis to fMRI data. *Neuropsychologia*. 2005;43:2084–2100.

Miller EK, Wallis JD. *The Prefrontal Cortex and Executive Brain Functions*. San Diego, CA: Academic Press; 2003.

Mitchell RLC, Crow TJ. Right hemisphere language functions and schizophrenia: the forgotten hemisphere? *Brain*. 2005;128:963–978.

Page S, Levine P. Forced use after TBI: promoting plasticity and function through practice. *Brain Inj*. 2003;17:675–684.

Page SJ, et al. Modified constraint–induced therapy in acute stroke: a randomized controlled pilot study. *Neurorehab Neural Repair*. 2005;19(1):27–32.

Root JC, et al. Left hemisphere specialization for response to positive emotional expressions: a divided output methodology. *Emotion*. 2006;6:473–483.

Rotenberg VS. The peculiarity of the right–hemisphere function in depression: solving the paradoxes. *Prog Neuropsychopharmacolog Biol Psychiatry*. 2004;28(1):1–13.

Schacter SC, Devinsky O, eds. *Behavioral Neurol and the Legacy of Norman Geschwind*. Philadelphia, PA: Lippincott, Williams, & Wilkins; 1997.

Schirmer A, Kotz SA. Beyond the right hemisphere: brain mechanisms mediating vocal emotional processing. *Trends Cogn Sci*. 2006;10:26–30.

Smith SD, et al. Anomaly detection in the right hemisphere: the influence of visuospatial factors. *Brain Cong*. 2004;55:458–462.

Thiel A, et al. Essential language function of the right hemisphere in brain tumor patients. *Ann Neurol*. 2005;57:128–131.

Treffert DA, Christensen DD. Inside the mind of a savant. *Sci Am*. 2005;6:108–113.

Vandenbulcke M, et al. Knowledge of visual attributes in the right hemisphere. *Nat Neurosci*. 2006;9:964–970.

Vigneau M, et al. Meta-analyzing left hemisphere language areas: phonology, semantics, and sentence processing. *NeuroImage*. 2006;30:1414–1432.

Section 26: Perceptual Functions and Dysfunctions of the Central Nervous System

Baier B, Karnath HO. Incidence and diagnosis of anosognosia for hemiparesis revisited. *J Neurol Neurosurg Psychiatry*. 2005;76:358–361.

Balasubramarian V. Dysphagia in two forms of conduction aphasia. *Brain Cong*. 2005;57(1):8–15.

Barrett AM, et al. Unawareness of cognitive deficits (cognitive anosognosia) in probable AD and control subjects. *Am Acad Neurol*. 2005;64:693–699.

Baynes K, Gazzaniga MS. Lateralization of language: toward a biologically based model of language. *Linguistic Rev*. 2005;22:303–326.

Bekkering H, et al. Goal-directed imitation in patients with ideomotor apraxia. *Cogn Neuropsychol*. 2005;22:419–432.

Berthier ML. Poststroke aphasia: epidemiology, pathophysiology, and treatment. *Drugs Aging*. 2005;22:163–182.

Carey DP, et al. Pointing to places and spaces in a patient with visual form agnosia. *Neuropsychologia*. 2006;44:1584–1594.

Coslett HB. Visual agnosia revisited. *J Int Neuropsychol Soc*. 2006;12:442–443.

Crosson B, et al. Role of the right and left hemispheres in recovery of function during treatment of intention in aphasia. *J Cogn Neurosci*. 2005;17:392–406.

Danckert J, Ferber S. Revisiting unilateral neglect. *Neuropsychologia*. 2006;44:987–1006.

Eriksen MK, et al. Sensitivity and specificity of the new international diagnostic criteria for migraine with aura. *J Neurol Neurosurg Psychiatry*. 2005;76:212–217.

Farah MJ. *Visual Agnosia*. 2nd ed. Cambridge, MA: MIT Press; 2005.

Gialanella B, et al. Functional recovery after hemiplegia in patients with neglect. The rehabilitative role of anosognosia. *Stroke*. 2005;36:2687–2690.

Gillen R, et al. Unilateral spatial neglect: relation to rehabilitation outcomes in patients with right hemisphere stroke. *Arch Phys Med Rehab*. 2005;86:763–767.

Hillis AE. Neurobiology of unilateral spatial neglect. *The Neuroscientist*. 2006;12:153–163.

Knibb JA, et al. Clinical and pathological characterization of progressive aphasia. *Ann Neurol*. 2006;59:156–165.

LaPointe LL. *Aphasia and Related Neurologic Language Disorders*. 3rd ed. New York: Thieme, 2005.

Malhotra P, et al. Spatial working memory capacity in unilateral neglect. *Brain*. 2005;128:424–435.

Natale E, et al. What kind of visual spatial attention is impaired in neglect? *Neuropsychologia*. 2005;43:1072–1085.

Pataraia E, et al. Organization of receptive language-specific cortex before and after left temporal lobectomy. *Neurology*. 2005;64:481–487.

Pizzamiglio L, et al. Development of a rehabilitative program for unilateral neglect. *Restorative Neurol Neurosci*. 2006;24:337–345.

Rumiati RI. Right, left or both? Brain hemispheres and apraxia of naturalistic actions. *Trends Cogn Sci*. 2005;9:167–169.

Sacks O. *A Leg to Stand On*. New York: Harper Perennial; 1993.

Sacks O. *An Anthropologist on Mars*. New York: Vintage Books; 1995.

Sacks O. *The Man Who Mistook His Wife for a Hat and Other Clinical Tales*. New York: Harper Perennial; 1990.

Silver MA, et al. Topographic maps of visual spatial attention in human parietal cortex. *J Neurophysiol*. 2005;94:1358–1371.

Vallar G, Ronchi R. Anosognosia for motor and sensory deficits after unilateral damage: a review. *Restorative Neurol Neurosci*. 2006;24:247–257.

Van Deusen J. *Body Image and Perceptual Dysfunction in Adults*. Philadelphia, PA: Saunders; 1993.

Vogel A, et al. Cognitive and functional neuroimaging correlates for anosognosia in mild cognitive impairment and Alzheimer's disease. *Int J Geriatr Psychiatry*. 2005;20:238–246.

Yang J, et al. Preserved implicit form perception and orientation adaptation in visual form agnosia. *Neuropsychologia*. 2006;44:1833–1842.

SECTION 27: BLOOD SUPPLY OF THE BRAIN: CEREBROVASCULAR DISORDERS

Ahn JY, et al. Aneurysm in the penetrating artery of the digital middle cerebral artery presenting as intracerebral hemorrhage. *Acta Neurochirurgica*. 2005;147:1287–1290.

Battaglia F, et al. Unilateral cerebellar stroke disrupts movement preparation and motor imagery. *Clin Neurophysiol*. 2006;117:1009–1016.

Broderick JP. Advances in the treatment of hemorrhagic stroke: a possible new treatment. *Cleve Clin J Med*. 2005;72:341–344.

Chhabra R, et al. Distal anterior cerebral artery aneurysms: bifrontal basal anterior interhemispheric approach. *Surg Neurol*. 2005,64:315–319.

de Sousa AA, et al. Unilateral pterional approach to bilateral aneurysms of the middle cerebral artery. *Surg Neurol*. 2005;63:S1–S7.

Eftekhar B, et al. Are the distributions of variations of circle of Willis different in different populations? Results of an anatomical study and review of literature. *BMC Neurol*. 2006;6:22.

Emsley HCA, et al. Circle of Willis variation in a complex stroke presentation: a case report. *BMC Neurol*. 2006;6:13.

Endo H, et al. Paraparesis associated with ruptured anterior cerebral artery territory aneurysms. *Surg Neurol*. 2005;64:135–139.

Hamada J, et al. Clinical features of aneurysms of the posterior cerebral artery: 15-year experience with 21 cases. *Neurosurgery*. 2005;56:662–670.

Kern R, et al. Stroke recurrences in patients with symptomatic vs asymptomatic middle cerebral artery disease. *Neurology*. 2005;65:859–864.

Lam WWM, et al. Early computed tomography features in extensive middle cerebral artery territory infarct: prediction of survival. *J Neurol Neurosurg Psychiatry*. 2005;76:354–357.

Musa F, Taguri A. Intermittent visual field loss associated with an anterior cerebral artery aneurysm. *Neuro-Ophthalmology*. 2006;30–63–67.

Ng Ys, et al. Clinical characteristics and rehabilitation outcomes of patients with posterior cerebral artery stroke. *Arch Phys Med Rehab*. 2005;86:2138–2143.

Park KC, et al. Deafferentation–disconnection neglect induced by posterior cerebral artery infarction. *Neurology*. 2006;66(1):56–61.

Sadatomo T, et al. The characteristics of the anterior communicating artery aneurysm complex by three dimensional digital subtraction angiography. *Neurosurg Rev*. 2006;29:201–207.

Saito H, et al. Treatment of ruptured fusiform aneurysm in the posterior cerebral artery with posterior artery–superior cerebellar artery anastomosis combined with parent artery occlusion: case report. *Surg Neurol*. 2006,65:621–624.

Thanh G, et al. A digital map of middle cerebral artery infarcts associated with middle cerebral artery trunk and branch occlusion. *Stroke*. 2005;36:986–991.

Uz A, Tekdemir I. Relationship between the posterior cerebral artery and the cisternal segment of the oculomotor nerve. *J Clin Neurosci*. 2006;13:1019–1022.

van Zandvoort, et al. Cognitive deficits and changes in neurometabolites after a lacunar infarct. *J Neurol*. 2005;252:183–190.

Section 28: Commonly Used Neurodiagnostic Tests

Anderson V, et al. Magnetic resonance approaches to brain aging and Alzheimer's disease-associated neuropathology. *Topics Magn Imag*. 2005;16:439–452.

Barber PA, et al. Imaging of the brain in acute ischaemic stroke: comparison of computed tomography and magnetic resonance diffusion–weighted imaging. *J Neurol Neurosurg Psychiatry*. 2005;76:1528–1533.

Benseler JS. *The Radiology Handbook: A Pocket Guide to Medical Imaging*. Athens, OH: Ohio University Press; 2006.

Ell PJ. The contribution of PET/CT to improved patient management. *Brit J Radiol*. 2006;79(937):32–36.

Frank Y, Pavlakis SG. Brain imaging in neurobehavioral disorders. *Pediatric Neurol*. 2001;25:278–287.

Harrigan M, et al. Computed tomographic perfusion in the management of aneurysmal subarachnoid hemorrhage: new application of an existent technique. *Neurosurgery*. 2005;56:304–317.

Hoeffner EG, et al. Cerebral perfusion CT: technique and clinical applications. *Radiol*. 2004;231:632–644.

Jacobs AH, et al. Human gene therapy and imaging in neurological diseases. *Eur J Nucl Med Mol Imag*. 2005;32: S358–S383.

Johannsen B. The usefulness of radiotracers to make the body biochemically transparent. *Amino Acids*. 2005;29:307–311.

Kingsley RE, Jones VF. Neuroimaging techniques. In: Kingsley RE, ed. *Concise Text of Neuroscience*. 2nd ed. Philadelphia, PA: Lippincott, Williams, & Wilkins; 2000:617–630.

Leondes CT. *Medical Imaging Systems Technology: Modalities*. Hackensack, NJ: World Scientific; 2005.

Lin A, et al. Efficacy of proton magnetic resonance spectroscopy in neurological diagnosis and neurotherapeutic decision making. *NeuroRx*. 2005;2:197–214.

Mathis CA, et al. Imaging technology for neurodegenerative diseases: progress toward detection of specific pathologies. *Arch Neurol*. 2005;62:196–200.

McVeigh ER. Emerging imaging techniques. *Circ Res*. 2006;98:879–886.

Morano GN, Seibyl JP. Technical overview of brain SPECT imaging: improving acquisition and processing of data. *J Nucl Med Technol*. 2003;31:191–195.

Rocca MA, et al. Imaging spinal cord damage in multiple sclerosis. *J Neuroimag*. 2005;15:297–304.

Strangman G, et al. Functional neuroimaging and cognitive rehabilitation for people with traumatic brain injury. *Am J Phys Med Rehab*. 2005;84(1):62–75.

Wesolowski J, Lev M. CT: history, technology, and clinical aspects. *Semin Ultrasound CT MRI*. 2005;26:376–379.

Zaidi H, Montandon ML. The new challenges of brain PET imaging technology. *Curr Med Imag Rev*. 2006;2(1):3–13.

Section 29: Neurotransmitters: The Neurochemical Basis of Human Behavior

American Psychiatric Association. *Diagnostic and Statistical Manual of Mental Disorders—Text Revision*. 4th ed. Washington, DC: American Psychiatric Association; 2001.

Ball SG, et al. Selective serotonin reuptake inhibitor treatment for generalized anxiety disorder: a double-blind, prospective comparison between paroxetine and sertraline. *J Clin Psychiatry*. 2005;66:94–99.

Bowery NG, Smart TG. GABA and glycine as neurotransmitters: a brief history. *Brit J Pharmacol*. 2006;147: S109–S119.

Clinton SM, et al. Dopaminergic abnormalities in select thalamic nuclei in schizophrenia: involvement of the intracellular signal integrating proteins calcyon and spinophilin. *Am J Psychiatry*. 2005;162:1859–1871.

Goff DC. Pharmacologic implications of neurobiological models of schizophrenia. *Harvard Rev Psychiatry*. 2005;13:352–359.

Kalia M. Neurobiological basis of depression: an update. *Metabolism*. 2005;54(5):24–27.

Kirwin JL, Goren JL. Duloxetine: a dual serotonin–norepinephrine reuptake inhibitor for treatment of major depressive disorder. *Pharmacotherapy*. 2005;25:396–410.

O'Shanick GJ. Update on antidepressants. *J Head Trauma Rehab*. 2006;21:282–284.

Parsey RV, et al. Lower serotonin transporter binding potential in the human brain during major depressive episodes. *Am J Psychiatry*. 2006;163:52–58.

Pearl PL, et al. Inherited disorders of GABA metabolism. *Future Neurol*. 2006;1:631–636.

Phelps NJ, Cates ME. The role of venlafaxine in the treatment of obsessive–compulsive disorder. *Ann Pharmacother*. 2005;39:136–140.

Reynolds G. The neurochemistry of schizophrenia. *Psychiatry*. 2005;4(10):21–25.

Sulzer D, et al. Mechanisms of neurotransmitter release by amphetamines: a review. *Prog Neurobiol*. 2005;75:406–433.

Synder S. Turning off neurotransmitters. *Cell*. 2005;125(1):13–15.

Thase ME, et al. Remission rates following antidepressant therapy with buproprion or selective serotonin reuptake inhibitors: a meta-analysis of original data from 7 randomized controlled trials. *J Clin Psychiatry*. 2005;66:974–981.

To SE, et al. The symptoms, neurobiology, and current pharmacological treatment of depression. *J Neurosci Nurs*. 2005;37:102–107.

Valenstein ES. *The War of the Soups and the Sparks: The Discovery of Neurotransmitters and the Dispute Over How Nerves Communicate*. New York: Columbia University Press; 2005.

Zhou F, et al. Co-release of dopamine and serotonin from striatal dopamine terminals. *Neuron*. 2005;46(1):65–74.

Section 30: The Neurologic Substrates of Addiction

American Psychiatric Association. *Diagnostic and Statistical Manual of Mental Disorders—Text Revision*. 4th ed. Washington, DC: American Psychiatric Association; 2001.

Bailey KP. The brain's rewarding system and addiction. *J Psychosoc Nurs*. 2004;42:14–18.

Bowirrat A, Oscar-Berman M. Relationship between dopaminergic neurotransmission, alcoholism, and reward deficiency syndrome. *Am J Med Genet*. 2005;132(1):29–37.

Bressan RA, Crippa JA. The role of dopamine in reward and pleasure behaviour—review of data from preclinical research. *Acta Psychiatrica Scandinavica*. 2005;111:S14–S21.

Dani JA, Harris RA. Nicotine addiction and comorbidity with alcohol abuse and mental illness. *Nat Neurosci*. 2005;8:1465–1470.

DeVries TJ, Shippenberg TS. Neural systems underlying opiate addiction. *J Neurosci*. 2002;22:3321–3325.

Everitt BJ, Trevor WR. Neural systems of reinforcement for drug addiction: from actions to habits to compulsion. *Nat Neurosci*. 2005;8:1481–1489.

Frank G, et al. Increased dopamine D2/D3 receptor binding after recovery from anorexia nervosa measured by positron emission tomography and [11C] Raclopride. *Biol Psychiatry*. 2005;58:908–912.

Franken IH, et al. The role of dopamine in human addiction: from reward to motivated attention. *Eur J Pharmacol*. 2005;526:199–206.

Hasler G, et al. Obsessive–compulsive disorder symptom dimensions show specific relationships to psychiatric comorbidity. *Psychiatry Res*. 2005;135:121–132.

Hyman SE, et al. Neural mechanisms of addiction: the role of reward-related learning and memory. *Annu Rev Neurosci*. 2006;29:565–598.

Kalivas P, et al. Unmanageable motivation in addiction: a pathology in prefrontal-accumbens glutamate transmission. *Neuron*. 2005;45:647–650.

Kalivas PW, Volkow ND. The neural basis of addiction: a pathology of motivation and choice. *Am J Psychiatry*. 2005;162:1403–1413.

Koob GF. Neurobiology of addiction: toward the development of new therapies. *Ann N Y Acad Sci*. 2000;909:170–185.

Koob GF, Le Moal M. Plasticity of reward neurocircuitry and the "dark side" of drug addiction. *Nat Neurosci*. 2005;8:1442–1444.

Lapish CC, et al. Glutamate–dopamine cotransmission and reward processing in addiction. *Alcohol: Clin Exp Res*. 2006;30:1451–1465.

Laviolette SR, van der Kooy D. GABAA receptors signal bidirectional reward transmission from the ventral tegmental area to the tegmental pedunculopontine nucleus as a function of opiate state. *Eur J Neurosci*. 2004;20:2179–2187.

Melis M, et al. The dopamine hypothesis of drug addiction: hypodopaminergic state. *Int Rev Neurobiol*. 2005;63:101–154.

Morgane PJ, et al. A review of systems and networks of the limbic forebrain/limbic midbrain. *Prog Neurobiol*. 2005;75:143–160.

Nestler EJ. Is there a common molecular pathway for addiction? *Nat Neurosci*. 2005;8:1445–1449.

Nestler EJ, Malenka RC. The addicted brain. *Sci Am*. 2004;90:78–85.

Spanagel R, Heilig M. Addiction and its sciences. *Addiction*. 2005;100:1813–1822.

Weiss F. Neurobiology of craving, conditioned reward, and relapse. *Curr Opin Pharmacol*. 2005;5(1):9–19.

Tamminga CA, Nestler EJ. Pathological gambling: focusing on the addiction, not the activity. *Am J Psychiatry*. 2006;163:180–181.

SECTION 31: LEARNING AND MEMORY

Arbib M. From monkey-like action recognition to human language: an evolutionary framework for neurolinguistics. *Behav Brain Sci*. 2005;28:105–167.

Azar B. How mimicry begat culture. *Monitor Psychol*. 2005;36:54–57.

Bayley P, et al. The neuroanatomy of remote memory. *Neuron*. 2005;46:799–810.

Bedwell JS, et al. Functional neuroanatomy of subcomponent cognitive processes involved in verbal working memory. *Int J Neurosci*. 2005;115:1017–1032.

Bright P, et al. Retrograde amnesia in patients with hippocampal, medial temporal lobe, or frontal pathology. *Learning Memory*. 2006;13:545–557.

Borenstein E, Ruppin E. The evolution of imitation and mirror neurons in adaptive agents. *Cogn Syst Res*. 2005;6:229–242.

Buckley MJ. The role of the perirhinal cortex and hippocampus in learning, memory, and perception. *Q J Exp Psychol*. 2005;58:246–268.

Budson AE, Price BH. Memory dysfunction. *N Engl J Med*. 2005;352:692–699.

Cipolotti L, Bird CM. Amnesia and the hippocampus. *Curr Opin Neurol*. 2006;19:593–598.

Gold JJ, Squire LR. The anatomy of amnesia: neurohistological analysis of three new cases. *Learning Memory*. 2006;13:699–710.

LaBar KS, Cabeza R. Cognitive neuroscience of emotional memory. *Nat Rev Neurosci*. 2006;7(1):54–64.

Moscovitch M, et al. Functional neuroanatomy of remote, episodic, semantic, and spatial memory: a unified account based on multiple trace theory. *J Anat.* 2005;207(1):35–66.

Muller NG, Knight RT. The functional neuroanatomy of working memory: contributions of human brain lesion studies. *Neurosci.* 2006;139(1):51–58.

Nolin P. Executive memory dysfunctions following mild traumatic brain injury. *J Head Trauma Rehab.* 2006;21(1):68–75.

Rajah MN, McIntosh AR. Overlap in the functional neural systems involved in semantic and episodic memory retrieval. *J Cogn Neurosci.* 2005;17:470–482.

Rizzolatti G. The mirror neuron system and its function in humans. *Anat Embryol.* 2005;210:419–421.

Rizzolatti G, Craighero L. The mirror–neuron system. *Annu Rev Neurosci.* 2004;27:169–192.

Schacter D. *Searching for Memory: The Brain, the Mind, and the Past.* New York: Basic Books; 1996.

Schlosser RG, et al. Assessing the working memory network: studies with functional magnetic resonance imaging and structural equation modeling. *Neuroscience.* 2006;139:91–103.

Schott BH, et al. Redefining implicit and explicit memory: the functional neuroanatomy of priming, remembering, and control of retrieval. *Proc Natl Acad Sci U S A.* 2005;102:1257–1262.

Stuss DT, Alexander MP. Does damage to the frontal lobes produce impairment in memory? *Curr Dir Psychol Sci.* 2005;14:84–88.

Valera E, et al. Functional neuroanatomy of working memory in adults with attention deficit hyperactivity disorder. *Biol Psychiatry.* 2005;57:439–447.

Wais P, et al. The hippocampus supports both the recollection and the familiarity components of recognition memory. *Neuron.* 2006;49:459–466.

Williams JHG, et al. Imitation, mirror neurons, and autism. *Neurosci Biobehav Rev.* 2001;25:287–295.

Xiao D, Barbas H. Pathways for emotions and memory I. Input and output zones linking the anterior thalamic nuclei with prefrontal cortices in the rhesus monkey. *Thalamus Related Syst.* 2002;2(1):21–32.

SECTION 32: THE NEUROLOGIC SUBSTRATES OF EMOTION

Anand A, et al. Activity and connectivity of brain mood regulating circuit in depression: a functional magnetic resonance study. *Biol Psychiatry.* 2005;57:1079–1088.

Braun M, et al. Emotion recognition in stroke patients with left and right hemispheric lesions: results with a new instrument—the FEEL Test. *Brain Cong.* 2005;58:193–201.

Cordon B, et al. Fear recognition ability predicts differences in social cognitive and neural functioning in men. *J Cogn Neurosci.* 2006;18:889–897.

Couturier JL. Efficacy of rapid-rate repetitive transcranial magnetic stimulation in the treatment of depression: a systematic review and meta-analysis. *J Psychiatry Neurosci.* 2005;30:81–82.

Damasio AR. *Descartes' Error: Emotion, Reason, and the Human Brain.* New York: Putnam; 1994.

Dapretto M, et al. Understanding emotions in others: mirror neuron dysfunction in children with autism spectrum disorders. *Nat Neurosci.* 2006;9(1):28–30.

Dolcus F, et al. Remembering one year later: role of the amygdala and the medial temporal lobe memory system in retrieving emotional memories. *Proc Natl Acad Sci U S A.* 2005;102:2626–2631.

Davidson RJ, et al. Depression: perspectives from affective neuroscience. *Annu Rev Psychol.* 2002;53:545–574.

Demarcee H, et al. Brain lateralization of emotional processing: Historical roots and a future incorporating "dominance." *Behav Cogn Neurosci Rev.* 2005;4(1):3–20.

Feldman-Barrett L, Wager TD. The structure of emotion. *Curr Dir Psychol Sci.* 2006;15:79–83.

Haupt M. Emotional lability, intrusiveness, and catastrophic reactions. *Int Psychogeriatr.* 1997;8:409–414.

Lee GP, et al. Neural substrates of emotion as revealed by functional magnetic resonance imaging. *Cogn Behav Neurol.* 2004;17(1):9–17.

Lee TMC, et al. Neural activities associated with emotion recognition observed in men and women. *Mol Psychiatry.* 2005;10:450–455.

Liddell B, et al. A direct brainstem–amygdala–cortical "alarm" system for subliminal signals of fear. *NeuroImage.* 2005;24:235–243.

Lyketsos CG, et al. Forgotten frontal lobe syndrome or "executive dysfunction syndrome." *Psychosomatics*. 2004;45:247–255.

McDonald S. Are you crying or laughing? Emotion recognition deficits after traumatic brain injury. *Brain Impairment*. 2005;6:56–67.

Ochsner KN, Gross JJ. The cognitive control of emotion. *Trends Cogn Sci*. 2005;9:242–249.

Phan K et al. Neural substrates for voluntary suppression of negative affect: a functional magnetic resonance imaging study. *Biol Psychiatry*. 2005;57:210–219.

Prigatano GP. Disturbance of self-awareness and rehabilitation. 20th anniversary issue: progress and future prospects in TBI rehabilitation. *J Head Trauma Rehab*. 2005;20(1):19–29.

Rolls, ET. *Emotion Explained*. Oxford, UK: Oxford University Press; 2005.

Sherer M. Neuroanatomic basis of impaired self-awareness after traumatic brain injury. Findings from early computed tomography. *J Head Trauma Rehab*. 2005;20:287–300.

Taylor JG, Fragopanagos NF. The interaction of attention and emotion. *Neural Networks*. 2005;18:353–369.

Tranel D, et al. Altered experience of emotion following bilateral amygdala damage. *Cogn Neuropsychiatry*. 2006;11:219–232.

Youngstrom EA. Brain lateralization of emotional processing: historical roots and a future incorporating "dominance." *Behav Cogn Neurosci Rev*. 2005;4(1):3–20.

Section 33: The Aging Brain

Allen JS, et al. The aging brain: the cognitive reserve hypothesis and hominid evolution. *Am J Hum Biol*. 2005;17:673–689.

Barzilai N, et al. Einstein's Institute for Aging Research: collaborative and programmatic approaches in the search for successful aging. *Exp Gerontol*. 2004;39:151–157.

Burke SN, Barnes CA. Neural plasticity in the ageing brain. *Nat Rev Neurosci*. 2006;7(1):30–40.

Conde JR, Streit WJ. Microglia in the aging brain. *J Neuropathol Exp Neurol*. 2006;65:199–203.

Erraji-Benchekroun L, et al. Molecular aging in human prefrontal cortex is selective and continuous throughout adult life. *Biol Psychiatry*. 2005;57:549–558.

Fillit HM, et al. Achieving and maintaining cognitive vitality with aging. *Mayo Clin Proc*. 2002;77:681–696.

Head D, et al. Frontal-hippocampal double dissociation between normal aging and Alzheimer's disease. *Cereb Cortex*. 2005;15:732–739.

Kempermann G, et al. Neuroplasticity in old age: sustained fivefold induction of hippocampal neurogenesis by long-term environmental enrichment. *Ann Neurol*. 2002;52:135–143.

Kertesz A, et al. The evolution and pathology of frontotemporal dementia. *Brain*. 2005;128:1996–2005.

Korolainen MA, et al. Proteomic analysis of glial fibrillary acidic protein in Alzheimer's disease and aging brain. *Neurobiol Dis*. 2005;20:858–870.

Mariani E, et al. Oxidative stress in brain aging, neurodegenerative and vascular diseases: an overview. *J Chromatography*. 2005;827:65–75.

Mocchegiani E, et al. Brain, aging and neurodegeneration: role of zinc ion availability. *Prog Neurobiol*. 2005;75:367–390.

Raz N, et al. Regional brain changes in aging healthy adults: general trends, individual differences and modifiers. *Cereb Cortex*. 2005;15:1676–1689.

Riley KP, et al. Early life linguistic ability, late life cognitive function, and neuropathology: findings from the Nun Study. *Neurobiol Aging*. 2005;26:341–347.

Scarmeas N, et al. Influence of leisure activity on the incidence of Alzheimer's disease. *Neurology*. 2001;57:2236–2242.

Snowdown DA, et al. Linguistic ability in early life and the neuropathology of Alzheimer's disease and cerebrovascular disease. Findings from the Nun study. *Ann N Y Acad Sci*. 2000;903:34–38.

Stadtman, ER. Protein oxidation and aging. *Free Radical Res*. 2006;40:1250–1258.

Sullivan PG, Brown MR. Mitochondrial aging and dysfunction in Alzheimer's disease. *Prog Neuropsychopharmacol Biol Psychiatry*. 2005;29:407–410.

Tuppo EE, Arias HR. The role of inflammation in Alzheimer's disease. *Int J Biochem Cell Biol*. 2005;37:289–305.

Verghese J, et al. (2003). Leisure activities and the risk of dementia in the elderly. *N Engl J Med.* 2003;348:2508–2516.

Wilson RS, et al. (2002). Participation in cognitively stimulating activities and risk of incident Alzheimer's disease. *J Am Med Assoc.* 2002;287:742–748.

Zlokovic BV. Neurovascular mechanisms of Alzheimer's neurodegeneration. *Trends Neurosci.* 2005;28:202–208.

Section 34: Sex Differences in Male and Female Brains

Achiron R, et al. Sex-related differences in the development of the human fetal corpus callosum: in utero ultrasonographic study. *Prenatal Diagn.* 2001;21:116–120.

Anokhin AP. Complexity of electrocortical dynamics in children: developmental aspects. *Dev Psychobiol.* 2000;36:9–22.

Baron-Cohen, S. The extreme male brain theory of autism. *Trends Cogn Sci.* 2002;6:248–254.

Bayliss AP, et al. Sex differences in eye gaze and symbolic cueing of attention. *Q J Exp Psychol.* 2005;58:631–650.

Becker JB, et al. Strategies and methods for research on sex differences in brain and behavior. *Endocrinology.* 2005;146:1650–1673.

Dessens AB, et al. Gender dysphoria and gender change in chromosomal females with congenital adrenal hyperplasia. *Arch Sexual Behav.* 2005;34:389–397.

Geschwind N. *Cerebral Lateralization: Biological Mechanisms, Associations, and Pathology.* Cambridge, MA: MIT Press; 1987.

Geschwind N, Galaburda AM, eds. *Cerebral Dominance: The Biological Foundations.* Cambridge, MA: MIT Press; 1988.

Guillem F, Mograss M. Gender differences in memory processing: evidence from event-related potentials to faces. *Brain Cong.* 2005;57(1):84–92.

Kilgore W, et al. Sex specific developmental changes in amygdala responses to affective faces. *NeuroReport.* 2001;12:427–433.

Lee TMC, et al. Neural activities associated with emotion recognition observed in men and women. *Mol Psychiatry.* 2005;10:450–455.

Luders E, et al. Mapping cortical gray matter in the young adult brain: effects of gender. *NeuroImage.* 2005;26:493–501.

McAlonan G, et al. Mapping the brain in autism. A voxel-based MRI study of volumetric differences and intercorrelations in autism. *Brain.* 2005;128:268–276.

Njemanze PC. Cerebral lateralization and general intelligence: gender differences in a transcranial Doppler study. *Brain Lang.* 2005;92:234–239.

Paus T. Mapping brain maturation and cognitive development during adolescence. *Trends Cogn Sci.* 2005;9:60–68.

Piefke M, et al. Gender differences in the functional neuroanatomy of emotional episodic autobiographical memory. *Hum Brain Mapping.* 2005;24:313–324.

Ronald A, et al. The genetic relationship between individual differences in social and nonsocial behaviours characteristic of autism. *Dev Sci.* 2005;8:444–458.

Singer T, et al. Empathic neural responses are modulated by the perceived fairness of others. *Nature.* 2006;439:466–469.

Spelke ES, et al. Sex differences in intrinsic aptitude for mathematics and science? *Am Psychol.* 2005;60:950–958.

Walla P, et al. Physiological evidence of gender differences in word recognition: a magnetoencephalographic (MEG) study. *Cong Brain Res.* 2001;12:49–54.

Clinical Test Questions

Sections 1 – 3

1. After his car accident, Jeff had difficulty with problem-solving, understanding the consequences of his actions, refraining from verbalizing inappropriate thoughts, and demonstrating insight. Which lobe of his brain was likely injured?
   a. occipital lobe
   b. frontal lobe
   c. parietal lobe
   d. temporal lobe

2. Jane was diagnosed with a tumor in her temporal lobe. Which functions was Jane likely experiencing difficulty with?
   a. cognition
   b. detection and perception of visual data
   c. detection and perception of auditory data
   d. detection and perception of somatosensory data

3. Joe is a patient who has been in a coma for 3 weeks after a motorcycle accident. Because of depressed respiration, he has been hooked up to a ventilator. The following reflexes are absent: cough and gag reflex, swallowing reflex, and pupillary reflex. Which neuroanatomical structure is likely damaged?
   a. hypothalamus
   b. thalamus
   c. basal ganglia
   d. brainstem

4. At 1 month post-injury, Alice displays poor regulation of temperature, acne (secondary to a hormonal imbalance of the pituitary gland), dysregulated sleep patterns, and emotional mood swings. Which neuroanatomical structure primarily controls all of these functions?
   a. thalamus
   b. hypothalamus
   c. basal ganglia
   d. limbic system

5. Upon examination, Mrs. Peters was found to present motor incoordination, decreased proprioception, bilateral ataxia, and dysarthric speech. It is likely that neurologic pathology lies in which structure?
   a. cerebellum
   b. basal ganglia
   c. limbic system
   d. thalamus

6. Jack, a 19-year-old male, was diagnosed with autism as a toddler. Today, he displays difficulty interpreting social cues, difficulty feeling empathy in response to another's pain, and difficulty understanding the norms of social conversation. Researchers have found that pathology in the _______ may account for some of these symptoms.
   a. claustrum
   b. hippocampus
   c. amygdala
   d. caudate nucleus

7. Sam enters his youngest child's elementary school class on Parents' Day and smells the odors of white paste, finger paints, and chalk. He recalls his first grade teacher, whom he had not thought of in many years. This phenomenon is mediated by the connection among which of the following structures?

   1. limbic system
   2. caudate nucleus
   3. olfactory bulb and tract
   4. midbrain
   5. hippocampus
   6. optic nerve

   a. 1, 2, 3
   b. 2, 3, 4
   c. 1, 3, 6
   d. 1, 3, 5

8. Pathology of which structures may play a role in the flashbacks and flooding of images associated with post-traumatic stress disorder?
   a. hypothalamus and thalamus
   b. hippocampus and amygdala
   c. caudate and putamen
   d. brainstem and reticular formation

9. Mr. Edwards has been diagnosed with Parkinson's disease and displays difficulty initiating and terminating automatic movements such as walking and writing. Which two neuroanatomical structures are likely involved?
   a. basal ganglia and substantia nigra
   b. amygdala and hippocampus
   c. cerebellum and thalamus
   d. hypothalamus and thalamus

10. Which brainstem center should be suspected when a patient's state of consciousness is diagnosed as stuporous?
   a. cerebral peduncles
   b. reticular formation
   c. inferior colliculi
   d. pyramidal decussation

SECTIONS 4 – 7

1. Edna's daughter relates that in the last month, Edna's gait has become increasingly unsteady. She has also lost considerable cognitive function and has urinary incontinence. Edna's physician suspects normal pressure hydrocephalus. This disorder results when:

   a. a blockage prevents the flow of cerebrospinal fluid
   b. the choroid plexus produce an excess of cerebrospinal fluid
   c. the arachnoid villae cannot absorb the cerebrospinal fluid

2. A 1-month-old infant has been diagnosed with hydrocephalus. Which of the below symptoms are likely present?

   1. compression of neural tissue
   2. skull expansion
   3. separation of cranial sutures
   4. compromised cognitive development

   a. all of the above
   b. 1, 2, 3
   c. 2, 3, 4

3. This artery commonly ruptures in the occurrence of a traumatic brain injury, causing fatal hemorrhages in the subdural space if not immediately addressed:

   a. basilar artery
   b. anterior cerebral artery
   c. middle meningeal artery
   d. posterior cerebral artery

4. Mr. Simon was brought to the emergency room after experiencing pain in his left arm and jaw. This pain is commonly known as _______ and results from _______ .

   1. radiating pain
   2. referred pain
   3. stimulation of a spinal nerve that innervates both a specific dermatomal region and a visceral organ
   4. the simultaneous innervation of both parasympathetic and sympathetic nervous systems

   a. 1, 3
   b. 2, 4
   c. 1, 4
   d. 2, 3

5. Mrs. Smith presents with pain that begins in her lower back and radiates down both legs. Her therapist tells her that this pain is likely due to a herniated lumbar disc that is impinging upon several spinal nerves innervating the lower extremities. What is the clinical name for this condition?

   a. sciatic nerve damage
   b. radial nerve damage
   c. peroneal nerve damage
   d. tarsal tunnel syndrome

6. Carol's deep tendon reflexes are hyper-reflexive when percussed. The therapist suspects:

   a lower motor neuron injury
   b. upper motor neuron injury
   c. injury to the cauda equina

7. Sally and Greg's baby was born with a severe spinal cord deformity in which the meninges, spinal cord, and the spinal nerves protrude through an incomplete closure of the vertebral column. This deformity is called:

   a. meningocele
   b. meningomyelocele
   c. Arnold Chiari malformation
   d. hydrocephalus

8. Tom's third and fourth cranial nerves were damaged in a head-on auto collision in which Tom was a passenger. When cranial nerves are damaged, such pathology is considered to be _______ motor neuron injury and presents as _______ at and below the lesion level.

   1. an upper
   2. a lower
   3. spasticity
   4. flaccidity

   a. 1, 3
   b. 2, 3
   c. 1, 4
   d. 2, 4

9. The meninges, protective glial cells, and capillary beds collectively form this structure, whose purpose is protection of the brain from exposure to toxins:

   a. blood brain barrier
   b. dura mater
   c. cisterna magna
   d. falx cerebri

10. This structure is a commonly used shunt placement in the treatment of hydrocephalus:

   a. foramen of Luschka
   b. foramen of Monro
   c. superior sagittal sinus
   d. cisterna magna

Section 8

1. Norman presents with the following symptoms: lateral strabismus, diplopia (double vision), ptosis (drooping) of the eyelid, and nystagmus. His therapist suspects pathology in:
    a. CN 4, trochlear nerve
    b. CN 2, optic nerve
    c. CN 3, oculomotor nerve
    d. CN 6, abducens nerve

2. Two weeks ago, Matt was sick with an upper respiratory infection. He has subsequently been hospitalized as the result of an acute onset of vertigo, vomiting, decreased balance, and nystagmus. His likely diagnosis is:
    a. Bell's palsy
    b. vestibular neuritis
    c. trigeminal neuralgia

3. The above clinical disorder results from pathology to cranial nerve:
    a. 7
    b. 5
    c. 8

4. Bill lost sensation on the left side of his face. He also reports that he no longer has the ability to chew food on the left side of his mouth. It is likely that Bill has:
    a. glossopharyngeal nerve palsy
    b. Bell's palsy
    c. facial nerve palsy
    d. trigeminal nerve palsy

5. One morning Bob could no longer voluntarily move the muscles of his face on the right side although sensation remained intact. He also found that sounds were experienced as abnormally heightened. Bob's neurologist diagnosed _______ .
    1. Bell's palsy
    2. trigeminal neuralgia
    3. facial nerve damage
    4. vestibulocochlear damage

    a. 1, 3
    b. 1, 4
    c. 4
    d. 2

6. After John's cerebral vascular accident, he displayed dysphagia and loss of the gag and swallowing reflexes. John's CVA may have damaged:
    1. the glossopharyngeal nerve
    2. the hypoglossal nerve
    3. the vagus nerve
    4. the vestibulocochlear nerve

    a. 1, 2
    b. 1, 3
    c. 1, 2, 3
    d. 1, 2, 3, 4

7. When asked to stick out her tongue, Mary's tongue deviates to the left side. There is also atrophy of the tongue muscles on the left side, and she is unable to push her tongue against her left cheek. The therapist suspects damage to which cranial nerve?
    a. CN 10 vagus nerve
    b. CN 12 hypoglossal nerve
    c. CN 11 accessory nerve
    d. CN 9 glossopharyngeal nerve

8. Martha has a medial strabismus. Which cranial nerve is likely weak?
    a. CN 6, abducens nerve
    b. CN 4, trochlear nerve
    c. CN 3, oculomotor nerve

9. After a motor vehicle accident, Ellen lost her sense of smell. This condition is referred to as:
    a. anomia
    b. agnosia
    c. anosmia

10. The above condition is caused by:
    a. bilateral damage to CN 1, olfactory nerve
    b. unilateral damage to CN 1, olfactory nerve
    c. bilateral damage to CN 2, optic nerve
    d. unilateral damage to CN 2, optic nerve

SECTIONS 9 – 11

1. Some children with sensory processing problems may have CNSs that may be hypersensitive to sensation from the environment. Such children may experience tactile defensiveness and/or gravitational insecurity. When sensory receptors are viewed from a developmental classification, it may be said that such children have a dominant _______ sensory system.

   a. epicritic
   b. protopathic

2. Over the last 3 years, Mrs. Johnson has experienced severe muscular weakness and fatigue that first affected the muscles of her eyes and head and then progressed to her extremities. What disease is a chronic progressive autoimmune disorder that affects the neuromuscular junction of voluntary muscles, destroying acetylcholine receptors?

   a. amyotrophic lateral sclerosis
   b. Huntingdon's disease
   c. multiple sclerosis
   d. myasthenia gravis

3. Mr. Alexander complains of numbness in his extremities, paresthesias, and Lhermitte's sign (causalgia that radiates down the back and lower extremities in response to neck flexion). He also presents an abnormal gait, tremors, and reports extreme fatigue after participation in minimal activity. These signs and symptoms are characterized by periods of exacerbation and remission. Mr. Alexander's physician makes a diagnosis of _______ .

   a. amyotrophic lateral sclerosis
   b. myasthenia gravis
   c. multiple sclerosis

4. The above disease is considered an upper motor neuron disorder characterized by:

   a. random demyelination of the PNS
   b. random demyelination of the CNS
   c. death of upper and lower motor neurons
   d. destruction of acetylcholine receptors at the neuromuscular junction

5. Jennifer has been diagnosed with a neurologic disorder that affects both the upper and lower motor neurons (in other words, both the CNS and PNS). Early signs involved muscle cramps in her legs followed by a slow progressive weakness and atrophy of the muscles in her right arm. Eventually, most of the muscles of her body became affected. This disease is known as ______ and is unique because it can involve both muscle flaccidity and spasticity in different muscle groups.

   a. amyotrophic lateral sclerosis
   b. multiple sclerosis
   c. myasthenia gravis

6. After a traumatic brain injury, Peter lost his sense of smell. This is called ______ and is caused by _____ .

   1. anomia
   2. anosmia
   3. a bilateral lesion to the olfactory nerve
   4. a unilateral lesion to the olfactory nerve

   a. 1, 3
   b. 1, 4
   c. 2, 3
   d. 2, 4

7. Because Peter (in the previous question) has lost the ability to process olfactory data, he has also lost which sense?

   a. gustation
   b. audition
   c. vision
   d. somatosensory processing

8. Amanda is diagnosed with a left contralateral homonymous hemianopsia. This results from:

   a. a lesion in the right optic nerve
   b. a lesion in the central that destroys the entire optic chiasm
   c. a lesion in the left optic tract
   d. a lesion in the right optic tract

9. Joe is blind but his external visual receptor anatomy is intact. This type of blindness if called _______ and results from _______ .

   1. cortical blindness
   2. primitive blindness
   3. pathology of the primary visual cortex in the occipital lobe
   4. pathology of the lateral geniculate nucleus in the thalamus

   a. 1, 3
   b. 1, 4
   c. 2, 3
   d. 2, 4

10. Mr. Stone can hear but cannot attach meaning to sounds. This type of hearing his called _______ and results from _______ .

    1. cortical deafness
    2. auditory agnosia
    3. a lesion to the association auditory cortices
    4. a lesion to the primary auditory cortex

    a. 1, 3
    b. 1, 4
    c. 2, 3
    d. 2, 4

SECTIONS 12 – 14

1. Tamara often becomes ill during long car rides and feels nauseous. In addition to the motion sickness drugs her doctor prescribed she has been advised to do which of the following to alleviate feelings of motion sickness?
   a. sit in the back seat of the car and read a book to distract her mind
   b. eat before making a car trip to increase activity of the parasympathetic nervous system
   c. sit in the front seat of the car and watch the upcoming traffic changes so that the visual system signals will match those of the vestibular system

2. Mr. Hernandez has been admitted to the hospital after a severe fall. He complains of vertigo and tinnitus and demonstrates a broad-based gait. He also reports that his balance becomes worse at night and when he is unfamiliar with an environment. It is likely that Mr. Hernandez has impairment of the:
   a. proprioceptive system
   b. reticular formation
   c. vestibular system
   d. visual system

3. In the above question, Mr. Hernandez is relying on which systems for balance?
   1. proprioception
   2. visual
   3. vestibular
   4. autonomic nervous system

   a. 1, 2
   b. 1, 2, 3
   c. 2, 3
   d. 2, 3, 4

4. Mr. Adams feels anxious before his scheduled magnetic resonance imaging (MRI) scan. His palms are visibly sweaty, his heart rate is accelerated, and his blood pressure is elevated. Which autonomic nervous system component has become activated?
   a. somatic
   b. sympathetic
   c. visceral
   d. parasympathetic

5. Sam has begun taking an antidepressant medication that increases the amount of serotonin in his CNS. Since starting the medication, Sam has experienced increased diarrhea and nausea. This is likely due to which of the following?
   a. inhibition of parasympathetic nervous system activity and depressed peristalsis
   b. stimulation of the sympathetic nervous system leading to bowel and bladder problems
   c. stimulation of the vagus nerve with high levels of serotonin
   d. stimulation of the GI tract with high levels of serotonin

6. When Wendy was 5 years old, she loved to spin around in a swivel chair. As an adult, such stimulation commonly causes her nausea. This is likely due to the relationship between the vestibular system and which cranial nerve?
   a. CN 12, hypoglossal nerve
   b. CN 10, vagus nerve
   c. CN 11, accessory nerve
   d. CN 3, oculomotor nerve

7. Andrew is asked by his therapist to stand with his eyes closed, his feet together, and his shoulders flexed to 90 degrees. Andrew's therapist then observes his degree of postural sway, balance, and arm stability. This clinical test is called the:
   a. Rhinne test
   b. Romberg test
   c. Kleiger test
   d. Hughston test

8. Mrs. Chong was diagnosed with an inoperable tumor in her brainstem. She was hospitalized five days ago after losing consciousness and becoming comatose. Mrs. Chong's tumor may be affecting which structure?
   a. reticular inhibiting system
   b. limbic system
   c. reticular activating system
   d. substantia nigra

9. Which structure is believed to alert the cortex to important incoming sensory data and acts as a screen to filter extraneous information so that cortical concentration can be enhanced?
   a. reticular activating system
   b. reticular inhibiting system
   c. vestibular system
   d. autonomic nervous system

10. Which system is responsible for the unconscious regulation of vegetative functions?
    a. reticular activating system
    b. reticular inhibiting system
    c. vestibular system
    d. autonomic nervous system

SECTIONS 15 – 17

1. Ed was admitted to the hospital with an acute inflammatory polyradiculopathy that was preceded by an infectious illness 3 weeks earlier. His condition is characterized by progressive ascending muscular weakness of the limbs—which produces a bilateral symmetric flaccid paralysis. Ed's diagnosis is likely _______ , which is a condition caused by demyelination of peripheral spinal roots.
   a. Guillain-Barre
   b. myasthenia gravis
   c. diabetes mellitus
   d. multiple sclerosis

2. Mrs. Albert has been experiencing an intense burning sensation on the left side of her face. In which tract is pathology likely?
   a. reticulospinal tract
   b. trigeminothalamic tract
   c. spinothalamic tract
   d. dorsal columns

3. Upon examination, Mr. Perez's skin on the left lateral aspect of his forearm appears smooth and glossy; hair on this region has also fallen off. These signs have occurred after Mr. Perez was stabbed in an attempted mugging in which his left radial nerve was severed. Such skin changes are referred to as:
   a. hypalgesia
   b. analgesia
   c. thermhyperesthesia
   d. autonomic trophic changes

4. After his left below-the-knee amputation, Greg reports feeling nonpainful sensations in his amputated body part. This is clinically referred to as:
   a. amputation pain
   b. stump pain
   c. phantom pain
   d. phantom limb

5. Greg's condition (in the above question) is likely due to which of the following?
   a. the formation of neurofibromas in the stump of the amputated body part
   b. the sensory homunculus in the cortex continues to perceive the amputated body part as present
   c. stimulation of the sensory nerves in the stump of the amputated body part

6. Upon evaluation of sensation in his upper extremities, John demonstrates decreased sensation to pin prick on his right first and second phalanges. Decreased sensation is referred to as:
   a. hypoesthesia
   b. hyperesthesia
   c. paresthesia
   d. allodynia

7. After Mr. Wong sustained cervical nerve impingement in a severe fall he reports the sensation of pins and needles in his left arm. Clinically, this condition is referred to as:
   a. thermesthesia
   b. hyperesthesia
   c. paresthesia
   d. allodynia

8. Ron has diabetes-related neuropathy. He presents a classic pattern of bilateral numbness and tingling, pain, and decreased proprioception in both feet and calves (due to bilateral nerve damage). The same sensations eventually spread to his fingers, hands, and forearms. This classic neuropathy pattern is known as:
   a. stocking and glove plexopathy
   b. stocking and glove radiculopathy
   c. stocking and glove polyneuropathy
   d. stocking and glove mononeuropathy

9. After Tom's neck injury in a motor vehicle accident, he reports intense burning pain radiating from his neck down both shoulders and arms. This type of burning pain is referred to as:
   a. allodynia
   b. thermesthesia
   c. hypoesthesia
   d. causalgia

10. Mr. Goodman presents with a classic wrist and finger drop after a severe injury to his brachial plexus. This condition results from:
    a. medial nerve compression
    b. radial nerve compression
    c. ulnar nerve compression
    d. peroneal nerve compression

Section 18

Matching

SC Tract	Function
1. ___ Corticospinal Tracts	A. Facilitation of Extensor Muscles
2. ___ Medial Longitudinal Fasciculus	B. Proprioception of Lower Extremities
3. ___ Dorsal Columns	C. Pain and Temperature
4. ___ Anterior Spinothalamic Tracts	D. Voluntary Movement
5. ___ Cuneocerebellar Tracts	E. Proprioception of Trunk and Upper Extremities
6. ___ Posterior Spinocerebellar Tracts	F. Coordination of Head and Neck Movements
7. ___ Lateral Spinothalamic Tracts	G. Discriminative Touch, Pressure, Vibration, Proprioception, Kinesthesia
8. ___ Vestibulospinal Tracts	H. Crude Touch, Light Touch

SECTION 19

1. What are the five most important SC tracts to clinically access and why?

2. Jack's SCI was at the level of C6. Which of the following abilities is he capable of?
   1. self-feeding with adaptive equipment
   2. independent self-feeding without adaptive equipment
   3. upper extremity dressing with adaptive equipment
   4. independent upper extremity dressing without adaptive equipment

   a. 1 and 3
   b. 2 and 3
   c. 1 and 4
   d. 2 and 4

3. After her SCI, Susan presents with the following:
   - spasticity below the lesion level
   - hyperactive reflexes below the lesion level
   - clonus (sustained rhythmic jerking in a muscle) below the lesion level
   - flaccidity at the lesion level

   Susan's signs and symptoms are indicative of _______ motor neuron injury.

   a. an upper
   b. a lower

4. Elwood sustained a stab wound in a gang fight that pierced his SC. He presents with the following signs and symptoms:
   - ipsilateral loss of motor below the lesion level
   - ipsilateral loss of discriminative touch, pressure, vibration, and proprioception—below the lesion level
   - contralateral loss of pain and temperature below the lesion level

   His diagnosis is:

   a. dorsal column syndrome
   b. anterior horn cell syndrome
   c. Brown-Séquard syndrome
   d. central cord syndrome

5. Immediately after his SCI, Mr. Foreman lost all spinal reflexes below the lesion level. He also experienced flaccid paralysis below the lesion level and lost autonomic function. This state lasted for several days (after which all muscles below the lesion level became spastic). This state is a temporary condition that occurs immediately after SCI and is called:

   a. orthostatic hypotension
   b. spinal shock or neurogenic shock
   c. autonomic dysreflexia
   d. poikilothermy

6. Mrs. Jones was diagnosed with neurosyphilis and presents with bilateral loss of tactile discrimination, vibration, pressure, and proprioception. Her diagnosis is:

   a. Brown-Séquard syndrome
   b. anterior cord syndrome
   c. anterior horn cell syndrome
   d. dorsal column disease

7. After his SCI, Dan's muscles are flaccid and atrophied below his lesion level. His reflexes are also hyporeflexive below the lesion level. These signs are indicative of _______ motor neuron injury.

   a. an upper
   b. a lower

8. Mark can no longer regulate his body temperature after his T1 SCI. Instead, his body tends to take on the temperature of his environment. This condition is known as:

   a. poikilothermy
   b. neurogenic shock
   c. autonomic dysreflexia
   d. orthostatic hypotension

9. After Janice's C6 SCI, she experienced spinal shock that has resolved. However, she is now experiencing an acute episode of exaggerated sympathetic reflexes characterized by severe hypertension, bradycardia, severe headache, and profuse sweating above the lesion level. This condition is called _______ and can occur or recur at any time during the patient's lifespan.

   a. poikilothermy
   b. orthostatic hypotension
   c. autonomic dysreflexia
   d. neurogenic shock

10. The above condition is a clinical emergency that requires immediate treatment. It can be caused by:
    1. full bladder or rectum
    2. decubitus ulcers
    3. ingrown toenails
    4. dressing changes

    a. 1 and 2
    b. 1, 2, 3
    c. 1 and 3
    d. all of the above

Sections 20 – 21

1. To reduce tone in a spastic muscle, all of the following can be used except for which one?
   a. performing a sustained stretch of the spastic muscle
   b. placing pressure on the tendon of a spastic muscle
   c. performing a quick stretch to the spastic muscle

2. When Mr. James' elbow is moved passively by the therapist into greater extension, the elbow joint is first severely spastic and cannot be moved. With sustained stretch, the spasticity suddenly gives way and the elbow can be moved more easily into extension. This phenomenon is known as:
   a. cogwheel rigidity
   b. clasp knife syndrome
   c. lead pipe rigidity
   d. clonus

3. Mrs. Edmonds is 2 weeks status post cerebral vascular accident. She displays:
   - scapular elevation and retraction
   - shoulder abduction and external rotation
   - elbow, wrist, and finger flexion
   - hip and knee extension
   - ankle and toe plantar flexion

   This phenomenon is called _______ and is a stereotyped set of movements that occur in response to a stimulus or voluntary movement.
   a. an associated reaction
   b. a synergy pattern
   c. a decorticate rigidity pattern

4. The term _______ refers to an uncontrolled oscillation of a muscle in a spastic muscle group. It can be reduced by _______.
   1. lead pipe rigidity
   2. clonus
   3. placing the spastic muscle on a sustained stretch and thereby facilitating the Golgi tendon organs
   4. performing a quick stretch of the spastic muscle to facilitate the muscle spindles

   a. 1 and 3
   b. 1 and 4
   c. 2 and 3
   d. 2 and 4

5. Mrs. Washington has been diagnosed with Parkinson's disease. When the therapist attempts to range her elbow joint, the resistance in the joint is jerky and characterized by a pattern of release/resistance. This condition is known as:
   a. clasp knife syndrome
   b. cogwheel rigidity
   c. lead pipe rigidity
   d. clonus

6. Splinting, serial casting, and seating and positioning are all therapeutic techniques used to decrease tone in a hypertonic muscle group. These techniques are based on which of the following principles:
   a. a quick stretch of a spastic muscle group facilitates the Golgi tendon organs which inhibit the spastic muscle
   b. placing pressure on the muscle belly of a spastic muscle group facilitates the Golgi tendon organs which inhibit the spastic muscle
   c. a sustained stretch of a spastic muscle group facilitates the Golgi tendon organs which inhibit the spastic muscle

7. Jim had a stroke three weeks ago that resulted in hemiparesis on his right side. He is now able to walk with a quad cane but demonstrates heightened spasticity in his right arm during walking and other effortful movements. This phenomenon is known as _______ and occurs because of an inability to selectively inhibit the interneurons that synapse on motor cell bodies of the opposing limb.
   a. an associated reaction
   b. a synergy pattern

8. After her stroke, Mrs. Mays' right upper extremity has become hypertonic and can be passively moved on only one side of the joint (into greater flexion) without resistance. This type of hypertonicity is referred to as _______ and is indicative of _______ motor neuron pathology.
   1. spasticity
   2. rigidity
   3. upper
   4. lower

   a. 1 and 3
   b. 1 and 4
   c. 2 and 3
   d. 2 and 4

9. Mr. Samuel was diagnosed with Parkinson's disease. When the therapist attempts to range his elbow joint, the movement is characterized by a uniform and continuous resistance to passive movement. This phenomenon is referred to as:
   a. cogwheel rigidity
   b. clasp knife syndrome
   c. decerebrate rigidity
   d. lead pipe rigidity

10. To facilitate tone in a muscle of a child with hypotonicity, all of the following can be used except for which one?
    a. performing a quick stretch to a hypotonic muscle
    b. placing pressure on the muscle belly of the hypotonic muscle
    c. performing a sustained stretch of a hypotonic muscle

SECTION 22

1. Mr. Chaudry was diagnosed with a brain tumor. As a result of the tumor's anatomical position, Mr. Chaudry has lost voluntary motor control on the left side of his body. His tumor is in which region?

   a. left primary motor area
   b. right primary motor area
   c. left premotor area
   d. right premotor area

2. Sarah presents the following:
   - rigidity in certain muscle groups
   - difficulty initiating, continuing, and terminating movements
   - involuntary, undesired movements such as tremor or chorea

   Her doctor suspects pathology of which structure?

   a. basal ganglia
   b. internal capsule
   c. cerebellum
   d. primary motor area

3. Pam has been taking medication for schizophrenia all of her adult life. She sometimes experiences involuntary movements such as facial grimacing, tongue protrusion, dystonia, and blepharospasm. This condition is called _______ and occurs after the chronic use of neuroleptic drugs that block dopamine receptors in the basal ganglia.

   a. tardive dystonia
   b. tardive chorea
   c. tardive dyskinesia
   d. tardive athetosis

4. Mr. Oberi sustained a stroke 3 weeks ago. It is found that when Mr. Oberi's balance is displaced, he does not extend his arms to prevent a fall. Extending one's arms to prevent a fall is a reflexive response referred to as _______ . Its absence indicates neurologic damage.

   a. a righting reaction
   b. protective extension
   c. an equilibrium reaction
   d. an associated reaction

5. When examined in the clinic, Mrs. Faye presents ataxia in both upper and lower extremities, dysmetria, and intention tremors. She also has difficulty judging the distance of an object and often over-reaches for objects (dysmetria). When asked by the therapist to pronate and supinate both forearms (diadochokinesia) as fast as she can, she demonstrates an inability to perform these movements at the same rate and speed (one arm lags and drifts). The therapist suspects pathology of the:

   a. basal ganglia
   b. cerebellum
   c. primary motor area
   d. premotor area

6. From birth, Nicholas experienced muscle contractions that produced twisting movements resulting in abnormal postures. These sustained muscle contractions can sometimes last hours at a time. This disorder is called _______ and is caused by pathology of the _______ .

   1. dyssynergia
   2. dystonia
   3. cerebellum
   4. basal ganglia

   a. 1, 3
   b. 1, 4
   c. 2, 3
   d. 2, 4

7. Mrs. Ahmed sustained a right hemisphere stroke. As a result, she can understand the correct use of a hairbrush but when offered one, she attempts to brush her teeth with it. Mrs. Ahmed has _______ . The area of brain damage is likely in the _______ .

   1. ideational apraxia
   2. ideomotor apraxia
   3. supplemental motor area
   4. premotor area

   a. 1, 3
   b. 1, 4
   c. 2, 3
   d. 2, 4

8. Roy is a 65-year-old man who sustained a severe head injury after falling from a roof he was repairing 2 weeks ago. When the outer border of the plantar surface of his foot is stroked, this elicits extension of the first toe and fanning of the other toes. This is indicative of a ______ and signifies damage to the corticospinal tracts.

   a. flexor withdrawal
   b. Hoffman's sign
   c. Babinski sign
   d. extensor thrust

9. Cheryl presents the following signs and symptoms:
   - hypertonicity
   - cogwheel rigidity
   - bradykinesia
   - tremors at rest (pill-rolling)

   In addition, she has difficulty beginning movement such as walking and equal difficulty stopping. She often bumps into walls as a result. Her handwriting has also become extremely small and of poor quality. Her therapist suspects:

   a. Huntingdon's chorea
   b. Parkinson's disease
   c. athetosis
   d. hemiballismus

10. Abraham reports that he has always had uncontrollable urges to carry out repetitive motor and vocal tics—such as shoulder shrugging, throat clearing, eye blinks, and head jerks. While he is able to suppress these urges briefly, an inner tension then builds, and he feels even more compelled to express these behaviors. This condition is called ______ and is likely due to pathology of ______ .

    1. Tourette's syndrome
    2. tardive dyskinesia
    3. dopamine system
    4. norepinephrine system

    a. 1, 3
    b. 1, 4
    c. 2, 3
    d. 2, 4

Sections 23 – 25

1. Seven days ago, Ed sustained a myocardial infarction that caused ischemia (lack of blood flow) and hypoxia to the brain for several minutes. This caused extreme damage to both cerebral hemispheres but left the brainstem and vegetative functions intact. Two days ago, Ed awakened in a condition of eyes-open unconsciousness in which he only responds to painful stimuli (he reflexively withdraws from a pinch or pin prick); however, he cannot communicate and does not respond to commands. It is likely that Ed is in:

   a. a persistent vegetative state
   b. brain death
   c. a stuporous state of unarousability
   d. a locked-in syndrome

2. Eight-year-old Jimmie has been diagnosed with a sensory processing disorder. He is hyperactive, cannot easily concentrate or filter extraneous noise, and becomes highly agitated by certain types of sensory stimulation in the environment (such as florescent lighting in the supermarket). Jimmie's therapist uses slow rocking, deep pressure, and vibration to enhance the _______ and calm the CNS.

   a. reticular activating system
   b. reticular inhibiting system

3. In the above case, Jimmie's _______ may be in overdrive and may be unable to filter extraneous information from reaching the cortex. Consequently, all sensory data may be perceived in a heightened manner.

   a. reticular activating system
   b. reticular inhibiting system

4. As a result of left hemisphere damage, which functions are more likely to be impaired?

   1. mathematical skills
   2. language skills
   3. motor planning skills
   4. visual-spatial skills

   a. 1, 2
   b. 1, 2, 3
   c. 2, 3
   d. all of the above

5 Mr. O'Neil has the ability to see objects but cannot interpret the visual data that enters his CNS. The umbrella term for this condition is _______ and indicates impairment in the _______ .

   1. visual agnosia
   2. blind sight
   3. primary visual cortex
   4. visual association areas

   a. 1, 3
   b. 1, 4
   c. 2, 3
   d. 2, 4

6. After Samantha's left hemisphere stroke, she can understand when others speak to her but is unable to communicate intelligible and meaningful sentences. This condition is known as _____ and occurs as a result of lesions to _____.

   1. expressive aphasia
   2. receptive aphasia
   3. Broca's area
   4. Wernicke's area

   a. 1, 3
   b. 1, 4
   c. 2, 3
   d. 2, 4

7. Mr. Marconi has a malignant brain tumor. As a result of the tumor's growth, Mr. Marconi has lost sensation on the right half of his body. The tumor is likely located in which structure?

   a. right primary somatosensory area
   b. left secondary somatosensory area
   c. left primary somatosensory area
   d. right secondary somatosensory area

8. Mrs. Janasky is being evaluated in the therapy clinic. The therapist asks Mrs. Janasky to close her eyes and then places three different objects—one at a time—in her hand. She then asks Mrs. Janasky to identify the objects by touch alone. This technique screens for _______ and evaluates the _______ area.

   1. stereognosis
   2. graphesthesia
   3. contralateral primary somatosensory
   4. contralateral secondary somatosensory

   a. 1, 3
   b. 1, 4
   c. 2, 3
   d. 2, 4

9. After a severe stroke, Bob remains in what appears to be a vegetative state with eyes open but no voluntary movement. Bob's therapist notices that he is able to move his eyeball muscles and can communicate using eye blinks. This condition is called:

   a. persistent vegetative state
   b. brain death
   c. locked-in syndrome
   d. stupor

10. Neuroplasticity occurs when other areas of the brain assume the functions once mediated by regions that have become damaged. Neuroplasticity is most viable in children because the central nervous system is not fully developed but is also dependent upon which of the following?

   1. severity of neurologic damage
   2. premorbid health status
   3. pre-injury use of damaged brain areas
   4. volume of gray matter and dendritic connections

   a. 1
   b. 1, 2
   c. 1, 2, 3
   d. all of the above

SECTION 26

1. As a result of neurologic damage to the right multi-modal association area, Mrs. Klein can no longer identify familiar faces. She recognizes her loved ones and friends by their voices and gait patterns. This phenomenon is known as:
   a. simultanagnosia
   b. metamorphopsia
   c. abarognosia
   d. prosopagnosia

2. The therapist gives Tim a box of knives, forks, and spoons that are all the same color and size. She asks Tim to pick out the forks from the other utensils. This skill is called _______ and is a form of _______ .
   1. form constancy discrimination
   2. figure–ground discrimination
   3. visual acuity
   4. visual-spatial perception

   a. 1, 3
   b. 1, 4
   c. 2, 3
   d. 2, 4

3. After her right hemisphere stroke Mrs. Fuentes cannot perceive information from the left side of the environment and the left side of her own body. She is often observed with her left arm hanging over the side of her wheelchair or in her food on the left side of her dinner tray. This condition is known as:
   a. bilateral neglect
   b. unilateral neglect
   c. double simultaneous extinction
   d. left field cut

4. After a tumor was removed from the left hemisphere language centers, Mr. Thompson cannot express emotions using words. When he is upset he breaks into tears or angry outbursts but cannot use words to describe his feelings. This phenomenon is called:
   a. agrommation
   b. alexithymia
   c. araphia
   d. anomia

5. The therapist asks Mr. Lambert to close his eyes. She then writes letters on the palm of his hand and asks him to identify each letter. This skill is called _______ and is a form of _______ .
   1. graphesthesia
   2. stereognosis
   3. visual–spatial perception
   4. tactile perception

   a. 1, 3
   b. 1, 4
   c. 2, 3
   d. 2, 4

6. Bill has the ability to see colors when he hears music. This skill is called _______ and is a perceptual phenomenon.
   a. synesthesia
   b. chromathesia
   c. graphesthesia
   d. harmothesia

7. Mrs. Kelly cannot dress independently after her stroke. While she understands that shirts are to be worn on the torso, she mistakenly places her head through the sleeves. Clinically this is referred to as _______ and can signify problems with _______ .
   1. dressing apraxia
   2. constructional apraxia
   3. body schema perception
   4. depth perception

   a. 1, 3
   b. 1, 4
   c. 2, 3
   d. 2, 4

8. Patients with receptive aphasia resulting from a right hemisphere lesion who can understand the literal meaning of words but cannot interpret tonal inflection have:
   a. aprosodia
   b. asymbolia
   c. alexia
   d. agrommation

9. After his head injury Steve can no longer comprehend the relationship of one location to another. In the community he cannot find his way from his house to the grocery store, even though the store is located one block from his house. This disorder is referred to as:
   a. position in space dysfunction
   b. depth perception dysfunction
   c. topographical disorientation
   d. right–left discrimination dysfunction

10. Dave sustained right parietal damage in a motor vehicle accident 3 years ago. As a result of the brain damage he sustained he has lost the ability to attach appropriate colors to specific objects. For example, when shown three apples (one green, one blue, and one red) he is unable to identify the green and blue apples as inappropriate. This condition is known as:
    a. color anomia
    b. color agnosia
    c. color achromatopsia
    d. color morphopsia

SECTIONS 27 – 29

1. After Mr. Li's stroke, he presents the following signs and symptoms:
   - contralateral hemiplegia
   - contralateral hemiparesthesia
   - aphasia
   - cognitive dysfunction
   - emotional lability

   Which cerebral artery was likely involved as a site of occlusion?

   a. left hemisphere posterior cerebral artery
   b. right hemisphere posterior cerebral artery
   c. right hemisphere middle cerebral artery
   d. left hemisphere middle cerebral artery

2. Mabel's daughter brought her to the doctor after she experienced numbness and weakness of one side of her body, blurred vision, dizziness, and confusion. These episodes occurred on two separate occasions, 2 days apart. It is likely that Mabel experienced a:

   a. hemorrhagic stroke
   b. berry aneurysm
   c. transient ischemic attack
   d. embolic stroke

3. In the above case, Mabel's doctor ordered a _______ to observe blood flow in the brain. This imaging technology is able to detect mini strokes in order to prevent a more serious CVA.

   a. CT scan
   b. MRI scan
   c. PET scan
   d. SPECT scan

4. Myasthenia gravis is a disease in which a specific neurotransmitter at the neuromuscular junction becomes ineffective, resulting in muscle weakness and fatigue. This neurotransmitter is:

   a. GABA
   b. glutamate
   c. acetylcholine
   d. norepinephrine

5. This neurotransmitter system has been shown to be impaired in people with anxiety and panic disorders.

   a. GABA
   b. glutamate
   c. acetylcholine
   d. Substance P

6. This neurotransmitter is believed to play a major role in learning and memory functions and has been used in the design of pharmaceuticals that inhibit the progression of Alzheimer's disease.

   a. GABA
   b. glutamate
   c. serotonin
   d. norepinephrine

7. Too much of this neurotransmitter substance has been linked to schizophrenia. Too little has been linked to Parkinson's disease. This neurotransmitter also plays a major role in addiction and stimulating the brain's reward system.

   a. endorphin
   b. norepinephrine
   c. serotonin
   d. dopamine

8. Imbalances in this neurotransmitter system has been linked to the following pathologies: depression, eating disorders, sleep disorders, and OCD.

   a. endorphin
   b. dopamine
   c. serotonin
   d. Substance P

9. This neurotransmitter is linked to arousal and vigilance, activation of the sympathetic nervous system, the fight/flight response, increased heart rate, and increased muscular preparedness.

   a. norepinephrine
   b. serotonin
   c. dopamine
   d. glutamate

10. This neurotransmitter acts in the nociceptor pathway and is involved in the transmission of pain receptors to the CNS.

    a. norepinephrine
    b. dopamine
    c. serotonin
    d. Substance P

SECTIONS 30 – 32

1. This term refers to the phenomenon in which people learn to associate the euphoria of a drug-induced state with the objects and people involved in the process of drug use. This phenomenon involves the neurotransmitter glutamate and occurs in response to the brain's addicted state.
   a. dependence
   b. tolerance
   c. priming
   d. sensitization

2. This term refers to the condition in which the brain becomes less responsive to a chronically used drug of addiction. The reward response of the mesolimbic system becomes diminished and people experience depressed mood and motivation.
   a. dependence
   b. tolerance
   c. priming
   d. sensitization

3. The neurotransmitter glutamate also plays a role in the phenomenon in which abstinence from an addictive substance produces a heightened neurologic response to the addictive substance. This state underlies the intense cravings that cause relapse.
   a. dependence
   b. tolerance
   c. priming
   d. sensitization

4. After Tom's head injury, he has difficulty remembering people and events encountered less than 1 hour ago. Tom cannot remember people to which he has recently been introduced or events that have transpired in the last day. If he wants to remember something of importance, he must write everything down. This type of memory problem involves _______ which is believed to be mediated by the _______ .
   1. short-term memory
   2. recent memory
   3. frontal lobes
   4. hippocampus

   a. 1, 3
   b. 1, 4
   c. 2, 3
   d. 2, 4

5. After Kim's auto accident, in which she lost consciousness, she could not remember her entire personal past history for several days. This type of memory problem is called _______ .
   a. anterograde amnesia
   b. retrograde amnesia
   c. post-traumatic memory loss

6. Mr. Chen was in a car accident 2 years ago in which he sustained moderate brain damage. Since that time, he has been unable to transfer short-term memories into long-term storage. This has resulted in his inability to remember ongoing day-to-day events since the accident—although Mr. Chen is able to remember his entire past history prior to the accident. This type of memory problem is called _______ .
   a. anterograde amnesia
   b. retrograde amnesia
   c. post-traumatic memory loss

7. Anthony sustained brain damage after falling from a ladder on the job site 1 week ago. As a result, he is emotionally labile, depressed, and despondent. This clinical phenomenon—which occurs in the acute stages of injury—is referred to as _______ and occurs as a result of _______ .
   1. euphoric reaction
   2. catastrophic reaction
   3. left prefrontal lobe damage
   4. right prefrontal lobe damage

   a. 1, 3
   b. 1, 4
   c. 2, 3
   d. 2, 4

8. Anthony's hospital roommate, Josh, also sustained severe brain damage but in a motor vehicle accident. Josh has experienced an opposite emotional syndrome, reporting that he has never felt better despite his severe physical and cognitive impairments. This clinical phenomenon that occurs in the acute stages of injury is referred to as _______ and occurs as a result of _______ .
   1. euphoric reaction
   2. catastrophic reaction
   3. left prefrontal lobe damage
   4. right prefrontal lobe damage

   a. 1, 3
   b. 1, 4
   c. 2, 3
   d. 2, 4

9. In a similar emotional syndrome resulting from neurologic damage, Paul exhibits impulsiveness, disinhibition, poor judgment, and risk-taking behaviors. He also easily becomes agitated and angry in response to perceived offenses. This emotional syndrome results from:
   a. orbitofrontal lobe lesions
   b. dorsolateral lobe lesions

10. Ken has been diagnosed with post-traumatic stress disorder (PTSD) after serving in the Gulf War. He reports feeling that the events of his military service are vividly occurring over and over again in real-life flashbacks. Researchers have found that the _______ become highly active in PET scans of people experiencing PTSD flashbacks—as though these structures are flooding the visual, auditory, and somatosensory cortices with the painful memories.
   a. prefrontal and dorsolateral frontal lobes
   b. hypothalamus and thalamus
   c. occipital and parietal lobes
   d. amygdala and hippocampus

SECTIONS 33 – 34

1. Researchers recommend that the most effective ways to reduce one's risk of age-related dementia involves engaging in which of the following?
   1. mentally stimulating activities such as reading
   2. leisure activities such as watching television
   3. mentally stimulating activities such as taking an adult education class
   4. leisure activities such as completing crossword puzzles

   a. 1, 3
   b. 1, 2, 3
   c. 1, 3, 4
   d. 2, 3, 4

2. Researchers have shown that at 1 day old, _______ infants demonstrate greater eye contact with others while _______ infants look longer at suspended mechanical mobiles. At 18 months old, _______ toddlers display a greater vocabulary than _______ toddlers.

   a. female, male, female, male
   b. male, female, male, female
   c. female, male, male, female
   d. male, female, female, male

3. The above gender-based difference is believed to be due to higher levels of _______ . The higher the level, the less eye contact and the smaller a child's vocabulary was measured to be.

   a. prenatal testosterone
   b. prenatal estrogen

4. Nine-year-old Jack displays the following:
   - difficulty developing social relationships
   - difficulty interpreting and responding to social cues
   - narrow interests, primarily in systems and ordering of objects
   - strong adherence to routines
   - lack of eye contact and poor language skills

   These signs are characteristic of:

   a. savant syndrome
   b. Asperger's syndrome
   c. Klinefelter's syndrome
   d. Turner syndrome

5. Simon Baron-Cohen suggests that Asperger's syndrome may reflect brain development that has been affected by high levels of testosterone in fetal development that slow the maturation of the _______ hemisphere and accelerate growth of the _______ hemisphere.

   a. left, right
   b. right, left

6. Baron-Cohen suggests that when too much testosterone accelerates fetal development of the right hemisphere (at the expense of the left) a range of problems can occur, such as:
   1. attention-deficit hyperactivity disorder (ADHD)
   2. autism
   3. congenital adrenal hyperplasia
   4. learning disability

   a. 1, 2
   b. 1, 2, 3
   c. 1, 2, 4
   d. 1, 2, 3, 4

7. Research has shown that women who sustain stroke are less likely than men to experience aphasia and that, if aphasia is experienced in women, it tends to resolve more quickly than in men. This may be due to which of the following?
   1. Women's cortical language centers are larger than men's.
   2. Women tend to use language centers in both the right and left hemispheres while men tend to use only left hemisphere language centers.
   3. Language centers have been shown to mature approximately 6 years earlier in girls than in boys.

   a. 1, 3
   b. 1, 2
   c. 1, 2, 3

8. Robin and Joe's baby Rachel was born with a normal 46 XY chromosomal pattern. In embryonic development, the baby's cells were unable to respond to testosterone. As a result, the internal anatomy of the fetus developed as a male but was incomplete. Because testosterone was present in fetal development, it prevented the fetus from developing internal female anatomy. External anatomy, however, developed as a normal female. When Rachel was born, she was pronounced to be a healthy baby girl. The condition, called _______ , was not diagnosed until Rachel reached adolescence and received medical attention to address the absence of menstruation.

   a. Klinefelter's syndrome
   b. DHT deficiency syndrome
   c. retained mullerian syndrome
   d. testicular feminizing syndrome (androgen insensitivity syndrome)

9. Researchers have found as early as childhood, _______ are more attentive to the social needs of others. When both male and female children were observed as they attempted to host or lead a group, female children _______ .
   1. females
   2. males
   3. were more sensitive to a newcomer and helped the newcomer to become part of group activities
   4. ignored the newcomer and continued participating in the activities in which they were already engaged

   a. 1, 4
   b. 1, 3
   c. 2, 3
   d. 2, 4

10. Researchers have found that the right and left inferior parietal lobules tend to be larger in males than in females. These brain regions also appear to mature approximately 4 years earlier in boys than in girls. The inferior parietal lobules are believed to be responsible for which of the following skills?
    1. abstract mathematical reasoning skills
    2. manipulation of objects in space
    3. navigating routes through the use of landmarks
    4. precision fine motor skills

    a. 1, 2
    b. 1, 2, 3
    c. 1, 2, 4
    d. 1, 2, 3, 4

Answers

Sections 1 – 3
1. b
2. c
3. d
4. b
5. a
6. c
7. d
8. b
9. a
10. b

Sections 4 – 7
1. c
2. b
3. c
4. d
5. a
6. b
7. b
8. d
9. a
10. d

Section 8
1. c
2. b
3. c
4. d
5. a
6. c
7. b
8. a
9. c
10. a

Sections 9 – 11
1. a
2. d
3. c
4. b
5. a
6. c
7. a
8. d
9. a
10. c

Sections 12 – 14
1. c
2. c
3. a
4. b
5. d
6. b
7. b
8. c
9. a
10. d

Sections 15 – 17
1. a
2. b
3. d
4. d
5. b
6. a
7. c
8. c
9. d
10. b

Section 18
1. D
2. E
3. G
4. H
5. E
6. B
7. C
8. A

Section 19

1.
 a. Lateral corticospinal tracts—voluntary motor control on the contralateral side
 b. Dorsal columns—conscious discriminative touch, pressure, vibration, and proprioception on the contralateral side
 c. Lateral Spinothalamic tracts—conscious pain and temperature on the contralateral side.
 d. Spinocerebellar tracts—unconscious proprioception
 e. Vestibulospinal tracts—facilitation of extensor tone (important to assess in neurologic injury)
2. a
3. a
4. c
5. b
6. d
7. b
8. a
9. c
10. d

Sections 20 – 21
1. c
2. b
3. b
4. c
5. b
6. c
7. a
8. a
9. d
10. c

Section 22
1. b
2. a
3. c
4. b
5. b
6. d
7. d
8. c
9. b
10. a

Sections 23 – 25
1. a
2. b
3. a
4. a
5. b
6. a
7. c
8. b
9. c
10. d

Sections 26 – 27
1. d
2. d
3. b
4. b
5. b
6. a
7. a
8. a
9. c
10. b

Sections 27 – 29
1. d
2. c
3. a
4. c
5. a
6. b
7. d
8. c
9. a
10. d

Sections 30 – 32
1. c
2. b
3. d
4. a
5. b
6. a
7. c
8. b
9. a
10. d

Sections 33 – 34
1. c
2. a
3. a
4. b
5. a
6. c
7. c
8. d
9. b
10. a

Glossary

A1: See *Primary Auditory Area.*

Abarognosis: The inability to accurately estimate the weight of objects—particularly in comparison to each other.

Abducens Nerve: CN 6. Responsible for extraocular eye movements. Lesion symptoms include medial strabismus, diplopia, and nystagmus.

Acalculia: A type of expressive aphasia that involves the inability to calculate mathematical problems.

Accessory Nerve: CN 11. Responsible for elevation of the larynx during swallowing, innervation of the sternocleidomastoid muscle (for head rotation and flexion/extension), and innervation of the upper trapezius muscle (for shoulder elevation and flexion above 90 degrees).

Accommodation: The ability of the eye to focus images of near or distant objects on the retina. The ciliary muscles are responsible for changing the thickness of the lens to focus images on the retina.

Acetylcholine (ACh): A neurotransmitter that acts at the neuromuscular junction to facilitate muscle movement.

Acetylcholinesterase (AChE): The enzyme that destroys ACh soon after it is released from its terminal boutons, thus terminating the postsynaptic potential.

ACh: See *Acetylcholine.*

AChE: See *Acetylcholinesterase.*

Achromatopsia: A condition that occurs when V4 is lost and the world appears in shades of gray. The memory of color is erased.

Action Potential: The brief electrical impulse that provides the basis for conduction of nerve signals along the axon. It results from the brief changes in the cell's membrane permeability to sodium and potassium ions. A strong enough action potential will cause the neuron to become excited and start the conduction process.

Adiadochokinesia: Inability to perform rapid alternating movements—as in supinating and pronating one's forearms and hands quickly and synchronously. Sign of cerebellar pathology.

Agnosia: Literally "not to know." Tactile Agnosia: An inability to interpret sensations through touch. Auditory Agnosia: An inability to interpret sounds. Visual Agnosia: An inability to interpret visual stimuli. All of the above agnosias result from lesions to the cortex; the sensory receptor anatomy remains intact.

Agraphesthesia: Loss of the ability to interpret letters written on the contralateral hand. Indicates damage to SS2.

Agraphia: A type of expressive aphasia resulting in the inability to write intelligible words and sentences.

Agrommation: A type of expressive aphasia that involves the inability to arrange words sequentially so that they form intelligible sentences.

Ahylognosia: The inability to discriminate between different types of materials by touch alone.

Akathisia: An inability to remain still, caused by an intense urge to move or fidget.

Akinesia: An inability to perform voluntary movement. Commonly seen in the late stages of Parkinson's disease. Patients report that a tremendous amount of mental concentration is required to perform basic motor tasks.

Akinetopsia: A cortical visual disorder in which people neither see nor understand the world in motion. Occurs as a result of lesions to V5.

Alexia: A type of receptive aphasia resulting in the inability to read and interpret written words. See also *Dyslexia* (difficulty interpreting the written word).

Alexithymia: A type of expressive aphasia. Inability to attach words to one's emotions; inability to express one's emotions using words. Dyslexithymia is difficulty attaching words to one's emotions.

Allodynia: A condition in which non-painful stimuli now produce pain.

Amorphognosia: The inability to discriminate between different forms/shapes by touch alone.

Amygdala: An almond-shaped nucleus in the anterior temporal lobe that attaches to the caudate nucleus. May have roles in the mediation of fear and anger and the perception of social cues.

Analgesia: Loss of pain sensation.

Anhedonia: An inability to experience pleasure. Often accompanies states of depression.

Anomia: A type of expressive aphasia that involves the inability to remember and express the names of people and objects.

Anosmia: Loss of smell (olfaction).

Anosognosia: Extensive neglect and failure to recognize one's own body paralysis. Results from lesions to the right hemisphere.

ANS: See *Autonomic Nervous System*.

Anterior: Also referred to as ventral. Refers to the front of an organism. Ventral means the belly of a four-legged animal.

Anterior Commissure: Located in the anterior thalamus; allows information to travel between both thalamic lobes.

Anterior Median Fissure: Divides the medulla into equal left and right halves. The fissure continues all the way down the spinal cord.

Anterior Spinocerebellar Tract: Ascending sensory spinal cord tract. Carries unconscious information from the lower extremities to the cerebellum regarding proprioception.

Anterior Spinothalamic Tract: Ascending sensory spinal cord tract. Carries conscious information about crude and light touch.

Anterograde Amnesia: A memory dysfunction resulting from brain damage, in which—after the injury or disease process—the person cannot remember ongoing day-to-day events, although memory of one's personal past remains intact. Anterograde amnesia is a dysfunction of the encoding process—the individual cannot transfer short-term memories into long-term memory storage. In effect, the individual cannot develop any new long-term memories.

Antitransmitter: A chemical substance that breaks down a neurotransmitter so that the postsynaptic neuron can repolarize in order to fire again. Antitransmitters terminate the postsynaptic neuron's response.

Aphagia: Inability to swallow. See also *Dysphagia* (difficulty swallowing).

Aphasia: Impairment in the expression and/or comprehension of language. Receptive aphasia (Wernicke's aphasia) is the impairment in the comprehension of language. Expressive aphasia (Broca's aphasia) is the impairment in the expression of language.

Aphonia: An inability to make sounds. Hypophonia refers to reduced vocal force.

Aphrenia: Stoppage of thought. The individual experiences poverty of thought.

Apnea: Arrest of breathing.

Apraxia: Inability or difficulty (dyspraxia) executing motor plans. Results from lesions to the motor cortices in the frontal lobe. Ideational apraxia involves an inability to cognitively understand the motor demands of the task. Ideomotor apraxia involves the loss of motor plans for specific activities; or the motor plan may be intact but the individual cannot access it.

Aprosodia: A receptive aphasia that involves difficulty comprehending tonal inflections used in conversation. Results from lesions to the right hemisphere language centers.

Arachnoid Mater: The middle meningeal layer located just below the subdural space. Has the appearance of a spider web.

Arachnoid Villi: Projections of the arachnoid mater into the dura mater. CSF is reabsorbed in the arachnoid villi.

Associated Reactions: Stereotyped movements in which effortful use of one extremity influences the posture and tone of another extremity (usually the opposite extremity). Can occur as a result of neurologic pathology or as part of normal movement (as a result of reflex stimulation). Associated reactions result from an overflow of activity into the opposite limb. This occurs because of an inability to selectively inhibit the interneurons that synapse with the motor cell bodies of the opposite limb.

Astereognosis: The inability to identify objects by touch alone. Results from damage to SS2.

Asthenia: Muscle weakness. A sign of cerebellar damage.

Asymbolia: A receptive aphasia that involves difficulty comprehending gestures and symbols.

Ataxia: Uncoordinated movements resulting from cerebellar lesions.

Atopognosia: The inability to accurately perceive the exact location of a sensation.

Audition: The detection and perception of sound.

Auditory Agnosia: The umbrella term for the inability to attach meaning to non-language sounds.

Auditory Association Areas: Responsible for the interpretation of auditory data. There are several auditory association areas located throughout the cortex. These areas have not as yet been mapped as precisely as the visual association areas.

Autonomic Nervous System (ANS): Composed of the parasympathetic nervous system and the sympathetic nervous system. Responsible for the innervation of visceral muscles, regulates glandular secretion, and controls vegetative functions (eg, temperature, digestion, heart rate).

Axon: Fiber emerging from the axon hillock and extending to the terminal boutons of a neuron. Axons transmit action potentials, or nerve signals, to the terminal boutons.

Axon Collaterals: Project from the main axon structure of a neuron and serve to transmit nerve signals to several parts of the nervous system simultaneously.

Axon Hillock: The region where a neuron's cell body and axon attach.

Basal Ganglia: An unconscious motor system that mediates stereotypic or automatic motor patterns such as those involved in walking, riding a bike, and writing. Composed of three primary structures: caudate nucleus, putamen, and globus pallidus. Some sources now include the subthalamus and substantia nigra as part of the basal ganglia system. Disorders of the basal ganglia often result in dystonia and dyskinesia.

Bitemporal Hemianopsia: Occurs when the temporal fields in both eyes have been lost. Results from a lesion to the central optic chiasm. Results in tunnel vision.

Blood-Brain Barrier: Consists of the meninges, the protective glial cells, and the capillary beds of the brain. Responsible for the exchange of nutrients between the CNS and the vascular system. Some molecules can cross the membrane while others cannot. This accounts for the inability of many pharmaceuticals to cross the blood-brain barrier.

Body Schema Perceptual Dysfunction: Body schema is a neural perception of one's body in space—formed by a synthesis of tactile, proprioceptive, and pressure sensory data about the body. Dysfunction occurs when there is a severe discrepancy between body schema and reality. Includes finger agnosia, unilateral neglect, anosognosia, and extinction of simultaneous stimulation.

Brachial Plexus: A network of peripheral spinal nerves that supply the upper extremities. Includes C5, C6, C7, C8, and T1. Common site of compression injuries.

Bradykinesia: Slowness of voluntary movement. Seen commonly in Parkinson's disease. Also seen in depression.

Bradyphrenia: Slowness of thought. Seen commonly in Parkinson's disease and depression.

Brainstem: Composed of the midbrain, pons, and medulla. Controls vegetative functions (eg, respiration, cough and gag reflex, pupillary response, swallowing reflex).

Broca's Aphasia: An expressive language disorder in which patients can understand what is spoken to them, but they cannot express their ideas in an understandable way.

Broca's Area: Located only in the left hemisphere, just above the lateral fissure in the premotor area. Mediates the motoric functions of speech and is responsible for the verbal expression of language.

Callosal Sulcus: Sulcus separating the corpus callosum and the cingulate gyrus.

Catastrophic Reaction: An acute emotional syndrome related to the site of brain injury. Patients with left prefrontal lobe damage tend to be emotionally labile, depressed, and despondent. Catastrophic reaction tends to occur during the acute stages of brain damage.

Cauda Equina: At the end of the spinal cord—the conus medullaris—the spinal cord sends off the remaining spinal nerves that have not yet exited the vertebral column. This mass of spinal nerves is called the cauda equina because it resembles a horse's tail.

Caudal: Refers to the tail of the organism. Also refers to structures that are below others.

Caudate Nucleus: A basal ganglial structure involved in the planning and execution of automatic movement patterns. The caudate acts like a brake on certain motor activities. When the brake is not working, extraneous, purposeless movements appear (eg, tics, dyskinesias).

Causalgia: An intense burning pain accompanied by trophic skin changes.

Cell Body: Contains the nucleus of the neuron, which stores the genetic code of the organism.

Central Canal: Passageway through which CSF flows. Begins in the caudal medulla and descends throughout the entire length of the spinal cord.

Central Nervous System (CNS): Composed of the brain and spinal cord.

Central Sulcus: (also called the sulcus of Rolando) Separates the primary motor cortex from the primary somatosensory cortex.

Cerebellar Peduncles: Carries sensorimotor information from the pons to the cerebellum about the body's position in space. There are three paired cerebellar peduncles: middle, inferior, and superior cerebellar peduncles.

Cerebellum: Responsible for proprioception or the unconscious awareness of the body's position in space. The cerebellum is a sensorimotor system; it receives sensory information from joint and muscle receptors concerning the body's position. The cerebellum uses this information to make decisions about how to adjust the body for the coordinated, precision control of movement and balance.

Cerebral Achromatopsia: A condition that occurs when V4 is lost and the world appears in shades of gray. The memory of color is erased.

Cerebral Aqueduct: Part of the ventricular system. A narrow channel that connects the third and fourth ventricles allowing CSF to flow through.

Cerebral Peduncles: Large fiber bundles located on the anterior surface of the midbrain. Carries descending motor tracts from the cerebrum to the brainstem. Have an inner coat (consisting of the red nucleus and substantia nigra—collectively called the tegmentum) and an outer coat (consisting of the crus cerebri).

Cerebromedullary Cistern: (also called cisterna magna) Largest cistern in the subarachnoid space; allows CSF to flow from the fourth ventricle to the subarachnoid space. Located between the medulla and the cerebellum. Often used as a shunt placement.

Cerebrospinal Fluid (CSF): A clear and colorless fluid that bathes and nourishes the brain and spinal cord. The composition of CSF is used for diagnostic purposes to identify disease processes.

Chemoreceptors: Sensory receptors that respond to the presence of a particular chemical; involved in olfaction and gustation.

Choroid Plexus: Vascular structures in the brain that protrude into the ventricles and produce cerebrospinal fluid.

Cingulate Gyrus: Most medial and deepest gyrus in the frontal and parietal lobes. Sits right above the corpus callosum. Shares vast connections with limbic system structures.

Cingulate Sulcus: The sulcus that separates the cingulate gyrus from other gyri in the fronto-parietal regions; located on the medial aspect of each hemisphere.

Circle of Willis: A circuit of five interconnecting arteries that function to prevent lack of blood flow to the brain due to occlusion.

Clasp Knife Syndrome: Involves severe spasticity at a joint. A sustained stretch will relax the muscle group and the spasticity will suddenly give way.

Claustrum: A group of nuclei located just lateral to the extreme capsule and just medial to the insula.

Clonus: An uncontrolled oscillation of a spastic muscle group that results from a quick muscle stretch. Occurs in UMN lesions.

CN: See *Cranial Nerves*.

CNS: See *Central Nervous System*.

Cogwheel Rigidity: Cogwheel rigidity is characterized by a pattern of release/resistance in a quick jerky movement. Commonly seen in Parkinson's disease.

Color Agnosia: A visual perceptual disorder in which individuals appear to forget the concept of color. They do not appear to know the color of common objects.

Color Anomia: A visual perceptual disorder in which individuals have lost the names for colors. However, they would still recognize that a blue banana was strange.

Commissure: Any collection of axons that connect one side of the nervous system to the other. An example is the corpus callosum.

Contractures: Limitation in joint movement due to shortening of muscles, tendons, and ligaments. Results from inactivity at a joint.

Contralateral Homonymous Hemianopsia: A loss of the visual field on the opposite side of the lesion. A left visual field cut, or left contralateral homonymous hemianopsia, results from a lesion in the right optic tract. A right visual field cut, or right contralateral homonymous hemianopsia, results from a lesion in the left optic tract.

Conus Medullaris: The end of the spinal cord at the L1 – L2 vertebral area.

Convolutions: The collective name for the gyri and sulci located on the surface of the cerebral hemispheres.

Coronal Plane: (also called frontal or transverse plane) The coronal planes run perpendicular to the sagittal planes. Coronal planes divide the anterior aspect of the brain from the posterior aspect.

Coronal Suture: The suture lines are areas where cranial bones have fused. The coronal suture runs along the coronal plane and connects the frontal bone with the parietal bones.

Corpus Callosum: Largest commissure in the brain. Allows the right and left cerebral hemispheres to communicate with each other.

Corpus Striatum: Collective name for the caudate, putamen, and globus pallidus (structures of the basal ganglia).

Cortex: A cortex is a layer of gray matter that contains nuclei, or nerve cell bodies. Humans have a cerebral cortex and a cerebellar cortex. The cortex sits on the surface of the cerebrum and cerebellum—underneath of which is white matter, or axons.

Corticobulbar Tract: Spinal cord tract that descends off of the corticospinal tract and projects to CN nuclei having a motor component.

Corticospinal Tracts: Descending motor tracts originating from the primary motor cortex. Responsible for voluntary movement on the contralateral side of the body.

Cranial Nerves (CNs): The CNs are 12 pairs of nerves that are considered to be part of the PNS. Their nuclei are located in the brainstem and are considered to be within the CNS. CNs carry sensory and motor information to and from the receptors of the head, face, and neck.

CSF: See *Cerebrospinal Fluid.*

Cunctation: Resisting or hindering; the opposite of festination—which means quickened. Together, the cunctating–festinating gait—characteristic of Parkinson's disease—describes difficulty initiating movement and inability to stop movement once started.

Cuneocerebellar Tract: Ascending sensory spinal cord tract. Carries unconscious information from the trunk and upper extremities to the cerebellum regarding proprioception.

Cutaneous Receptors: Respond to pain, temperature, pressure, vibration, and discriminative touch. Found in the layers of the skin.

DA: See *Dopamine.*

Decerebrate Rigidity: Damage to any tract that originates in the brainstem may result in decerebrate rigidity. Involves spastic extension of both the upper and lower extremities. The occurrence of decerebrate rigidity indicates a much poorer prognosis than does decorticate rigidity.

Decorticate Rigidity: Results from damage to the corticospinal tracts. Decorticate rigidity presents as spastic flexion of the upper extremities; spastic extension of the lower extremities.

Deep Tendon Reflexes: A reflex arc in which a muscle contracts when its tendon is percussed. Deep tendon reflexes work on the principle of the spinal reflex arc. In an UMN injury, deep tendon reflexes become hyper-reflexive—because the spinal reflex arc remains intact below the lesion level causing the reflex arc to run unmodified by cortical input. In a LMN injury, deep tendon reflexes become hyporeflexive—because the reflex arc is lost.

Dendrites: The treelike processes that attach to the cell body and receive messages from the terminal boutons of a presynaptic neuron. Dendrites can bifurcate, or produce additional dendritic branches. Bifurcation increases the neuron's receptor sites.

Dentate Ligaments: A projection of the pia mater of the spinal cord. The dentate ligaments are a series of 22 triangular bodies that anchor the spinal cord.

Dentate Nuclei: One of fours pairs of cerebellar nuclei.

Depth Perception Dysfunction: Involves difficulty determining whether one object is closer to the individual than another object. Also referred to as stereopsis.

Dermatome: Skin segment innervated by a specific peripheral nerve.

Diencephalon: Collective name for the thalamus, hypothalamus, epithalamus, and subthalamus.

Diplopia: Double vision.

Dopamine (DA): The DA system has major effects on the motor system and on cognition and motivation. Loss of DA from the substantia nigra is the primary cause of Parkinson's disease. Too much DA has been implicated in schizophrenia. The DA system also plays a role in addictive behaviors.

Dorsal: (also referred to as posterior) Refers to the back of an organism.

Dorsal Columns: Ascending sensory spinal cord tract. Carries conscious information about discriminative touch, pressure, vibration, proprioception, and kinesthesia.

Dorsal Horn: The dorsal horn is considered to be part of the CNS. It contains the cell bodies of the sensory spinal cord tracts. In the dorsal horn, the dorsal rootlets may synapse on interneurons. These interneurons then synapse with spinal cord tracts. Or the dorsal rootlets may synapse directly on the cell bodies of spinal cord tracts.

Dorsal Intermediate Sulcus: Sulci that are located just lateral to the dorsal median sulcus.

Dorsal Median Sulcus: A sulcus that divides the posterior medulla into equal left and right halves.

Dorsal Root and Rootlets: Dorsal roots are axon bundles that emerge from an ascending spinal nerve. The dorsal root leads into the dorsal rootlets—thin stringlike axons that emerge from the dorsal root and synapse in the dorsal horn of the spinal cord. The dorsal root and rootlets are considered to be part of the PNS.

Dorsal Root Ganglion: Contains the cell bodies of sensory nerves that are part of the somatic PNS. Each sensory nerve has its own dorsal root ganglion. The dorsal root emerges from the dorsal ganglia.

Double Simultaneous Extinction: The inability to determine that one has been touched on both the involved side and the uninvolved side—the neural sensation of the uninvolved side overrides the ability to perceive touch on the involved side. Also called extinction of simultaneous stimulation.

Down Regulation: A decrease in the number of receptors for a neurotransmitter on the postsynaptic neuron, often due to long-term exposure to the neurotransmitter.

Dressing Apraxia: A form of ideomotor apraxia involving an inability to dress oneself due to impairment in either (a) body schema perception or (b) perceptual motor functions. Example: a patient may attempt to put his arm through pant legs or dress only one half of his body.

Dura Mater: The outermost meningeal layer. The dura has two projections that extend into the brain: falx cerebri and tentorium.

Dural Sinuses: Openings for blood vessels and nerves in the dura. Located above the frontal and parietal lobes. The sinuses function as a circulatory system: cerebral veins empty into the sinuses; they also receive CSF from the subarachnoid space.

Dysarthria: Difficulty articulating words clearly. Slurring words.

Dysesthesia: Unpleasant sensation such as burning.

Dyslexia: The impaired ability to read. Dyslexia is a language problem in which the ability to break down words into their most basic units—phonemes—is impaired. See also *Alexia*.

Dysmetria: Inability to judge distance. Past-pointing or over-shooting one's reach for objects. Occurs as a result of cerebellar lesions.

Dysphagia: Difficulty swallowing.

Dysphonia: Difficulty projecting one's voice audibly.

Dystonia: Also sometimes used interchangeably with dyskinesia. Both terms refer to abnormalities in muscle tone and movement. Includes athetosis, chorea, Parkinson's disease, and idiopathic torsion dystonia, etc.

Emboliform Nuclei: One of four pair of cerebellar nuclei.

Endorphins: Work conjointly with Substance P to act as pain modulators. The primary action of endorphin is the inhibition of nociceptive information.

Enteric Nervous System: An independent circuit that is loosely connected to the CNS but can function alone without instruction from the CNS. Located in sheaths of tissue that line the esophagus, stomach, small intestines, and colon. Composed of a network of neurons, neurotransmitters, and proteins.

Epicritic Sensory Receptors: Can detect sensation with precision, accuracy, and acuteness. Discriminative touch, sharp pain, exact joint position, and the exact localization of a stimulus are within the functions of the epicritic system. Evolutionary function: allows the organism to explore the environment with precise detail, thus allowing the ability to detect imminent danger.

Episodic Memory: A type of memory involving significant events that happened to an individual—the first day of school, one's wedding, the birth of a child.

Epithalamus: Very small structure located just posterior to the thalamus and just anterior to the pineal gland. A principal structure of the epithalamus is the habenula—a nucleus at the posterior of the epithalamus.

Esotropia: Internal or medial strabismus. Results from lesions to the abducens nerve (CN 6).

Euphoric Reaction: An acute emotional syndrome related to the site of brain injury. Right prefrontal lobe damage often leaves patients with an indifference to their impairment (referred to as anosognosia). Patients commonly report states of euphoria and well-being—despite severe impairment. Euphoric reaction tends to occur during the acute stages of brain damage.

Exotropia: External or lateral strabismus. Results from lesions to the oculomotor nerve (CN 3).

Explicit Memory: A type of memory involving knowledge that an individual consciously knows she has acquired. For example, an anatomist knows that he has acquired knowledge of the human body's musculoskeletal system.

Expressive Aphasia: A language perceptual problem involving difficulty expressing clear, meaningful language.

External Capsule: White matter located just lateral to the putamen (of the basal ganglia) and medial to the claustrum.

Exteroceptor: A sensory receptor that is adapted for the reception of stimuli from the external world (eg, visual, auditory, tactile, olfactory, and gustatory receptors).

Extinction of Simultaneous Stimulation: The inability to determine that one has been touched on both the involved side and the uninvolved side—the neural sensation of the uninvolved side overrides the ability to perceive touch on the involved side. Also called double simultaneous extinction.

Extrapyramidal System: Motor structures and spinal cord tracts that do not use the pyramids to send motor messages to the skeletal muscles.

Extreme Capsule: White matter located just lateral to the claustrum and medial to the insula.

Facial Nerve: CN 7. Responsible for taste on the anterior of the tongue. Also responsible for innervating the muscles of facial expression and eyelid closing.

Falx Cerebri: A projection of dura mater that extends into the medial longitudinal fissure.

Fasciculations: Brief contractions of motor units; can be observed in skeletal muscle and detected on clinical exam.

Fastigial Nuclei: One of four pair of cerebellar nuclei.

Festinating: Quickened. Festinating is the opposite of cunctation—resisting or hindering. Together, the cunctating–festinating gait—characteristic of Parkinson's disease—describes difficulty initiating movement and inability to stop movement once started.

Fibrillations: Fine twitches of single muscle fibers that usually cannot be detected on clinical exam but can be identified on an electromyogram.

Figure–Ground Discrimination Dysfunction: Involves difficulty distinguishing the foreground from the background.

Filum Terminale: A projection of the pia mater of the spinal cord. The filum terminale is a slender median fibrous thread that attaches the conus medullaris to the coccyx. It anchors the end of the spinal cord to the vertebral column.

Finger Agnosia: Involves an impaired perception concerning the relationship of the fingers to each other. It also involves an impaired identification and localization of one's own fingers.

Fissure: A deep groove in the surface of the brain—deeper than a sulcus.

Flaccidity: Loss of muscle tone resulting from denervation of specific peripheral nerves. Flaccidity occurs in LMN injuries. In an UMN injury, flaccidity occurs only at the lesion level; spasticity occurs below the lesion level.

Fontanels: Non-ossified spaces or soft spots located between the cranial bones of fetuses and newborns. These allow the skull to expand to accommodate the growing brain (anterior, posterior, sphenoid, and mastoid fontanels).

Foramen Magnum: The largest foramina in the skull—specifically the occipital bone. The opening through which the brainstem connects with the spinal cord.

Foramen of Luschka: (also called lateral aperture) Opening in the fourth ventricle through which CSF flows into the subarachnoid space. There are two foramen of Luschka in the fourth ventricle; located in the pons. Often a site of CSF blockage.

Foramen of Magendie: (also called median aperture) Opening in the fourth ventricle through which CSF flows into the subarachnoid space; located in the rostral medulla. Often a site of CSF blockage.

Foramen of Monro: A channel located in the ventricular system; allows CSF to flow from the lateral ventricles to the third ventricle. Often a site of CSF blockage.

Foramina: Openings in the skull for the passage of blood vessels and nerves.

Form Constancy Dysfunction: Involves difficulty attending to subtle variations in form or changes in form, such as size variation of the same object.

Fornix: The fornix bodies are a pair of arch-shaped fibers that begin in the uncus and wrap around to the mammillary bodies. The fornix is a relay system for messages generated by the limbic system.

Fossa: (or cranial fossa) The undersurface of the brain sits in three cranial sections or fossa. The anterior cranial fossa primarily supports the frontal lobes. The middle cranial fossa supports the anterior-inferior temporal lobes and the diencephalon. The posterior cranial fossa supports the cerebellum.

Fourth Ventricle: Region of the ventricular system through which CSF flows. Located between the pons and the cerebellum. Connected to the third ventricle via the cerebral aqueduct. The fourth ventricle connects to the spinal cord via the central canal.

Frontal Eye Field: Located in the middle frontal gyrus, anterior to the premotor area. Responsible for visual saccades.

Frontal Lobe: Responsible for cognition, expressive language, motor planning, mathematical calculations, and working memory.

Frontal Plane: (also called coronal or transverse plane) The frontal planes run perpendicular to the sagittal planes. Frontal planes divide the anterior aspect of the brain from the posterior aspect.

GABA: See *Gamma-Aminobutyric Acid.*

Gamma-Aminobutyric Acid (GABA): A major inhibitory neurotransmitter that turns off the function of cells. GABA deficiency is implicated in anxiety disorders, insomnia, and epilepsy. GABA excess is implicated in memory loss and the inability for new learning.

Ganglia: A collection of neural cell bodies (or nuclei) usually located outside of the CNS.

Globose Nuclei: One of four pair of cerebellar nuclei.

Globus Pallidus: A basal ganglia structure involved in stereotypic or automatic movement patterns. While the caudate nucleus works like a brake on motor activity, the globus pallidus is an excitatory structure.

Glossopharyngeal Nerve: CN 9. Responsible for taste on the posterior of the tongue. Also innervates muscles of swallowing.

GLU: See *Glutamate.*

Glutamate (GLU): One of the major excitatory neurotransmitters of the CNS. Responsible for cell death when the brain experiences a major traumatic event. May also play a role in learning and memory.

Golgi Tendon Organ: Proprioceptors that are embedded in the tendons, close to the skeletal muscle insertion. Detect tension in the tendon of a contracting muscle.

Gray Matter: Sits on the surface of the cerebrum and cerebellum. Consists of nerve cell bodies.

Gustation: The detection and perception of taste.

Gyri (s., Gyrus): The wrinkles or folds on the surface of the cerebral hemispheres.

Hemianopsia: Field cut. See also *Contralateral Homonymous Hemianopsia* and *Bitemporal Hemianopsia.*

Hemiparesis: Partial paralysis or muscular weakness of limbs on one side of the body. Occurs on the contralateral side (or opposite side) of the lesion site.

Hemiparesthesia: Loss of sensation of limbs on one side of the body. Occurs on the contralateral side (or opposite side) of the lesion site.

Hemiplegia: Complete paralysis of limbs on one side of the body. Occurs on the contralateral side (or opposite side) of the lesion site.

Hippocampus: Located within the parahippocampal gyrus. One of the major storehouses in the brain for long-term memory.

Homunculus: Cortical representation for each body part's motor and sensory function. See also *Motor Homunculus* and *Sensory Homunculus.*

Horizontal Plane: The horizontal planes divide the superior aspect of the brain from the inferior aspect.

Hypalgesia: A decrease in the ability to perceive pain.

Hyperalgesia: An increase in the ability to perceive pain (ie, pain sensation becomes heightened).

Hyperesthesia: An increase in sensory perception. Heightened perception of pain and temperature.

Hyperkinesia: Disorders involving speeded movement (eg, chorea, athetosis). Opposite of hypokinesia.

Hyperreflexia: An increase in deep tendon reflexes.

Hypertonia: Excessive muscle tone due to spasticity or rigidity. Opposite of hypotonia.

Hypoesthesia: A decrease in sensory perception.

Hypoglossal Nerve: CN 12. Responsible for innervating the tongue muscles.

Hypokinesia: The slowing of movement just short of complete loss of movement or akinesia.

Hyporeflexia: Decreased deep tendon reflexes.

Hypothalamus: Two lobes (one in each hemisphere) that contain nuclei responsible for the regulation of the autonomic nervous system, release of hormones from the pituitary gland, temperature regulation, hunger and thirst, and sleep/wake cycles.

Hypotonia: A lack of muscle tone. Opposite of hypertonia.

Ideational Praxis: The ability to cognitively understand the motor demands of a task. For example, a patient must understand that shirts are articles of clothing to be worn on the torso. This type of praxis is largely a function of the motor association area.

Ideomotor Planning I: The ability to access the appropriate motor plan. For example, this would involve the ability to sort through all stored motor plans and identify the specific one for shirt donning. Such motor plans are commonly stored in the premotor area.

Ideomotor Planning II: The ability to implement the appropriate motor plan—or put it into action. For example, after identifying the appropriate motor plan for donning a shirt, the patient must put that plan into action. Implementing motor plans commonly involves M1.

Implicit Memory: A type of memory involving knowledge that the individual does not consciously know he has acquired. For example, someone with amnesia may not consciously know she has acquired knowledge of how to play the piano. But when seated at the piano, she can play fluently.

Inferior: Refers to the direction "below."

Inferior Colliculi: A pair of relay centers for audition that communicate directly with the medial geniculate nuclei of the thalamus. Located on the posterior region of the midbrain.

Inferior Olives (or Olivary Nuclei): Relay nuclei that carry ascending sensory information to the cerebellum. The sensory data pertain to the body's position in space.

Infundibulum: The stalk that extends from the hypothalamus and holds the pituitary gland.

Insula: A portion of the cerebral cortex that lies deep in the lateral fissure; covered from view by the frontal, parietal, and temporal lobes. Possible roles in (a) the perceptual processing of gustatory information and (b) the interpretation of music.

Intention Tremor: Occurs during voluntary movement of a limb and tends to increase as the limb nears its intended goal. Tremor is the rhythmic oscillation of joints caused by alternating contractions of opposing muscle groups. Intention tremors tend to diminish or stop when the patient's limbs are at rest. A sign of cerebellar damage.

Internal Capsule: A large fiber bundle that connects the cerebral cortex with the diencephalon. All descending motor messages from the cortex travel through the internal capsule to the thalamus, brainstem, spinal cord, and to the skeletal muscles. All sensory information travels through the internal capsule before reaching the cortex.

Interoceptors: Receive sensory information from inside the body—such as stomach pain, pinched spinal nerves, or inflammatory processes in the viscera.

Interpeduncular Fossa: The indentation between the pair of cerebral peduncles (on the anterior of the midbrain) that contains the mammillary bodies.

Interthalamic Adhesion: A commissure that connects the two thalamic lobes. Runs through the third ventricle.

Intervertebral Discs: Dense cushionlike structures that lie between each vertebrae. Each disc consists of the nucleus pulposus (a soft, pulpy, highly elastic tissue in the center of the disc) and the annulus fibrosus (the more fibrous outer-covering of the disc).

Joint Receptors: Proprioceptors that are located in the connective tissue of a joint capsule. Respond to mechanical deformation occurring in the joint capsule and ligaments.

Kinesthesia: The ability to sense one's body movement in space.

Lambdoid Suture: The suture lines are areas where cranial bones have fused. The lambdoid suture connects the two parietal bones to the occipital bone.

Lateral: Refers to structures that are further from the midline of the body.

Lateral Fissure: (also called fissure of Sylvius) Separates the temporal lobe from the frontal lobe.

Lateral Spinothalamic Tract: Ascending sensory spinal cord tract. Carries conscious information about pain and temperature.

Lateral Ventricle: One of four ventricles that contain cerebrospinal fluid. There are two lateral ventricles—one in each hemisphere. The lateral ventricle is an arch shaped structure that has its anterior horn located in the frontal lobe; its body located in the parietal lobe; its posterior horn located in the occipital lobe; and its inferior horn located in the temporal lobe.

Lead Pipe Rigidity: Characterized by a uniform and continuous resistance to passive movement as the extremity is moved through its range of motion (in all planes).

Lenticular Nucleus: Collective name for the globus pallidus and putamen (of the basal ganglia). Also called lentiform nucleus.

Limbic System: Located deep within the core of the brain, the limbic system appears to be the source of human emotions before they are modulated by the frontal lobes. The limbic system is also a storehouse for long-term memory—particularly memories that have strong emotional significance.

LMN: See *Lower Motor Neuron.*

Locus Ceruleus: Located on the floor of the fourth ventricle in the brainstem. Secretes NE.

Long-Term Memory: The ability to remember one's past, familiar others, and events encountered more than several hours ago. Believed to be a function of the hippocampus and temporal lobe.

Lower Motor Neuron (LMN): Carries motor messages from the motor cell bodies in the ventral horn to the skeletal muscles in the periphery. A LMN is considered to be part of the PNS. Includes the CNs, peripheral spinal nerves, conus medullaris, cauda equina, and the ventral horn. Lesions to these structures will cause LMN signs.

Lumbar Plexus: A network of peripheral spinal nerves that supply the lower extremities. Includes L1, L2, L3, L4, and L5. Common site of compression injuries.

M1: See *Primary Motor Area.*

Mammillary Bodies: Two protrusions that sit within the interpeduncular fossa on the anterior surface of the midbrain. The mammillary bodies are nuclei groups that form attachments with the hypothalamus and fornix and may play a role in the processing of emotion.

Mechanoceptors: Sensory receptors that are stimulated by mechanical deformity (eg, hair cell receptors of the skin).

Medial: Refers to structures that are close to the midline of the body.

Medial Longitudinal Fasciculus: Mixed spinal cord tract with both ascending and descending fibers. Responsible for the coordination of head, neck, and eyeball movements.

Medial Longitudinal Fissure: Runs along the midsagittal plane. Separates the right and left cerebral hemispheres.

Medulla: Carries descending motor messages from the cerebrum to the spinal cord, and ascending sensory messages from the spinal cord to the cerebrum. Located between the pons and the spinal cord.

Medullary Reticulospinal Tract: Descending extrapyramidal motor tract that inhibits antigravity muscles (inhibits extensor muscles).

Membrane Potential: The electrical charge that travels across the cell membrane. It is the difference between the chemical composition inside and outside the cell—in other words, the sodium potassium balance inside and outside the cell. If the membrane potential is strong enough, it causes an action potential.

Meninges: The meninges are located between the skull and brain, and cover the spinal cord. They form a protective seal around the CNS. There are three layers of meninges: dura mater, arachnoid mater, and pia mater.

Mesolimbic System: The brain's reward system is called the mesocorticolimbic, or mesolimbic system. This system consists of a complex circuit of neurons that evolved to encourage people to repeat pleasurable behavior that supports survival. The primary neural pathway of the mesolimbic system originates in the ventral tegmental area of the midbrain. This area sends projections to the nucleus accumbens, which is located deep beneath the frontal cortex. The primary neurotransmitter used by this system is DA. Almost all drugs of abuse have the potential to become addictive because of their ability to increase DA levels in the brain's reward system.

Metamorphopsia: Involves a visual-perceptual distortion of the physical properties of objects so that objects appear bigger, smaller, or heavier than they really are.

Micrographia: A sign of Parkinson's disease in which handwriting becomes small and of poor quality to due movement decomposition.

Midbrain: The midbrain is the most rostral structure of the brainstem that sits atop of the pons and is just inferior to the thalamus. Has a role in automatic reflexive behaviors dealing with vision and audition. Site of the reticular activating system—has roles in wakefulness and consciousness.

Midsagittal Plane: The midsagittal plane divides the left and right cerebral hemispheres. This plane divides the brain in half and runs along the medial longitudinal fissure.

Mononeuropathy: Involves damage to a single peripheral nerve; usually due to compression or entrapment.

Motor Association Area: Also called the prefrontal lobe. Role in the cognitive planning of movement.

Motor Homunculus: The cortical map that represents each body part's area for motor function. Location of M1—the precentral gyrus.

Movement Decomposition: Movement is performed in a sequence of isolated parts rather than as a smooth, singular motion. A sign of cerebellar damage. Also referred to as dyssynergia.

Multimodal Association Area: Cortical areas where sensory information merges for integration and interpretation. Two multimodal association areas: (a) anterior multimodal association area—located in prefrontal lobe, (b) posterior multimodal association area—located where the parietal, occipital, and temporal lobes converge.

Muscle Spindle: Proprioceptors that are located in the skeletal muscle. Provide a constant flow of information regarding length, tension, and load on the muscles.

Myelin: Axons are covered by a cellular sheath called myelin—an insulating substance composed of lipids and proteins. Myelin serves to conduct nerve signals. The more myelin the axon has, the faster its conduction rate.

Myotome: A group of muscles innervated by a specific single spinal nerve. The myotomes of the human body correspond closely to the dermatomes.

NE: See *Norepinephrine*.

Neostriatum: Collective name for the caudate and putamen (of the basal ganglia).

Neuron: The electrically excitable nerve cell and fiber of the nervous system. Composed of a cell body with a nucleus, dendrites, a main axon branch, and terminal boutons.

Neuropathy: A general term for pathology involving one or more peripheral nerves.

Neuroplasticity: Occurs when other areas of the brain assume the functions once mediated by regions that have been damaged. Human brains appear to possess a vast amount of brain matter that does not become active until damage occurs to other areas. At this time, those previously unused regions may become active and take over the function of damaged areas. It is possible that the same brain function may be shared by several, separate brain regions that lie dormant until injury/disease occurs. This may be the brain's evolutionary attempt at compensation. Neuroplasticity is most viable in children because the CNS is not fully mature, but rather is still developing.

Neurotransmitter: Chemical stored in the terminal boutons. Released into the synaptic cleft to transmit messages to another neuron.

Nociception: The detection and localization of pain.

Nodes of Ranvier: Spaces between the myelin—on an axon—where nerve signals jump from one node to the next in the process of conduction.

Norepinephrine (NE): A neurotransmitter essential in the production of the fight/flight response, fear, and panic.

Nystagmus: Involuntary back and forth movements of the eye in a quick, jerky, oscillating fashion when the eye moves laterally or medially to either the temporal or nasal extreme visual fields.

Occipital Lobe: Responsible for the detection and interpretation of visual stimuli.

Occipital Pole: Most posterior region of the occipital lobe. Site of V1, responsible for the detection of visual stimuli.

Oculomotor Nerve: CN 3. Responsible for extraocular eye movements. Lesion symptoms can include nystagmus and lateral strabismus.

Olfaction: Detection and perception of smell.

Olfactory Bulb and Tract: Also known as CN 1. The olfactory tract travels directly to the hippocampus—which accounts for the deep association between specific odors and long-term memories that have emotional significance.

Olfactory Nerve: CN 1. Responsible for smell.

Oligodendrites: Compose the myelin in the CNS. Because oligodendrites do not produce nerve growth factor, most nerve damage in the CNS cannot resolve.

Optic Chiasm: A cross-shaped connection between the optic nerves. Carries visual information from the optic nerves to the optic tracts.

Optic Nerve: CN 2. The optic nerves (one in each hemisphere) carry visual information from the retina, through the optic disk, to the optic chiasm. Visual information then travels from the optic chiasm to the optic tracts. Responsible for visual acuity.

Paleostriatum: Another name for the globus pallidus (of the basal ganglia)—because of its striped appearance.

Parahippocampal Gyrus: Most medial and deepest gyrus in the temporal lobes. Folds back on itself at its anterior end to become the uncus. Relays information between the hippocampus and other cerebral areas, particularly the frontal lobes. The parahippocampal gyrus may function when humans compare a present event to an event stored in long-term memory to decide how to handle a present situation.

Parasympathetic Nervous System: A division of the autonomic nervous system. Responsible for peristalsis and the maintenance of homeostasis.

Paresis: Partial paralysis.

Paresthesia: The occurrence of unusual feelings, such as pins and needles.

Parietal Lobe: Responsible for sensory detection and interpretation.

Perception: The ability to interpret or attach meaning to sensory data from the external and internal environments. Perceptual impairment more often results from damage to the right hemisphere—which produces a distortion in the patient's perception of his physical body and environment.

Peripheral Nervous System (PNS): Composed of the CNs, autonomic nervous system, and the SNS.

Phantom Limb Phenomenon: The sensation that an amputated body part still remains. If the sensation is painful it is referred to as phantom pain. The cortical map of the body still retains the anatomical image of the amputated body part.

Photoreceptors: Sensory receptors that detect light on the retina of the eye.

Pia Mater: The deepest meningeal layer, located on the surface of the gyri and sulci of the brain and on the surface of the spinal cord. The pia mater of the spinal cord sends off two projections: the filum terminale and the dentate ligaments.

Pineal Gland: Innervated by the autonomic nervous system. Has a role in sexual hormonal functions and sleep-wake cycles.

Pituitary Gland: An endocrine gland that secretes hormones regulating growth, reproductive activities, and metabolic processes.

Plexopathy: A form of neuropathy involving damage to one of the plexus—brachial or lumbar. Involves multiple peripheral nerve damage.

PNS: See *Peripheral Nervous System*.

Polyneuropathy: A form of neuropathy involving bilateral damage to more than one peripheral nerve. An example is stocking and glove polyneuropathy. Usually caused by a disease process, such as diabetes.

Pons: Brainstem structure that acts as a relay system between the spinal cord, cerebellum, and cerebrum. It largely mediates sensorimotor information on an unconscious level—shifting weight to maintain balance and making fine motor adjustments in one's muscles to perform precise coordinated limb movement. Located between the midbrain and medulla. Site of the reticular inhibiting system—has roles in sleep states and unconsciousness.

Pontine Reticulospinal Tract: Descending extrapyramidal motor tract that facilitates antigravity muscles (ie, facilitates extensor muscles).

Position in Space Dysfunction: Involves difficulty with concepts relating to positions such as up/down, in/out, behind/in front of, and before/after.

Postcentral Gyrus: Located just posterior to the central sulcus; known as SS1 where sensory information from the contralateral side of the body is detected. Also referred to as the sensory homunculus—the cortical representation for each body part's sensory function.

Postcentral Sulcus: Sulcus located just posterior to the postcentral gyrus.

Posterior: (or dorsal) Refers to the back of an organism.

Posterior Commissure: Located just superior to the superior colliculi. Allows information to travel between both diencephalic hemispheres.

Posterior Spinocerebellar Tract: Ascending sensory spinal cord tract. Carries unconscious information from the lower extremities to the cerebellum regarding proprioception.

Postpontine Fossa: Shallow depression located between the anterior aspects of the pons and medulla.

Postsynaptic Neuron: Second order neuron. Receives the presynaptic neuron's neurotransmitter substance from the synaptic cleft.

Post-Tetanic Potentiation: Occurs in synapses that are frequently used. When the presynaptic bouton becomes excited, it releases greater amounts of neurotransmitter substance. The postsynaptic neuron then has prolonged and repetitive discharge after firing due to too much neurotransmitter release or too slow antitransmitter work.

Praxis: Motor planning. The ability to understand and implement the movements required for a specific activity.

Precentral Gyrus: Located just anterior to the central sulcus; considered to be M1 where voluntary movement (on the contralateral side of the body) is initiated. Also referred to as the motor homunculus—the cortical map that represents each body part's area for motor function.

Precentral Sulcus: Sulcus located just anterior to the precentral gyrus.

Premotor Area: Located just anterior to M1. Has a role in motor planning or praxis.

Prepontine Fossa: Shallow depression located between the anterior aspects of the midbrain and pons.

Presynaptic Neuron: First order neuron. Releases its neurotransmitter into the synaptic cleft.

Primary Auditory Area (A1): Located within the insula in the temporal lobe. Responsible for detecting sounds from the environment.

Primary Motor Area (M1): Location of the precentral gyrus. Where voluntary movement (on the contralateral side of the body) is initiated. Also referred to as the motor homunculus—the cortical map that represents each body part's area for motor function.

Primary Somatosensory Area (SS1): Location of the postcentral gyrus. Where sensory information from the contralateral side of the body is detected. Also referred to as the motor homunculus—the cortical map that represents each body part's area for motor function.

Primary Visual Area (V1): Responsible for the detection of visual stimuli. Located at the most posterior region of the occipital lobe.

Primitive Reflexes: Reflexes which humans are born with or which develop and become integrated by the CNS in infancy and toddlerhood. These reactions facilitate gross motor patterns of flexion and extension. Primitive reflexes exhibited by an adult are signs of neurologic damage.

Procedural Memory: A type of memory involving the recall of steps involved in specific tasks. For example, knowing the steps involved in fabricating a wrist cock-up splint. Or knowing the steps involved in changing a flat tire.

Proprioception: The ability to sense one's body position in space. Proprioception occurs mostly on an unconscious level because it is primarily mediated by the cerebellum.

Proprioceptor: Sensory receptors located in the muscles, tendons, and joints of the body, and in the utricles, saccules, and semicircular canals of the inner ear.

Prosopagnosia: The inability to identify familiar faces because the individual cannot perceive the unique expressions of facial muscles that make each human face different from each other.

Protopathic Sensory Receptors: Adapted to identify gross bodily sensation, rather than precise regions of sensation. Detect crude touch and dull pain rather than discriminative touch and sharp pain.

Putamen: A basal ganglial structure involved in stereotypic or automatic movement patterns. While the caudate nucleus works like a brake on motor activity, the putamen is an excitatory structure.

Pyramidal Decussation: The crossing-over point where motor fibers from the left cortex cross to the right side of the spinal cord. Motor fibers from the right side of the cortex cross to the left side of the spinal cord. This is why the right cerebral hemisphere controls the left side of the body and the left cerebral hemisphere controls the right side of the body.

Pyramids: Fiber bundles that carry descending motor information from the cortex to the spinal cord. The pyramids are two large structures on the anterior region of the medulla that are divided by the anterior median fissure.

Radiculopathy: Nerve root impingement that results from a lesion affecting the dorsal or ventral roots. Can result from herniated vertebral discs.

Rebound Phenomenon: The inability to regulate the action of opposing muscle groups. The patient is asked to resist the therapist's attempt to pull the patient's flexed elbow into extension. The therapist then releases the patient's forearm. Normally, the elbow would remain in approximately the same position due to the ability to regulate the speed of muscular contraction. Patients with cerebellar lesions are unable to regulate their opposing muscle groups and their limb suddenly hits their torso. This occurs as a result of impaired proprioceptive feedback. The patient cannot regulate the speed and force of opposing muscle groups quickly enough to prevent the arm from hitting the torso.

Recent Memory: Considered to be a component of long-term memory. Recent memory refers to the ability to recall events/information that occurred hours to weeks ago.

Receptive Aphasia: A perceptual problem involving impairment in the comprehension of language.

Receptor Field: A body area that contains specific types of sensory receptor cells. Small receptor fields are located on body areas with the greatest sensitivity—lips, hands, face, soles of feet. Large receptor fields are located on body areas with less sensitivity—legs, abdomen, arms, back.

Referred Pain: Occurs when a specific body region shares its spinal nerve innervation with a separate dermatomal skin segment. The pain experienced by the body part is misinterpreted by the cortex as pain coming from a separate dermatomal skin segment.

Remote Memory: A component of long-term memory; refers to the ability to recall events/information that occurred in the distant past (eg, several years ago). Believed to be a function of the hippocampus and temporal lobe.

Resting Tremor: (also called nonintention tremors). Characteristic of Parkinson's disease. Involuntary oscillating movements that occur in an extremity at rest. Resting tremors decrease with the intention of voluntary movement.

Reticular Formation: Two systems diffusely located in the brainstem: reticular activating system and reticular inhibiting system. The activating system alerts the cortex to attend to important sensory stimuli and is involved in states of wakefulness. The inhibiting system is involved in states of unconsciousness such as sleep, stupor, or coma.

Retrograde Amnesia: A type of memory problem that involves the loss of one's entire personal past and occurs after injury or trauma. Memory of one's past is often recovered at some point after injury.

Reuptake Process: Process by which a neurotransmitter is reabsorbed into the presynaptic neuron's terminal boutons.

Right–Left Discrimination Dysfunction: Involves difficulty understanding and using the concepts of right and left.

Rigidity: Involves an inability to passively and/or actively move a joint on both sides of the joint. Example: cogwheel rigidity (commonly seen in Parkinson's).

Rostral: Refers to the head of an organism. Also refers to structures that are above others.

Rostral Spinocerebellar Tract: Ascending sensory spinal cord tract. Carries unconscious information from the trunk and upper extremities to the cerebellum regarding proprioception.

Rubrospinal Tracts: Descending extrapyramidal motor tract. Facilitates antagonist of antigravity muscles (the flexors).

Saccule: One of the receptors of equilibrium in the inner ear. Responds to changes in head position. Part of the vestibular system.

Sagittal Plane: The sagittal planes run parallel to the midsagittal plane.

Sagittal Suture: The suture lines are areas where cranial bones have fused. The sagittal suture runs along the midsagittal plane and connects the two parietal bones.

Schwann Cell: The myelin in the PNS is composed of Schwann cells that produce nerve growth factor. This allows peripheral nerve damage to resolve (unlike damage in the CNS).

Secondary Somatosensory Area (SS2): Responsible for the interpretation of sensory stimuli from the contralateral side of the body.

Semantic Memory: A type of memory involving the recall of facts—dates of people's birthdays, state capitols, the names of presidents. Semantic memory includes the definitions of words and how to use the rules of grammar.

Semicircular Canals: Receptors of equilibrium that form a system of canals called the bony labyrinth in the inner ear. There are three semicircular canals—each responds to movement of the head. Part of the vestibular system.

Sensory Homunculus: The cortical representation for each body part's sensory function. Location of the postcentral gyrus, or SS1.

Sensory Receptor: A specialized nerve cell that is designed to respond to a specific sensory stimulus (eg, touch, pressure, pain, temperature, light, sound, position in space).

Septal Area: The region on either cerebral hemisphere that comprises the subcallosa area and the corresponding half of the septum pellucidum.

Septum Pellucidum: A sheathlike cover that extends over the medial wall of each lateral ventricle. May have a role in the processing of emotion.

Serotonin (5-HT): A neurotransmitter implicated in sleep, emotional control, pain regulation, and carbohydrate feeding behaviors (eating disorders). Low levels of serotonin are associated with depression and suicidal behavior.

Short-Term Memory: The ability to remember others and events encountered less than 1 hour ago.

Simultanagnosia: Involves difficulty interpreting a visual stimulus as a whole. Patients often confabulate to compensate for what they cannot interpret visually.

SNS: See *Somatic Nervous System.*

Somatic Nervous System (SNS): Part of the PNS. Responsible for the innervation of skeletal muscles.

Source Memory: A type of memory involving the recall of how information is learned. For example, an anatomist may recall that he learned specific information about the human shoulder joint from a specific dissection class. Often, people remember facts but cannot recall how they came to learn those facts.

Spasticity: Involves the inability to move a joint on one side of the joint. Usually, either the flexors or the extensors are spastic, but not both. Results from UMN lesions.

Spinal Reflex Arc: A spinal reflex arc is mediated at the spinal cord level; there is no cortical involvement. A sensory receptor in the PNS sends a message along an ascending sensory spinal nerve that travels to the dorsal horn and synapses on an interneuron. The interneuron synapses with a motor cell body located in the ventral horn. The motor cell body in the ventral horn relays the message to a motor spinal nerve in the PNS, which sends the motor message to a skeletal muscle group for action in response to the initial sensory message.

SS1: See *Primary Somatosensory Area.*

SS2: See *Secondary Somatosensory Area.*

Staccato Voice: Broken speech; a sign of cerebellar damage. Because the modulation of speech is a proprioceptive function, patients with cerebellar damage may be unable to modulate the fluidity of speech.

Stereopsis: Depth vision.

Strabismus: Deviation of the eyeball laterally (lateral strabismus, also called exotropia) or medially (medial strabismus, also called esotropia).

Subarachnoid Space: Area beneath the arachnoid mater and above the pia mater. Contains CSF.

Substance P: Acts as a neurotransmitter in the nociceptive pathway, although it is classified as a peptide. The nociceptive pathway mediates the experience of pain.

Substantia Nigra: Located in the midbrain. The red nucleus and the substantia nigra of the midbrain form the inner coat of the cerebral peduncles. The substantia nigra produces DA—a neurotransmitter that functions in movement and mood regulation. The axons of the substantia nigra form the nigrostriatal pathway, which supplies DA to the striatum.

Subthalamus: A thalamic nuclei group located caudal to the thalamus. Contains cells that use DA. A key structure connecting feedback and feedforward circuits of the thalamus and basal ganglia.

Sulci (s., Sulcus): The valleys or crevices between the gyri.

Superior: Refers to the direction "above."

Superior Colliculi: Are a pair of relay centers for vision that communicate directly with the lateral geniculate nuclei of the thalamus. Located on the posterior region of the midbrain.

Supplemental Motor Area: Considered to be part of the premotor area. Located inside the medial longitudinal fissure. Has a role in the bilateral control of posture.

Sympathetic Nervous System: A division of the autonomic nervous system. Responsible for the fight/flight response and gearing the body up for action.

Synaptic Cleft: Space between a presynaptic neuron's terminal boutons and a postsynaptic neuron's dendrites. The terminal boutons release their neurotransmitter substances into the synaptic cleft.

Synaptic Delay: The time required for the neurotransmitter to diffuse across a postsynaptic neuron's membrane.

Synaptic Fatigue: Occurs as a result of a neurotransmitter depletion due to the repetitive simulation of a presynaptic neuron.

Synergy Pattern: A stereotyped set of movements that occur in response to a stimulus or voluntary movement. Involves pathology of muscle tone affecting joint position. Synergies are described as patterns because the involved joint positions occur consistently as a result of specific neurologic damage. Specific flexor and extensor synergies can be observed in the upper and lower extremities. Synergy patterns can change as the patient experiences stages of recovery. Or they may continue if recovery of damaged brain structures cannot occur.

Synesthesia: A perceptual phenomenon involving the ability to combine senses in response to specific stimuli. For example, the ability to see colors when one hears music.

Tachykinesia: Speeded movement. Commonly seen in Tourette's syndrome.

Tachyphrenia: Speeded thought. Commonly seen in mania.

Tactile Agnosia: The umbrella term for the inability to attach meaning to somatosensory data.

Tardive Dyskinesia: A movement disorder related to treatment with DA receptor antagonists (neuroleptics and antiemetics). The term tardive refers to the fact that this movement disorder occurs after chronic use of these drugs. Characterized by choreiform movements, dystonia, tics, and/or myoclonus. Example: tongue protrusions (orobucco-lingual movements), chewing-type movements, facial grimacing, blepharospasm, lip smacking. Tardive dyskinesia is different from most disorders in that the discontinuation of the causative agent (the neuroleptic) does not result in the amelioration of the movement disorder.

Tectum: The tectum is the collective name for the superior and inferior colliculi (of the midbrain).

Tegmentum: The tegmentum is the inner coat of the cerebral peduncles (of the midbrain). The tegmentum is the collective name for the substantia nigra and the red nucleus.

Temporal Lobe: Responsible for the detection and interpretation of sounds and long-term memory.

Tentorium: The projection of dura mater that extends as a horizontal shelf between the occipital lobe and the cerebellum.

Terminal Boutons: Emerge from the end branches of the axon and contain the neurotransmitter substances.

Thalamus: An egg-shaped lobe (one in each hemisphere) that contains 26 pairs of nuclei which act as a relay system for sensory and motor information traveling to and from the cortex.

Thermal Receptors: Sensory receptors that detect changes in temperature.

Thermesthesia: The ability to perceive temperature (hot and cold).

Thermohyperesthesia: An increase in temperature perception (ie, hot and cold sensations become heightened).

Thermohypesthesia: A decrease in temperature perception.

Third Ventricle: Part of the ventricular system. The walls of the third ventricle are created by the thalamus and hypothalamus. The lateral ventricle connects to the third ventricle via the foramen of Monro. The third ventricle connects to the fourth ventricle via the cerebral aqueduct.

Tics: Repetitive, brief, rapid, involuntary movements involving single muscles or multiple muscle groups. Tics are caused by an increased sensitivity to DA in the basal ganglia. With increased sensitivity to DA, the caudate, which normally acts like a brake on extraneous movements, cannot suppress movements like tics.

Topographical Disorientation: Involves difficulty comprehending the relationship of one location to another.

Transverse Plane: (also called coronal or frontal plane) The transverse planes run perpendicular to the sagittal planes. Transverse planes divide the anterior aspect of the brain from the posterior aspect.

Tremor: Involuntary oscillating movement resulting from alternating or synchronistic contractions of opposing muscles. See also *Resting Tremor* and *Intention Tremor*.

Trigeminal Nerve: CN 5. Responsible for sensation of face, head, and inner oral cavity. Also innervates the muscles of the jaw for chewing.

Trochlear Nerve: CN 4. Responsible for extraocular eye movements. Lesion symptoms include nystagmus, diplopia, and vertical, medial strabismus.

Two- and Three-Dimensional Constructional Apraxia: A type of apraxia involving an inability to copy two- and three-dimensional designs or models.

Two-Point Discrimination: The loss of the ability to determine whether one has been touched by one or two points. An aesthesiometer is the instrument used to assess two-point discrimination.

UMN: See *Upper Motor Neuron.*

Uncus: The bulblike anterior end of the parahippocampal gyrus.

Unilateral Neglect: Involves the inability to integrate and use perceptions from one side (the affected side) of the body or environment.

Upper Motor Neuron (UMN): An UMN carries motor messages from the primary motor cortex to (a) the CN nuclei (in the brainstem), or (b) interneurons in the ventral horn. An UMN travels up to but does not actually enter the ventral horn. An UMN is considered to be part of the CNS.

Utricle: One of the receptors of equilibrium in the inner ear. Responds to changes in head position. Part of the vestibular system.

V1: See *Primary Visual Area.*

V2 and up: See *Visual Association Areas.*

Vagus Nerve: CN 10. Responsible for taste on the palate and epiglottis. Carries parasympathetic information to and from the heart, pulmonary system, esophagus, and gastrointestinal tract. Also responsible for innervating the muscles of the larynx, pharynx, and upper esophagus (the muscles of swallowing and speaking).

Ventral: (also anterior) Refers to the front of an organism. Ventral means the belly of a four-legged animal.

Ventral Horn, Root, and Rootlets: The ventral horn contains the cell bodies of the motor spinal nerves that innervate skeletal muscle. Descending motor spinal tracts travel from the cortex down to the spinal cord. In the ventral horn, the motor spinal cord tracts synapse on interneurons. These interneurons then synapse with motor spinal nerves that travel to skeletal muscles in the PNS. The motor spinal nerve exits the ventral horn through the ventral rootlets. The ventral rootlets then merge into the ventral root. The ventral horn, rootlets, and root are all considered to be within the PNS.

Ventricular System: Hollow spaces in the brain through which CSF flows. There are four ventricles: two lateral ventricles, one third ventricle, and one fourth ventricle.

Vermis: The midline structure of the cerebellum that may have a role in the integration of information used by the right and left cerebellar hemispheres. Some have also suggested that the vermis may have a role in emotion and the timing of appropriate affective responses.

Vestibular System: Functions to maintain equilibrium and balance, maintains the head in an upright vertical position, coordinates head and eye movements, and influences muscle tone through the alpha and gamma motor neurons and the vestibulospinal tracts.

Vestibulocochlear Nerve: CN 8. Responsible for hearing or audition, balance, and the position of the head in space.

Vestibulospinal Tracts: An extrapyramidal tract (does not use the pyramids). Descending motor spinal cord tract. Facilitates antigravity muscles (extensors) that are responsible for posture and stance.

Visceral Sensory Receptors: Respond to pressure and pain from the internal organs.

Visual Acuity: The ability to see with accuracy.

Visual Agnosia: An umbrella term for the inability to identify and recognize familiar objects and people although the visual anatomy remains intact. Visual agnosia is a perceptual disorder involving the cortical visual association areas.

Visual Association Areas (V2 and up): Cortical areas in the occipital lobe that are responsible for the interpretation of visual stimuli.

Visual Perception: The ability to attach meaning to visual data.

Wernicke's Aphasia: A language perceptual problem involving difficulty comprehending the literal interpretation of language.

Wernicke's Area: Located only in the left hemisphere within the superior temporal gyrus. Responsible for the comprehension of the spoken word.

White Matter: Located beneath the gray matter in the internal regions of the cerebrum and cerebellum. Consists of myelinated fiber tracts or neuronal axons.

Withdrawal Reflex: A spinal reflex that works similarly to the spinal reflex arc. This reflex is a protective mechanism that allows reflexive withdrawal of a body part from physical danger, while simultaneously adjusting posture to avoid imbalance. The reflex works at the level of the spinal cord without cortical processing.

Working Memory: A subcomponent of short-term memory; involves moment-to-moment awareness. Also plays a role in the search and retrieval of archived information. Mediated by the frontal lobes.

Index